Digital Computers in Scientific Instrumentation

Applications to Chemistry

D1235743

Digital Computers in Scientific Instrumentation

Applications to Chemistry

SAM P. PERONE

Chemistry Department
Purdue University
Lafayette, Indiana

DAVID O. JONES

Creative Industrial Design
Chelan, Washington

McGraw-Hill Book Company

New York St. Louis San Francisco Düsseldorf Johannesburg
Kuala Lumpur   London   Mexico   Montreal   New Delhi
Panama   Rio de Janeiro   Singapore   Sydney   Toronto

Library of Congress Cataloging in Publication Data

Perone,   S   P   1938–
    Digital computers in scientific instrumentation.
    Includes bibliographies.
    1. Electronic data processing – Chemistry.
2. Chemical laboratories – Electronic equipment.
I. Jones, David O., Joint author. II. Title.
QD42.P44   540'.28'54   72–13679
ISBN 0–07–049319–7

**Digital Computers in Scientific
Instrumentation: Applications to Chemistry**

1 2 3 4 5 6 7 8 9 0 DODO 7 9 8 7 6 5 4 3

This book was set in Press Roman. The editors were
William P. Orr and M. E. Margolies; the
production supervisor was Ted Agrillo. The drawings
were done by Oxford Illustrators Limited.
The printer and binder was R. R. Donnelley & Sons
Company.

To all our girls

Contents

Preface

A scientist today finds himself in the midst of a dramatic revolution in scientific instrumentation. Digital electronics and minicomputers appear destined soon to dominate the whole area of laboratory experimentation. This revolution has come about so abruptly that the practicing scientist recognizes a very real technological gap in his background. He realizes that computerized instrumentation can provide benefits in routine and research work — but he does not know how to get involved. To compound his frustration, he even finds it difficult to communicate with the experts and computer manufacturers. Not only is the *jargon* of digital-computer technology quite foreign to the uninitiated, but the fundamental concepts of digital instrumentation are basically unfamiliar. The scientist is accustomed to thinking of experimental data and instrumentation from an *analog* viewpoint, and the introduction to the digital world is not without some difficulty. Not that the material is difficult — it is really quite simple for the scientist trained in logic — but mastery of the material necessitates the generation of drastically different instrumental and experimental concepts than the ones to which the scientist is generally accustomed. The principles, methodology, and jargon of digital instrumentation are so unique that even the instrument-oriented scientist has difficulty mastering the area independently. Moreover, the scientist's ability to use the large data processing computer is of little help in understanding *on-line* applications of the digital computer.

Because scientists have recognized this technological gap, many have taken steps to overcome it. This has required herculean efforts on their part, generally, since very little detailed information is available. Most of what is available is provided by computer manufacturers and may require thorough familiarization with a particular computer for any appreciation. The most fundamental problem, however, is that an adequate appreciation of this new technology cannot be acquired through reading or attendance at technical meetings. Moreover, it is not sufficient to have a computer available in the laboratory with which the scientist can play in his spare time, nor is it sufficient to develop programming skills which are not oriented toward on-line experimental applications. The most efficient way to develop the technological skills necessary for the incorporation of the digital computer into laboratory experimentation is by total involvement in an intensive training program which includes substantial *hands-on* experience with digital instrumentation and the digital computer in an experimental on-line environment. This text is intended as a companion guide for the scientist who undertakes such a program — formally or informally.

It should be pointed out that this text was developed for use in a graduate-level one-semester analytical chemistry course at Purdue University. The course consists of two lectures and the equivalent of three hours of laboratory per week. The lectures cover the material presented in the chapters of this text. The laboratory provides the student with hands-on experience with a digital-computer system, digital electronics, and interfacing to a variety of laboratory instrumentation. This same course has been presented as a 3-week summer short course annually since 1968. The summer program has been supported by the National Science Foundation.

The material presented in this text is confined primarily to discussions pertinent for dedicated applications of the small digital computer in the laboratory. It is the authors' belief that a sound introduction to the principles of computer applications in the laboratory at the dedicated level will provide the scientist with the background necessary to cope with more complex computerized laboratory systems. Moreover, Chapter 10 provides an introduction to considerations related to large laboratory computer systems utilizing time sharing.

Because the authors are chemists, the illustrative examples in this text — and in particular the material in Chapter 12 — are taken from this field. However, the tutorial material presented in this text is generally applicable to scientists of all disciplines, and constitutes about 85 percent of the presentations.

One primary objective of this text is to introduce the fundamental principles of small-computer programming. Because of the limited applicability of high-level languages to small computers, the emphasis here is on developing a familiarity with assembly-language or machine-language programming. On the other hand, Chapter 11 discusses the increasingly important role of high-level programming languages for on-line experimentation. It is our experience that the most effective use of high-level languages for on-line applications can be made when the user has a thorough appreciation of the intricacies of assembly-language programming.

A second objective of this text is to introduce the fundamentals of digital logic and experimental interfacing to digital computers. It has been our observation that even when the laboratory scientist has the services of a capable electrical engineer at his disposal for interfacing, that user must be familiar with the elements of interface design and computer input/output characteristics in order to take full advantage of the electronics specialist. It is not required that the user be skilled in the subtleties of optimization of digital circuitry, but he must at least be able to communicate with the engineer who can implement his desired interface.

A third objective of this text is to provide a summary of fundamental principles and illustrative examples for dedicated on-line computer applications in the chemistry laboratory. Detailed discussions of several actual laboratory studies (Chapter 12) are included for this purpose. In addition, many hypothetical examples are included.

Finally, it is desired to provide some insight into the whole field of laboratory computer instrumentation, including time sharing, systems design, and computer architecture.

The major dilemma in preparing this text was the fact that not every small-computer user employs the same type of computer. There is a large variety of small digital computers on the market, and each has its own unique hardware and software characteristics. To write a text which would be meaningful to each individual user posed a formidable problem to the authors. The approach taken here was decided upon after much deliberation and discussion with other workers in the field. The effectiveness of this approach has been tested in the educational program at Purdue University.

The fundamental aspect of our approach here is simply that we have selected arbitrarily a set of computer hardware and software characteristics upon which all the detailed discussions are based. These characteristics are essentially identical to the type of computer equipment employed in our research and educational program at Purdue University. The heart of these systems is the Hewlett-Packard 2100 family of computers. Thus, the software and hardware features are essentially consistent with that particular computer.

This approach was convenient for us, of course. However, we have found by experience that students who are trained in the fundamentals of computer programming and interfacing for on-line applications on one type of computer can very quickly switch over to another type of computer and carry over those principles acquired on the learning system.[1] In addition, the authors have provided in Chapter 13 a general discussion of a wide variety of computer hardware and software features. Also, Appendix D is included to allow the translation of text material to other types of small laboratory computers which are currently available. For example, the detailed programming discussions based on Hewlett-Packard assembly language can be made meaningful to the user of a Digital Equipment Corporation PDP-8 by reference to Appendix D. All the discussions have general applicability, of course, regardless of the specific computer system employed by the reader.

In this text, the authors have attempted to restrict their discussions to include only that type of computer system which is most likely to be used by the laboratory scientist. That is, we have considered only relatively small systems with limited core memory space, paper-tape input/output, and no expensive high-speed or high-capacity peripheral devices like a magnetic disc, magnetic tape, line printer, etc. On the other hand, the applicability and characteristics of some of these peripherals are considered in Chapter 10.

In Chapter 12 the authors have included several detailed discussions of laboratory computer applications carried out in their own and other research programs. The objective of this chapter is to demonstrate a wide variety of ways in which the computer can be used to enhance experimental measurement capabilities. The fundamental principles and detailed descriptions included should be generally useful to most readers.

In the body of this text, the authors have presented only those electronics principles and technological descriptions which are necessary to provide meaningful discussions of interface design and the development of computerized digital instrumentation. These discussions are confined primarily to Chapters 7 and 8. Here, the reader will find that he can generate a working understanding of the functional characteristics of digital data-acquisition devices and digital instrumentation useful for interface design. For the reader who desires a more detailed study of digital logic, the authors have included Appendixes B and C.

The individuals and groups who made possible the completion of this text are identical with those who have contributed to the success of the summer short course and graduate course on which the text is based.

Certain groups and individuals made particularly noteworthy contributions. These include Jack Frazer, Roger Anderson, and Jackson Harrar of the Lawrence Livermore Laboratory, who have provided, respectively, the inspiration, technical guidance, and critical appraisal for the educational program leading to this text. The support of the National Science Foundation and of the entire analytical chemistry faculty at Purdue University and the assistance of many analytical graduate students over the past few years for this same

[1] S. P. Perone, *J. Chem. Educ.*, **47**:105 (1970).

educational program have played a vital role. In particular, the authors are indebted to L. B. Rogers who suggested the whole project — many times — and who did not despair before we began to listen to him. Jon Amy provided invaluable assistance in obtaining the hardware necessary to support the educational program — back in 1968, when laboratory computers were a relative rarity and considerably more expensive than today. Harry Pardue collaborated with the authors in developing the course on which this text is based, and we are greatly indebted to his contributions to the success of that educational program. We are especially grateful for the truly dedicated contributions of each teaching assistant who has ever been involved in the computer-instrumentation courses. In particular, we must mention those five "pioneers" who spent the entire spring semester of 1968 learning enough computer technology to help put together and *teach* the first summer short course program. John Patterson, who has contributed Chapter 10 of this text, was a member of that pioneer group as a first-year graduate student. Another pioneer was Gerry Kirschner, who later assembled for the short course the first volume of reading materials from which this text eventually evolved. Perhaps the most indispensable person involved in the preparation of this text has been Connie Dowty, whose secretarial expertise included the ability to translate two different sets of often-misspelled, and occasionally incoherent, scratchings into nearly flawless prose.

Sam P. Perone
David O. Jones

1
Introduction to Computer Technology[1]

1-1 THE DIGITAL COMPUTER

Extensive, detailed descriptions of computer hardware and operations can be found in available texts [1-9]. A brief introduction to these topics will be presented here.

Figure 1-1 provides a block diagram of the essential components for a typical digital-computer configuration. (The reader is referred to Chap. 13 for a discussion of other possible architectures.) The *memory* is a component capable of storing many thousands of binary-coded (digital) *packets* of information. Each packet is composed of n binary digits (*bits*) and is called a *word* of information. Each of these n-bit words has an *address* associated with it, and the information contained within can be fetched or stored randomly by specification of the address. This internal memory is usually of the magnetic-ferrite-core type [10]. The information stored in memory can be of three types: data, instruction, or address. The speed with which information may be fetched or

---

[1] Taken in part from S. P. Perone, *J. Chromatogr. Sci.*, 7:714 (1969), by permission of the copyright owner.

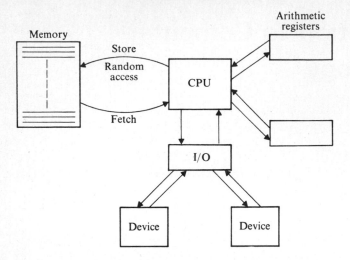

Fig. 1-1   Digital-computer configuration. *From S. P. Perone, J. Chromatogr. Sci., 7:714 (1969).*

stored in memory is the *memory cycle time* and is usually on the order of 1 μsec for core memory.

The *central processing unit* (CPU) is the workhorse of the computer. It controls the overall operation. Specifically, the CPU is made up of electronic registers and logic circuits which execute the simple logical and arithmetic operations of which the computer is capable. When the arithmetic and/or logical operations are executed in appropriate sequences, the computer can accomplish complex mathematical or data processing functions. Moreover, if one provides the appropriate electronic *interface*, these simple operations can be used to control experimental systems, acquire data, or print results on a Teletype printer, line printer, oscilloscope, or other peripheral device.

The sequence of instructions to be executed by the computer is called a *program*. In actuality, the program is a set of binary-coded instructions stored in memory. The CPU fetches each instruction from memory, interprets and executes it, and then moves on to the next instruction. The CPU fetches instructions sequentially from memory unless told to do otherwise by one of the instructions.

The *arithmetic registers* are high-speed electronic *accumulators* (ACs). That is, each is a set of n electronic two-state devices (like flip-flops), which can be used to accumulate intermediate results of binary arithmetic involving n-bit data. Nearly all the arithmetic and logical operations of the CPU are carried out on data contained in the arithmetic registers. Binary information can be transferred to or from memory and the arithmetic registers by the execution of appropriate machine instructions.

The *console switch register and displays* are used for instant communication between the operator and the computer. The *switch register* (SR) consists of an array of n switches and can be used to select an address in memory, to deposit binary instructions or data in memory or an arithmetic register, or to communicate directly with the central processor. The console display will usually provide a binary representation (n-bit word with 1 light per bit) of current contents of various registers within the computer, e.g., the memory reference register, the program counter, arithmetic registers, and any selected memory location.

The *I/O (input/output) bus* allows transfer of binary-coded information to/from peripheral devices. The particular device selected on input or output will be indicated by including a *select code*, which will direct the I/O data from/to the one appropriate device. The number of peripheral devices that can be connected to the I/O bus is limited primarily by the sophistication of the device-selection hardware and software. For small computers, as many as 64 peripheral devices may not be unrealistic. However, a single system usually includes less than 10.

A small digital computer (of the types currently being used for laboratory automation and experimental control) may have an instruction set which includes on the order of 50 to 100 different instructions. Each of these instructions corresponds to a specific binary coding which when decoded by the CPU results in the execution of a fairly simple arithmetic or logical step. Examples of some simple machine operations are binary addition of a datum in some memory location to the contents of an AC, the transfer of the contents of an AC to a memory location (and vice versa), rotation of the binary digits of the AC contents to the left or right, and the application of logical tests such as determining whether the AC is zero, nonzero, odd, even, positive, negative, etc.

Obviously, the repertoire of machine instructions includes some rather elementary operations. However, by developing programs composed of appropriate sequences of many of these elementary operations, the most sophisticated mathematical computations can be carried out. Since the computer can execute instructions so rapidly—on the order of 10^6 instructions per second—it can complete complex computations with fantastic speed. For example, a program to multiply two n-bit integer values might require a program including 50 or 60 statements (instructions plus data and address information) and may require 200 or 300 μsec for completion. Thus, 3,000 to 5,000 multiplications per second could be accomplished. (Although it is usually possible to incorporate hard-wired multiply and divide circuitry in the CPU requiring only a single instruction for operation and resulting in a completed multiply or divide computation in the order of a few microseconds, most standard computations like square root, exponentiation, logarithms, sine, cosine, etc., are accomplished by appropriate program segments.)

Thus, the digital computer is really a very simple-minded device, which must be told how to accomplish even the most fundamental computations but which can accomplish these operations with blinding speed. Moreover, it is a tireless machine, which will be content to calculate endlessly and consistently. It is also a very versatile device since it is programmable and capable of accomplishing an infinite variety of computational, logical, or control operations. Finally, it is a device which can (in fact, *must*) communicate in a variety of ways with the outside world. It is this characteristic which defines the computer as a general-purpose experimental device.

Programming the digital computer   The programming of a computer is usually accomplished with some sort of symbolic language. That is, readily recognized symbols are used to represent simple machine operations or groups of machine operations. Translating programs are supplied by the computer manufacturer to convert symbolic programs into the *binary-coded machine-language* programs. The simplest of these symbolic languages is the *assembly language*, where there is nearly a one-for-one conversion from symbolic statements to machine language. A program for translating these programs into machine language is called an *Assembler*. The relationship between assembly and machine languages is shown in Fig. 1-2.

Because programming in assembly language can become very tedious, higher-level languages have been developed where single statements can be translated into large blocks of machine-language program segments. Such a high-level language is FORTRAN, and the translating program is called a *Compiler*. The relationship to machine language is also shown in Fig. 1-2.

Obviously, it is much simpler to prepare programs in FORTRAN than in assembly language. However, Compiler-generated programs may be very inefficient in the utilization of available memory space. Moreover, speed of execution and synchronization of computations with outside events are relatively difficult to control with these programs. These considerations are particularly important for *on-line* computer applications in the laboratory. For these applications,

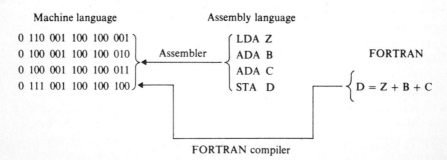

Machine language   Assembly language

0 110 001 100 100 001 ⎫   ⎧ LDA Z
0 100 001 100 100 010 ⎪ Assembler   ⎨ ADA B   FORTRAN
0 100 001 100 100 011 ⎬   ⎪ ADA C
0 111 001 100 100 100 ⎭◄─────   ⎩ STA D   ─── { D = Z + B + C

FORTRAN compiler

Fig. 1-2   Comparison of programming languages. *From S. P. Perone, J. Chromatogr. Sci., 7:714 (1969).*

Table 1-1 Computer size

Category	Word size	Core memory size (No. of words)	Bulk data storage	Standard peripherals	Cost
Small	8–18 bits	4K–8K	None	Teletype printer paper tape I/O	$10,000–$30,000
Medium	12–32 bits	8K–32K	Some (magnetic disc, tape, or drum)	Some (convenient I/O)	$30,000–$250,000
Large	≥32 bits	>32K	Extensive (multiple storage banks)	Extensive high-speed and convenient I/O	>$250,000

assembly-language programming must be used extensively so that the programmer can exercise the detailed control of computer operations required. Thus, the emphasis in this text is on assembly-language programming, and it is introduced in Chaps. 2 and 3.

Computer size   Digital computers are available in all sizes, shapes, colors, and prices. Computers are usually categorized with respect to size; although this is a fairly ambiguous term. Computer "size" has relatively little to do with physical dimensions in this day of microelectronics, but rather it is concerned with overall computer capability. Moreover, the classification should consider the complete system, including standard peripheral equipment. Thus, the pertinent criteria for computer size categorization should include word length (i.e., the number of bits used for each unit of binary information in the computer), core memory size, bulk data storage capacity, standard peripheral capability, and cost. Using these criteria, computer systems can be categorized as in Table 1-1.

There are, of course, many other characteristics of computer operation. Moreover, the manufacturing technology is changing so rapidly that the capabilities of what would have been called a large computer a few years ago we might now assign to medium or small systems. This trend seems to be continuing, and the rough lines drawn in Table 1-1 will be out of date probably by the time this text appears in print.

Off-line computers   The computer configuration with which most scientists are familiar is the *off-line* system. This configuration is diagramed in Fig. 1-3. To use the computer in this configuration, the scientist typically will write a data processing program in FORTRAN or some other high-level computer language, run the experiment(s), manually tabulate the data from the strip chart recorder or oscilloscope trace, transfer the tabulated data to punched cards, add the data cards to the deck of program cards, transport the combined card deck to the

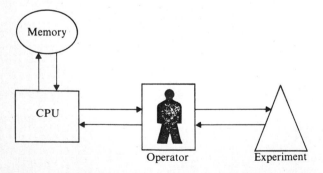

Fig. 1-3   Off-line computer configuration. *From S. P. Perone, J. Chem. Educ., 47:105 (1970).*

computer center for processing, and then wait until the program has been executed and the results printed. Turn-around times may vary from a few minutes to a few days, depending on the capacity of the computer facility, the number of users, and the backlog and priorities of work to be processed.

There are many variations of the above description of off-line computer usage. For example, the experimental system may include automatic data-acquisition and digitizing devices, which might store data on punched-card, punched-tape, or magnetic-tape buffers which can then be transmitted more conveniently to the remote computer center. Alternatively, the laboratory may be equipped with a remote terminal (such as a Teletype terminal) through which the investigator may enter his programs and data as well as receive a printout of results from the computer.

The important common characteristics of all off-line computer systems, however, are that the experimental data are transmitted to the computer through some intermediate storage medium, the data are processed after some finite time delay has occurred, and there is no direct feedback from computer to experiment. Depending on the modes of data acquisition and transmission to the computer, the turn-around time of the computer facility, and the speed with which the investigator can interpret the result printout, the time delay for experimental modifications based on the results of previous experiments can be excessive. Should this *reaction time* be a critical factor, an off-line computer facility may be inadequate for the particular experimental studies.

On-line computers For the investigator who requires very rapid or instantaneous results from his computer system—for whatever reason—the solution may be to employ an on-line computer system. Figure 1-4 presents a block diagram for a typical on-line computer configuration. The most important

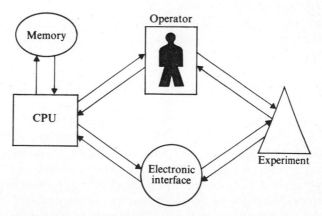

Fig. 1-4 On-line computer configuration. *From S. P. Perone, J. Chem. Educ., 47:105 (1970).*

distinction of this configuration is that there is a *direct* line of communication between the experiment and the computer. The line of communication is through an electronic interface. (This interface includes control logic, electronic elements to provide timing and synchronization, and conversion modules—such as digitizing devices—which translate real-world data into information which the digital computer recognizes.) Data are acquired under computer control or supervision, and the program for data processing is either resident in memory or immediately readable into memory to provide for very rapid completion of the computational tasks. Results may be made available to the investigator quickly by means of Teletype display, line-printer, oscilloscope, or other forms of printout. In addition, the computer may be programmed to communicate directly with the experiment by controlling electronic or electromechanical devices—such as solid-state switches, relays, stepping motors, servomotors—or any other devices which can be activated by voltage level changes.

The advantages of on-line computer operation can be summarized as the following: *elimination of the middle man* and the concomitant substitution of an electronic interface between the computer and the experimental system, which is much more compatible with the computer's characteristics and which does not suffer from the inherent inadequacies of the human as a communication link. For example, the computer can accept input data at rates on the order of 10^5 to 10^6 per second and can nearly instantaneously transmit control information or commands back to the experimental system. If this line of communication were handled through the human investigator, as in the typical off-line configuration, the response time of the overall experimental and data processing system would be many orders of magnitude greater. Moreover, the electronic communication link is generally a more reliable, tireless, objective, and accurate information transfer medium.

The possibility of *direct computer control of the experiment* is a second advantage associated with on-line computer operation. This facility allows the complex logical and decision-making capabilities of the computer to be implemented for an infinite variety of laboratory automation or experimental design problems.

A third advantage of the on-line computer is that *real-time interaction between computer and experiment* is possible. That is, because the computer can make computations and decisions at speeds exceeding most ordinary data-acquisition rates, it is possible for the computer to execute experimental control modifications *before* a given experiment has reached completion. These real-time operations allow the computer to be used to monitor the progress of an experiment and modify that experiment to follow a pathway which provides optimum conditions and which could only have been arrived at as a result of some observation on the initial course of the experiment.

A *breakdown of the logistic barriers of the remote computer system* is an additional advantage of on-line computer operation. Because of the direct communication link between the experiment and the computer, the mechanical and logistic roadblocks typically imposed by a computer facility toward the off-line introduction of data are irrelevant.

1-2 COMPUTER AUTOMATION FOR CHEMICAL ANALYSIS

Strictly speaking, computer automation of analytical instrumentation involves computer control of all phases of experimental operation. Specifically, these operations include sample introduction, scheduling, bookkeeping, experimental start/stop/control, data acquisition, data processing, and report preparation. However, the following discussions will pertain to any analytical system interfaced to a digital computer for on-line data acquisition, data processing, and some experimental control.

Initial steps The first step in providing computer automation for analytical instrumentation is the recognition that the services of an on-line computer are needed. Generally, the three most important reasons for developing an on-line system are (1) the need for very rapid response from the computer system, (2) the need for sophisticated experimental control operations, and (3) the desire to eliminate tedious, inefficient, manual operations in the analytical laboratory. These needs can be generated by an analytical laboratory with many instruments performing routine, repetitive functions, as well as by the research laboratory with a single challenging experimental application.

Once the decision has been made to incorporate the digital computer into laboratory instrumentation, the next problem faced is the decision of what type of system to choose. The choices range from a general-purpose small computer for dedicated operation with a single instrument to a large time-sharing computer servicing several smaller satellite computers and/or several different experimental systems while simultaneously executing off-line programs. Most applications will fall somewhere in between.

The most basic choice to be made is between a dedicated small computer system and time-sharing access to a larger computer system. The considerations involved in such a decision are too numerous and varied to be completely considered here; these are covered in detail in Chap. 10.

A third category of computer system, which includes both types mentioned above, is the *package* system. A package system may be purchased—including both hardware and software—which is designed to fit a specific laboratory situation. For example, several different packages are currently available for automation of multiple analytical gas chromatographic (GC) units in a given laboratory. (See Chap. 12.) These range from systems where one small computer is dedicated to each instrument to systems where a single medium-sized computer may service 10 or more chromatographs in a time-sharing fashion.

Other packages have been developed which will service nuclear magnetic resonance spectroscopy, mass spectrometry, microwave spectroscopy, and other large analytical instruments. Again, the package may provide for either dedicated or time-sharing operation. (See Chap. 12 for references to manufacturers currently furnishing a variety of package systems.)

Finally, some time-sharing packages exist which allow the user(s) to develop their own on-line service programs within the restrictions of the package time-sharing software. One distinct advantage is that higher-level languages may be used.

Operational problems Although many completed, successfully operating, automated systems now exist in analytical laboratories, many practical problems arise in the implementation of these systems or in the selection or design of new systems. For example, one of the most serious problems faced when a computer-automated analytical system (such as a group of gas chromatographs serviced by an on-line, multiple-input data processing computer) is put into operation is the conformity which the system must necessarily impose on the analytical investigations. That is, the package system must include specific restrictions and limitations on the type of input data which it can successfully handle and process. For example, with GC data the system may not handle adequately input data which include overlapping peaks or which have super-imposed noise in excess of some specific limit, where the range of peak sizes exceeds some particular bounds, where the experimental uncertainty in retention times exceeds some fixed limits, or where the data-acquisition rate must exceed some specific limit (e.g., 100 data points per second).

These limitations may be very reasonable and allow a great deal of latitude in the experiments processed by the automated system. However, for the experiment which lies outside those limits, the prospects for successfully using the automated system may be very slight. Certainly, it would be no small task to modify the computer-automated system to accommodate the "oddball" experiments. If the number and importance of such situations are small, the package system will provide very adequate automation of the bulk of the analytical investigations. However, if a laboratory finds that most of their routine analyses do not fall within the limitations specified by any of the commercially available package systems, that laboratory is faced with the decision of forgoing automation altogether, purchasing and modifying an existing package system which most nearly approaches their needs, or paying for the development of a completely new system which is exactly tailored to their needs. None of these alternatives is very desirable, and each can be very expensive.

The point of the preceding discussion is simply that it is virtually impossible to develop automated systems for laboratory instrumentation which will satisfy all users. Thus, it becomes necessary for each of the various laboratories to develop some in-house skills which will allow the intelligent modification of existing systems or the development of entirely new systems. This appears to be a fact of life with which many users must learn to live.

Another serious problem which the user must tackle is how the experimental data are to be transmitted. For example, should the digitizing devices be located at the computer facility, with analog experimental data transmitted from remote stations for digitization? Or should the data be digitized at the experimental source and transmitted digitally? This question is generally resolved on the basis of the frequency and level of the information to be transmitted. That is, for data in which the dominant frequency is somewhat less than 1 Hz, it is possible with filtering to transmit millivolt-level analog signals over long distances through well-shielded cable. If the signal was first amplified to a level in the range of a few volts to minimize problems of electronic pickup, analog signals up to about 1 kHz may be transmitted.

One of the most vital, and yet most subtle, facts which must be recognized when placing laboratory instrumentation on-line is that the introduction of digitized data into the computer does not automatically lead to more satisfactory data processing. Regardless of the fidelity and dynamic range of digitizing devices, the digital data can be no more accurate nor more precise than the original analog signal. In fact, a significant *loss* of information can occur if the digitization rate is not sufficiently rapid. Moreover, high-frequency background noise from the instrument, which might not even be observed with readout on a potentiometric recorder, may show up strongly in data obtained with a high-speed analog-to-digital converter. Thus, not only must a user be concerned with establishing appropriate data-acquisition parameters for his experiments, but he may also be faced with the task of "cleaning up" his analog instrumentation to provide useful digitized signals.

1-3 OBSERVATIONS

The most important observation that one should make here is that the digital computer with appropriate interfacing to laboratory instrumentation is really a very versatile, general-purpose device for scientific investigations. The incorporation of this device in a laboratory environment will soon be as common as the current usage of potentiometric recorders. Moreover, the on-line computer opens up whole new vistas for experimental measurements. Not only can tedious, time-consuming, routine tasks be automated, but more importantly the full instrumental capabilities of the digital computer can be focused on a single, challenging, nonroutine measurement problem.

One very significant question raised by the preceding discussions is, "How can the practicing scientist who is unskilled in computer technology use a laboratory computer in his own experimental work?" The answer to this question depends on the type of work involved. If what is desired is the automation of routine laboratory measurement functions, it may be possible to purchase a package system requiring no technological contribution from the scientist except for a statement of specifications. On the other hand, if the scientist requires digital-computer instrumentation for nonroutine experimental problems, he will be forced to develop some skill in computer technology. At the very least, he must be able to communicate with a programmer and an electronics specialist so that the specific experimental tasks can be implemented. In this instance, it is very important that the scientist have some detailed awareness of what can and cannot be done computationally with the digital computer, of the problems of interfacing, of the intricacies of the machine language, and of the timing and synchronization problems associated with computer communications.

In other words, the extent to which the scientist takes advantage of the digital computer's capabilities in his work depends on his mastery of the technology involved.

BIBLIOGRAPHY

1. Chu, Y.: "Digital Computer Design Fundamentals," McGraw-Hill, New York, 1962.
2. Nashelsky, L.: "Digital Computer Theory," Wiley, New York, 1966.
3. Flores, J.: "Computer Design," Prentice-Hall, Englewood Cliffs, N.J., 1967.
4. Baron, R. C., and A. T. Piccirilli: "Digital Logic and Computer Operations," McGraw-Hill, New York, 1967.
5. Hellermen, H.: "Digital Computer System Principles," McGraw-Hill, New York, 1967.
6. Rosen, S. (ed.): "Programming Systems and Languages," McGraw-Hill, New York, 1967.
7. Wegner, P.: "Programming Languages, Information Structures, and Machine Organization," chaps. 1 and 2, McGraw-Hill, New York, 1968.
8. Knuth, D. E.: "Art of Computer Programming," vol. 1, chap. 1, Addison-Wesley, Reading, Mass., 1968.
9. Gear, C. W.: "Computer Organization and Programming," chaps. 2 and 3, McGraw-Hill, New York, 1969.
10. Nashelsky, op. cit., p. 238 ff.

2
Digital-Computer Number Systems, Machine Language, and Assembly Language

2-1 BINARY, OCTAL NUMBER SYSTEMS

All information handled or generated by the central processing unit (CPU) must be binary or binary-coded. (Refer to Chap. 1.) This includes instructions, memory addresses, and data. Thus, the small-computer user must quickly become familiar with this number system. It would be well to review here the binary number system and binary arithmetic.

The decimal number 369_{10} can be broken down into $3 \times 10^2 + 6 \times 10^1 + 9 \times 10^0$. Similarly, the binary number 10101 represents $2^4 + 2^2 + 2^0$ ($= 21_{10}$). For large binary numbers, e.g., 101101110010101, it becomes convenient to represent them in a shorthand fashion for easy recall or reference. The usual shorthand reference to binary numbers is the *octal* system. The applicability of the octal system can be seen from the binary representation of the numbers 0 to 7:

000	0	011	3	110	6
001	1	100	4	111	7
010	2	101	5		

This sequence illustrates the normal binary counting sequence, which can be extended to an infinite number of binary digits (bits). It also shows the octal digits equivalent to all 3-bit combinations. To convert any large binary number to octal, group binary digits in groups of three, starting at the rightmost digit, e.g.,

$$\underbrace{..1}_{1} \quad \underbrace{001}_{1} \quad \underbrace{101}_{5} \qquad = 115_8$$

$$\underbrace{101}_{5} \quad \underbrace{101}_{5} \quad \underbrace{110}_{6} \quad \underbrace{010}_{2} \quad \underbrace{101}_{5} = 55625_8$$

$$201_8 = 010 \quad 000 \quad 001$$
$$356_8 = 011 \quad 101 \quad 110$$

Counting in octal: 1, 2, 3, 4, 5, 6, 7, 10, 11, 12, 13, 14, 15, 16, 17, 20, 21, 22, 23, . . . , 75 76, 77, 100, 101, 102, . . . , 766, 777, 1000, 1001, . . . , 7776, 7777, 10000, . . .

Often it is necessary to convert numbers from one base to another. The above examples illustrate the ease of converting from octal to binary and vice versa. Consider the following examples:

DECIMAL-TO-BINARY CONVERSION

For example, $876_{10} = ?$ in binary?

$$
\begin{array}{rl}
876 & \\
-512 & \quad 512 = 2^9 = \text{largest power of 2 to fit in } 876_{10} \\
\hline
364 & \\
-256 & \quad 256 = 2^8 = \text{largest power of 2 to fit in } 364_{10} \\
\hline
108 & \\
-\ 64 & \quad 64 = 2^6 = \text{largest power of 2 to fit in } 108_{10} \\
\hline
44 & \\
-\ 32 & \quad 32 = 2^5 \\
\hline
12 & \\
-\ 8 & \quad 8 = 2^3 \\
\hline
4 & \\
-\ 4 & \quad 4 = 2^2 \\
\hline
0 &
\end{array}
$$

Thus, the binary representation of 876_{10} must include 2^9, 2^8, 2^6, 2^5, 2^3, and 2^2, and the binary number is

$$
\begin{array}{ccccc}
1 & 101 & 101 & 100 \\
2^9 & 2^8 & 2^6 & 2^5 \ 2^3 \ 2^2
\end{array}
$$

The octal equivalent of the binary number is 1554_8.

An alternative conversion approach is shown below. It involves repetitively dividing by the new base 2.

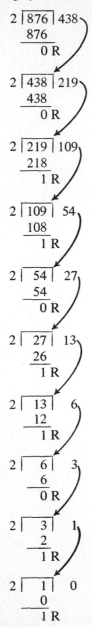

$$2\,\overline{|\,876\,|}\,438$$
$$\underline{876}$$
$$0\,R$$

$$2\,\overline{|\,438\,|}\,219$$
$$\underline{438}$$
$$0\,R$$

$$2\,\overline{|\,219\,|}\,109$$
$$\underline{218}$$
$$1\,R$$

$$2\,\overline{|\,109\,|}\,54$$
$$\underline{108}$$
$$1\,R$$

$$2\,\overline{|\,54\,|}\,27$$
$$\underline{54}$$
$$0\,R$$

$$2\,\overline{|\,27\,|}\,13$$
$$\underline{26}$$
$$1\,R$$

$$2\,\overline{|\,13\,|}\,6$$
$$\underline{12}$$
$$1\,R$$

$$2\,\overline{|\,6\,|}\,3$$
$$\underline{6}$$
$$0\,R$$

$$2\,\overline{|\,3\,|}\,1$$
$$\underline{2}$$
$$1\,R$$

$$2\,\overline{|\,1\,|}\,0$$
$$\underline{0}$$
$$1\,R$$

Taking the *remainders* in reverse order generates the binary number 1 101 101 100. Note that the result is the same as above.

DECIMAL-TO-OCTAL CONVERSION

For example, 576_{10} = ? in octal?

$$8_{10} \overline{\smash{\big)}576_{10}} \; 72$$
$$\underline{56}$$
$$16$$
$$\underline{16}$$
$$0 \; R$$

$$8 \overline{\smash{\big)}72} \quad 9$$
$$\underline{72}$$
$$0 \; R$$

$$8 \overline{\smash{\big)}9} \quad 1$$
$$\underline{8}$$
$$1 \; R$$

$$8 \overline{\smash{\big)}1} \quad 0$$
$$\underline{0}$$
$$1 \; R$$

Taking the remainders in reverse order gives 1100_8 = 576_{10}.

OCTAL-TO-DECIMAL CONVERSION

For example, 1000_8 = ? in decimal? (*Note*: Multiplication, division, addition, and subtraction are done in the base of the original number.)

$$12_8 \overline{\smash{\big)}1000_8} \; 63$$
$$\underline{74}$$
$$40$$
$$\underline{36}$$
$$2 \; R$$

$$12_8 \overline{\smash{\big)}63_8} \quad 5$$
$$\underline{62}$$
$$1 \; R$$

$$12_8 \overline{\smash{\big)}5_8} \quad 0$$
$$\underline{0}$$
$$5 \; R$$

Taking the remainders in reverse order gives 512_{10} = 1000_8.

Conversion tables for binary, octal, and decimal values within generally useful ranges are presented in Appendix A. However, the computer user must quickly develop the ability to convert binary to octal, and vice versa.

2-2 BINARY ARITHMETIC

Because all the arithmetic operations in the computer are carried out in binary, we should review those basic operations here.

Addition

```
    1010          1010          111011
 + 0100        + 1010        + 001111
 _____        _____        _____
   1110         10100         1001010
```

Note: Each binary digit is referred to as a bit (contraction of "binary digit"). Thus, the examples above yield 4-bit, 5-bit, and 7-bit answers.

Subtraction The subtraction operation is carried out by negation followed by addition. Thus, the question arises as to how a negative binary number is represented since the computer hardware has no provision for "+" and "−" signs. To accomplish this, *complementary arithmetic* is used. For example, the value 5 is represented as a 6-bit binary word 000101. One way we could represent −5 would be as the *1's complement* of the 6-bit word—where the complement is generated by replacing 1's with 0's, and vice versa. Thus,

$$-5 = 111010 \text{ (1's complement)}$$

However, if we add together the binary number for 5 and the 1's complement of 5, which we would like to correspond to subtracting 5 from 5, we get

```
   000101
 + 111010
 _____
   111111
```

However, if we complement this result, we get all 0's; therefore, 111111 must be −0 in 1's-complement notation!

If we subtract 7 from 5 using 1's-complement notation we get

```
   000101   (+5)
   111000   (−7)
 _____
   111101
```

which is an acceptable answer. That is, the 1's complement of 111101 = 000010 = 2. But if we subtract 5 from 7, we get

```
   000111   (+7)
   111010   (−5)
 _____
(1)  000001
```

which is *not* correct. However, if the bit carried out at the left can be carried around and added to the result, the correct answer is generated:

(1)   000001 → 000010

Thus, an *around-end carry* is needed to get the correct answer when using 1's-complement arithmetic for binary subtraction. If the computer hardware can tolerate −0 and can generate the around-end carry, 1's-complement arithmetic is perfectly adequate. However, there is a better approach, which is used by most computer manufacturers. That is to use *2's-complement* notation.

Two's-complement arithmetic involves, essentially, generating the around-end carry first. That is, the negative of a binary number is represented by *complementing* and *incrementing* the number. The result is the 2's complement. That is, one has, for instance,

$$
\begin{array}{ll}
+5 \rightarrow 000101 & \\
 111010 & \text{(complement)} \\
+ \quad\quad\ 1 & \text{(increment)} \\
\hline
 111011 & (-5 \text{ in 2's-complement notation)}
\end{array}
$$

Examples using 2's-complement arithmetic are

$$
\begin{array}{llll}
000101 & (+5) & & 000101 & (+5) \\
111011 & (-5) & \text{and} & 111001 & (-7) \\
\hline
000000 & & & 111110 & (-2)
\end{array}
$$

To check this last result, complement and increment to negate the answer:

$$
\begin{array}{ll}
111110 & (-2) \\
000001 & \text{(complement)} \\
+ \quad\quad\ 1 & \text{(increment)} \\
\hline
000010 & (+2)
\end{array}
$$

Thus, the result is correct in 2's-complement notation.

Note that if we always use one more bit than we need to represent the largest number we want, the most significant bit (leftmost bit) will always be a 1 if the number is negative, and 0 if the number is positive. We can use it as the *sign bit*. Thus, 2's-complement arithmetic is extremely useful and efficient for computer implementation. This is the arithmetic method which will be used in all future discussion here.

Binary multiplication   Binary multiplication is carried out in conventional fashion, e.g.,

$$
\begin{array}{r}
10101 \\
\times \quad 10100 \\
\hline
1010100 \\
101010000 \\
\hline
110100100
\end{array}
$$

also,

$$10101 \quad \times 2^3 \ = 10101000$$

$$101010 \times 2^{-1} = 10101$$

Thus, multiplication can be generated by the appropriate combination of *shifting* and *adding*, each of which the computer can handle. Also, note that any multiplication by a power of 2 can be carried out simply by shifting bits left or right.

Binary division also can be carried out with a familiar procedure:

$$
\begin{array}{r}
1011 \quad \text{(quotient)} \\
100 \overline{)\, 101101} \\
100 \\
\hline
110 \\
100 \\
\hline
101 \\
100 \\
\hline
1 \quad \text{(remainder)}
\end{array}
$$

Thus, computer division can be generated by shifting and subtracting binary values. A detailed consideration of binary multiplication and division programs is given in Chap. 5.

2-3 MACHINE-LANGUAGE PROGRAMMING

In order to describe the fundamentals of machine-language computer programming, it will be necessary, first, to define the computer system with which we will be working. The computer system selected as a model for all further specific discussions is the Hewlett-Packard 2100 family (2116A, B, C; 2115A; 2114A, B; 2100). This system has characteristics representing a composite picture of many of the small and medium-sized computers currently available. Thus, most of the discussions will be readily transferable to the particular computer system with which the reader is involved. In those cases where large fundamental differences in machine philosophy exist (such as for I/O hardware), alternative design characteristics will be discussed. (See Chap. 13 and Appendix D.)

Our typical computer system has the following basic characteristics (refer to the Glossary for definitions of unfamiliar terms):

HARDWARE

16-bit word length
8,192-word core memory
Two 16-bit arithmetic registers [accumulators (ACs)] —Labeled A and B
One 1-bit Link, Extend, or Carry register
2.0-μsec complete read/write memory cycle time
Multichannel hardware I/O
Multichannel hardware priority interrupt

CONSOLE FEATURES

Displays

Memory display (16 bit)
Memory address register display (16 bit)
Dual AC display (16 bits each)
Note: Whenever referring to a 16-bit word for this computer system, the
following format is used for numbering each bit:

    15 14 13 12 ... 2   1   0
     1   1   1   1 ... 1   1   1

Controls

Power on/off
Start (program initiation)
Halt (program stop)
Single cycle (program execution one cycle at a time)

Switch register

16-bit register (used in conjunction with input features below)

Input

Load AC: load the binary value contained in the switch register (SR) into the
    A or B accumulator
Load address: load the SR into the memory address register
Load memory: load the SR into the memory location whose address is in the
    memory address register
Display memory: display the contents of the memory location whose address
    is in the address register

STANDARD PERIPHERAL DEVICES FOR INPUT/OUTPUT

Teletype terminal
High-speed paper-tape reader
High-speed paper-tape punch

We will not be concerned with all the computer's basic characteristics at first. (For example, we will not consider using both ACs in the early discussion.) In later chapters, however, all these features will be incorporated into the discussions. Moreover, some optional features peculiar to a wide variety of available computer systems will be discussed.

MACHINE LANGUAGE

It is appropriate at this point to consider the machine language associated with our typical computer. (A complete summary and description of this language is given in Appendix A. Also, the reader is referred to "A Pocket Guide for Hewlett-Packard Computers," Hewlett-Packard, Cupertino, Calif.) A *machine language* includes that set of binary-coded instruction words which when interpreted by the CPU cause the computer to execute the various arithmetic and logical operations of which it is capable. A computer can be programmed by storing an appropriate sequence of these machine-language instruction words in memory and directing the processor to fetch, interpret, and execute these instructions. The computer will start where told and continue to fetch and execute instructions *sequentially* in memory unless told otherwise by one of the instructions in the sequence.

Each of the $8,192_{10}$ core memory locations has a specific address varying from 0 to $8,192_{10}$ or from 0 to 17777_8. Octal representation of memory addresses, because of its simple conversion to binary, is preferred since console specification of memory addressing is done in binary and the memory address register is displayed in binary. The address of a memory location might be included as part of a machine-language instruction which references a particular location, or it might be stored in some other memory location, where the program can fetch it to obtain address information.

Thus, we see that the computer can contain three types of binary-coded information in memory or the working registers—*instruction* words, *data* words, or *address* words. Ordinarily, the computer differentiates between the three by the *context* in which it accesses a given memory location or working register.

Example 2-1 illustrates machine-language programming.

Although the information stored in locations 1441_8 to 1443_8 is intended to be data, the computer does not know a priori that they are data words. If the program were started at location 1441_8, for example, the CPU would attempt to interpret at least the first datum as an instruction. What happens after that is indeterminate. The specific value stored originally in location 1500_8 is irrelevant because it gets wiped out when the final sum is stored in that memory location.

MNEMONIC STATEMENTS

The program as written above could be loaded into memory through use of the SR and load-memory console feature. It could then be executed to carry out the

Example 2-1   Add three numbers, and store the sum in memory location 1500_8. Start the program at memory location 200_8.

Memory location	/ ₃	Contents	Computer interpretation of binary information
200_8		0 110 001 100 100 001 (061441_8)	Instruction: load AC with contents of location 1441_8
201_8		0 100 001 100 100 010 (041442_8)	Instruction: add contents of location 1442_8 to contents of AC; result in AC
202_8		0 100 001 100 100 011 (041443_8)	Instruction: add contents of location 1443_8 to contents of AC; result in AC
203_8		0 111 001 101 000 000 (071500_8)	Instruction: store contents of AC in location 1500_8
204_8		1 000 010 000 000 000 (102000_8)	Instruction: halt, stop program execution
1441_8		0 000 000 000 001 000 (000010_8)	Data
1442_8		0 000 000 001 000 000 (000100_8)	Data
1443_8		0 000 001 000 000 000 (001000_8)	Data
1500_8		– – – – – – before – – – – – – 0 000 001 001 001 000 – – – – – – after – – – – – –	Storage location

Note: For the instructions stored in locations 200_8 to 203_8, the rightmost (least significant) 10 bits provide the address of the memory location containing the piece of data to be fetched. These are called *memory reference instructions*. The operation executed in these particular instructions is determined by the coding of the leftmost (most significant) 6 bits. The Load Accumulator instruction, for example, has bits 14 and 13 set; whereas the Store Accumulator instruction has bits 14 to 12 set. (Remember that the bits are numbered 0 to 15 from right to left. Also note that we will only be considering using one AC corresponding to the A register of our computer.) Other instructions use other bits to determine the operation performed. For example, the instruction word stored at location 204_8 is decoded by the CPU and recognized as *not* being a memory reference instruction because bits 14 to 12 are clear 0's. The unique combination of the other bits identifies it as a Halt instruction.

desired operation. However, it would be extremely tedious to program a computer by employing binary-coded machine instructions. A more efficient programming approach involves using a symbolic language which can be translated automatically into the computer's machine language. Such a language might use *mnemonic* symbols to represent machine-language instructions. The mnemonic symbols can be brief three- or four-letter combinations designed to recollect the machine operation for which they stand. For example, the machine instruction 102000_8 stops program execution and can be represented by the mnemonic symbol HLT. The instruction 061441_8 directs the CPU to fetch the information in memory location 1441_8 and load it into the AC; this can be represented by the mnemonic statement LDA   1441. The other instructions in Example 2-1 can be represented similarly as shown in Example 2-2. In addition, memory addresses referred to in program statements may be represented symbolically; e.g., location 1441_8 may be referred to as location NUM1, 1442_8 as location NUM2, 1443_8 as location NUM3, and 1500_8 as location SUM. The obvious advantage of symbolic addressing is that one need not keep track of location numbers. These modifications allow the simplified program statements of Example 2-2.

Example 2-2   Add numbers and store them in location 1500_8. Use mnemonic program statements.

Memory location (octal)	Mnemonic statement	Memory contents (octal)
200	LDA NUM1	061441
201	ADA NUM2	041442
202	ADA NUM3	041443
203	STA SUM	071500
204	HLT	102000
1441	NUM1 10	10
1442	NUM2 100	100
1443	NUM3 1000	1000
1500	SUM 0	0
		(before execution)

Note the distinction between an LDA operation and an ADA operation in Example 2-2. LDA causes the previous contents of the AC to be *replaced* by the contents of the memory location referenced; whereas the ADA instruction causes the contents of the referenced memory location to be *added* to the contents of the AC. Also note that the initial content of location SUM is not critical since the STA instruction causes the previous contents of the memory location to be *replaced* by the current contents of the AC. Thus, the statement SUM   0 is used only to set aside a memory location in which information is stored by the program.

A complete list of the machine-language instruction set by the correspond-ing mnemonics for our computer system is given in Appendix A. We will eventually define and incorporate all the machine's instructions into our programming discussions in subsequent chapters.

These mnemonic representations of machine-language instructions are much more convenient to use than the basic binary coding. Thus, the computer manufacturer provides a program called the *Assembler* which will interpret programs written in terms of well-defined mnemonic instructions and translate these into a binary-coded machine-language program. The set of mnemonic instructions is called the assembly language. The assembly-language program provides about a one-for-one correspondence between mnemonic statements and machine-language statements. For this reason, the two terms are often used (incorrectly) interchangeably.

2-4   ASSEMBLY-LANGUAGE PROGRAMMING

The most complex numerical computations can be carried out with the appropriate combination of simple machine instructions. The following examples will illustrate the use of the assembly language to accomplish simple computational tasks.

Program to clear a block of core memory   Often it is desired to set the contents of each location in a block of core memory to 0. The short assembly-language program of Example 2-3 can be written to perform this task. Assume that the block of memory to be cleared includes locations 100_8 to 140_8, or a total of 41_8 locations.

Example 2-3   Direct approach to Clear Core program

Statements	*Comments*
ORG 200B	/STARTING ADRSS = 200 (OCT.)
CLA	/CLR ACCUMLATOR (AC)
STA 100B	/STORE AC IN LOCNS 100 to 140 (OCT.)
STA 101B	
STA 102B	
.	
.	
.	
STA 140B	
HLT	
END	/END OF PROGRM

In the program of Example 2-3 note the introduction of two new types of statements: the origin statement ORG   200B, which specifies the starting address of the machine-language program to be stored in memory, and the END statement, which indicates that there are no more statements in the program. Each of these statements provides necessary information to the Assembler program for proper assembling into binary-coded machine language and the eventual storage of the program in core memory, but they do not result in the generation of corresponding machine instructions and do not, therefore, use up core memory. Thus, we will not include these types of Assembler instruction statements as program statements when specifying program length. (*Note*: Numerical values which are to be interpreted as *octal* are tagged with a B. If the B is absent, the number is interpreted as *decimal* by the Assembler.)

The execution of the above program (after it has been converted to machine language and stored in memory) will indeed result in setting the contents of locations 100_8 to 140_8 to 0's. Note, however, that 35 programming statements are required, occupying 35 memory locations.

Generating the machine-language program   It is instructive at this point to consider the mechanics by which a machine-language program might be generated. For this purpose, included below are some detailed instructions which pertain to the particular computer system defined here. The details would be similar for other systems using paper-tape I/O.

Machine-language programming through the switch register   The program given in Example 2-3 could be placed into core memory by manually converting it to machine-language (binary) coding and entering it instruction by instruction through the SR. The operations involved are these: (1) Set the SR to the binary pattern corresponding to the address of the first location of the program (location 200_8). The pattern will be 0   000   000   010   000   000. (2) Press the LOAD ADDRESS button on the console front. This sets the memory address register to the value set on the SR. (3) Set the SR to the appropriate binary pattern for the first instruction of the program (2400_8 or 0   000   010   100   000   000). (4) Press the LOAD MEMORY button on the console front. The SR value will be deposited in memory location 200_8, and the memory address register *automatically* increments itself to the next address value 201_8. (5) Set the SR to the appropriate binary pattern for the next sequential instruction of the program (072100_8 or 0   111   010   001   000   000). Press the LOAD MEMORY button. (6) Continue until all instructions are stored in memory. (7) Set the SR to the value of the address of the first location of the program, location 200_8, press LOAD ADDRESS, and then press the RUN button. The program will execute.

Machine-language programming using the Assembler program   The Clear Core program can also be generated in core memory by using the more convenient programming and I/O facilities available. (One might argue that for a short program like the Clear Core program the most convenient procedure is to "toggle" the program into core memory as discussed above. However, for longer and more involved programs, such a procedure would be extremely tedious.)

The most direct conversion of programming concepts to machine language is provided by assembly-language programming. This involves writing programs with mnemonic symbols representing machine-language instructions. The Assembler is a program which looks at the mnemonic-coded program and translates it into the appropriate machine-language (binary-coded) statements. The mnemonic-coded program can be prepared initially by typing the program on a Teletype terminal in the LOCAL mode, with the paper-tape punch on. The resultant punched paper-tape segment will contain the mnemonic-coded assembly-language program, with 8-bit ASCII[1] coding representing each alphanumeric character typed.

The mechanism by which the generation of a binary-coded machine-language program is then accomplished is summarized here: (1) The Assembler program—which is provided as a binary-coded machine-language program recorded on punched paper tape—is first read into the computer's memory, using the high-speed paper-tape reader and the Basic Binary Loader[2]. (2) The mnemonic-coded program prepared earlier on punched paper tape (called the *source tape*) is placed in the high-speed paper-tape reader. (3) The Assembler program is started by loading the appropriate starting address and pressing RUN. (4) The Assembler program will then cause the source tape to read in, and it will read the whole tape through once, looking simply for symbolic address information and checking for format errors. The Assembler will detect any format errors or illegal symbols on this first pass and will output diagnostic messages on the Teletype terminal. When the first pass is completed, the Assembler will print out all symbolic addresses and their assigned locations. (5) The source tape is again loaded in the high-speed tape reader, and a second pass is initiated. On this second pass, the Assembler will translate the mnemonic-coded program into binary-coded machine language and punch out the binary-coded program on paper tape. This tape will include all required address information for allocation of the machine-language statements in memory. (6) The binary-coded tape produced by the Assembler can now be loaded into core memory through the high-speed tape reader using the Basic Binary Loader program, just as the Assembler program was loaded initially. (*Note*: This is true only for *absolute* programs, i.e., those where the programmer specifies the absolute location of all parts of the program in core memory by statements within his program. These are the only types of programs we will be discussing here.)

---

[1] ASCII stands for American Standard Code for Information Interchange.
[2] Basic Binary Loader is a core-resident utility program supplied by the manufacturer which, upon execution, reads binary-coded punched paper tapes into the appropriate areas of core memory. In the computer system defined here, this program resides in the last 64_{10} memory locations.

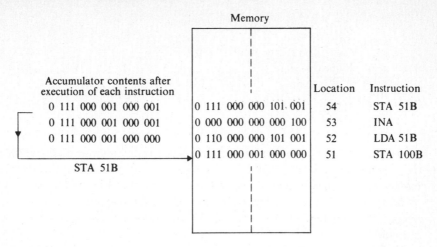

Fig. 2-1   Address modification by incrementing a memory reference instruction.

Address modification approach for Clear Core program   Example 2-3 is not very efficient in terms of amount of coding required. That is, if a different instruction is required for every time a different memory location is accessed during the execution of a program, core memory will be filled to overflowing with instructions, and there will be no room for data storage. Thus, alternative, more efficient techniques for addressing must be used.

   One alternative approach involves taking advantage of the fact that memory reference statements generate, as part of the binary-coded instruction, the particular address to be referenced. In our computer the address is contained in the last (least significant) 10 bits. (Refer to Fig. 2-1.) Thus, the address portion of the instruction word can be altered by arithmetic operations on the instruction itself. For example, Fig. 2-1 illustrates how one might take into the AC the contents of location 51_8, which is originally the binary-coded instruction 0   111   000   001   000   000 or STA   100B, increment that binary value with an INA (increment accumulator) instruction, and then store the modified instruction back in location 51_8, where it will now be executed as STA   101B. If the instruction is then reexecuted, location 101_8 will be cleared. However, a way of returning to location 51_8 is needed. That is, the normal sequential access of instructions from memory must be altered. To accomplish this, another type of memory reference instruction—the JMP instruction—is used. The JMP instruction can transfer program control to an address specified in the address portion of the instruction. Thus, a particular section of a program may be executed over and over again by providing a JMP statement to repetitively return to the origin of the program segment.

   These repetitively executed segments are called *loops*. For example, the loop of Example 2-4 might be written.

Example 2-4   Program segment with loop

```
ORG   50B
CLA
STA   100B
LDA   51B
INA
STA   51B
JMP   50B
```

The program segment of Example 2-4 illustrates address modification and a program loop, which can be combined for more efficient programming. However, the program segment, as written, will not do what we want since there is no provision for exiting from the loop. Thus, the computer will continue to execute the loop beyond the desired point where the core memory block is to be cleared.

To provide an exit from the program loop, a test must be made during each loop to determine if it has been executed a sufficient number of times. One convenient test can be generated by using the ISZ memory reference instruction. The ISZ (increment, skip if zero) instruction can cause the contents of a memory location to be incremented and then test to see if it has been set to 0; if it is 0, the next sequential instruction is skipped. Thus, some memory location, which we will call CNTR, can be set to contain, initially, *negative* the number of times a given loop is to be executed. Within the loop, just before the JMP instruction, an ISZ CNTR instruction can be included. And when the loop has been executed the specified number of times, CNTR will have been set to 0, the ISZ instruction will cause the JMP instruction to be skipped, and the loop will be exited. (Refer to Example 2-5.)

In addition to providing a means of counting out of a loop, the ISZ instruction can be used simply to add 1 to the contents of any given memory location. For example, it may be used to provide a more direct means of address modification, or it might be used to accumulate a count of the number of times a particular event occurs. Usually in these applications, the "test" part of the instruction is not desired. However, as long as the contents of the memory location is never incremented to 0, the instruction can be used as if it were a simple increment instruction. This is always a potential source of error in a program, and the programmer should never forget that he is not using a simple increment instruction.[1]

If one uses the instructions and techniques just described, the Clear Core program example can be rewritten as is in Example 2-5.

The program of Example 2-5 will work. However, it represents a potentially dangerous approach since arithmetic operations on an instruction

---

[1] Some machines do provide a simple increment memory reference instruction, but it is not common in the small digital computers.

Example 2-5 Clear Core Program with address modification and loop

```
        ORG     50B
        CLA
LOOP    STA     100B
        ISZ     LOOP
        ISZ     CNTR
        JMP     LOOP
        HLT
CNTR    DEC     -33
        END
```

Note: The loop must be executed 33_{10} times in order to clear locations 100_8 to 140_8. Also, a DEC statement is used for any data which the programmer wants the Assembler to interpret as decimal. Octal data are specified with an OCT statement.

might inadvertently modify the sense of the instruction. This is particularly true if the altered reference crosses a *page* boundary, as discussed below.

Memory reference limitations The memory reference instruction has only 10 bits available for address specification. (See Example 2-1.) Thus, the maximum number of memory locations which can be referenced directly is $1,024_{10}$, and the value which the address portion can assume varies from 0 to $1,023_{10}$. With $8,192_{10}$ (or greater) core memory locations, some additional information is necessary to allow access to all locations. (The specific approaches used by the various small-computer manufacturers for accomplishing this vary, and some discussion is provided in Appendix D.) For the computer system defined here, memory address specification is accomplished by arbitrarily organizing memory into pages of $1,024_{10}$ locations each. An $8,192_{10}$-word core memory, then, contains eight pages of $1,024_{10}$ locations each. Thus, the 10-bit address portion of a memory reference instruction could be interpreted by the Assembler as referring to a location found on the same page as the instruction itself is located.

Since only 4 bits of our memory reference instruction are used to generate the *op code*, two more bits could be used for address specification, raising to $4,096_{10}$ the total number of locations which could be addressed directly. However, this approach would effectively confine programs and associated data to $4,096_{10}$ words since nothing outside that block could be referenced.

A much better use of those two remaining bits is to use them to generate a more versatile addressing scheme, which is applicable to larger core memories. One bit can be used to specify whether the address referred to is to be found on the same, or *current*, page or on some other page—usually the first, or *zero*, page. Thus, the zero page could be referred to directly from anywhere in core memory, providing a useful common storage area.

The second remaining bit in the memory reference instruction can be used to specify whether addressing is to be *direct* or *indirect*. Indirect addressing instructs the CPU to fetch address information from the location whose address is referred to in the instruction and then go to the location whose address is fetched and carry out the indicated operation (such as LDA, JMP, ISZ, etc.). A 15-bit address can be stored in a given location. Thus, by a two-step indirect addressing sequence, up to $32,768_{10}$ core locations can be referenced. (The sixteenth bit in an address word is used for *multilevel* indirect addressing. That is, if bit 15 in an address word is set to 1, the memory location referred to by the lower 15 bits is interpreted as containing another address word. This indirect addressing sequence can continue indefinitely until an address word is found without bit 15 set.)

The illustration in Fig. 2-2 shows the complete assignment of bit information in a memory reference instruction. The programmer specifies whether addressing is to be direct or indirect in the memory reference statement as shown in Example 2-6. He also specifies the particular page to be addressed by specifying somewhere within the program where named locations reside in core memory. If the program should call for direct addressing of a location not on the zero or current pages, the Assembler cannot generate a workable code and, therefore, outputs an error diagnostic indicating an illegal reference. If the memory reference statement refers to a number address less than $1,024_{10}$, the Assembler assumes it is on the zero page.

The example of Figs. 2-3 and 2-4 illustrates how illegal references can be generated and how they can be corrected. Because the hypothetical program in

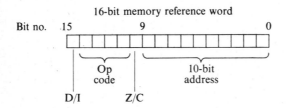

D/I = 0/1 = direct or indirect addressing
Z/C = 0/1 = zero or current page address

Some op codes	4-bit value
ADA	100 0
LDA	110 0
STA	111 0
JSB	001 1
JMP	010 1
ISZ	011 1
AND	001 0

Fig. 2-2 Construction of a 16-bit memory reference instruction.

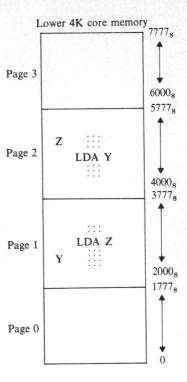

Fig. 2-3 Program containing two illegal references.

Fig. 2-3 has crossed a page boundary, two illegal references have developed. The programmer has three choices: (1) Revise the program so that page boundaries are never crossed. (2) Put locations Y, Z, and any others which may be referred to from across page boundaries on the zero page. (3) Use indirect addressing to cross page boundaries.

The first alternative is not always a possible choice and is tedious at best. The second is useful whenever a location is to be referenced from several different pages in core. However, since zero-page space is relatively special, a programmer does not want to assign referenced locations indiscriminately to that page. The third alternative is a generally applicable and recommended approach in the development of a program. These three alternatives are illustrated in Fig. 2-4.

Note the statements used to generate indirect addressing. Also, note that a statement such as PZ   DEF   Z tells the Assembler to place the 16-bit address of location Z in location PZ. The memory reference instruction followed by a ,I is interpreted by the Assembler as an indirect addressing operation. That is, LDA   PZ,I, when converted to binary machine language, will tell the CPU to go to location PZ, fetch the word contained there, interpret this word as an address, and then go to this new address and carry out the LDA operation on the contents of that location (location Z).

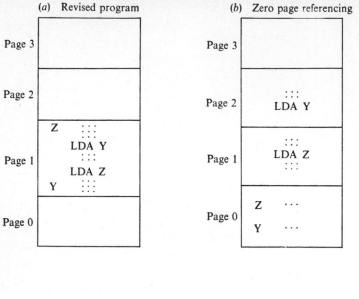

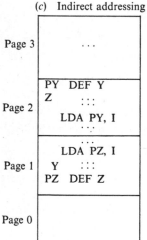

Fig. 2-4   Various ways to remove illegal references.

Address modification with indirect addressing   The technique of indirect addressing can also be very useful for operations requiring address modification. To illustrate this application, consider how the Clear Core operation outlined in the examples above is carried out with indirect addressing and address modification, as in Example 2-6.

Because of the more reliable method of address modification employed, Example 2-6 represents the most acceptable approach to generating the Clear Core program. It is interesting, however, to compare this last program with the

Example 2-6 Clear a block of core memory using indirect addressing and address modification.

```
            ORG    50B
            CLA
LOOP        STA    PBLCK,I
            ISZ    PBLCK
            ISZ    CNTR
            JMP    LOOP
            HLT
CNTR        DEC    -33
PBLCK       DEF    BLOCK
            ORG    140B
BLOCK       OCT    0
            END
```

first one written (Example 2-3). The first program represented a very inefficient programming effort (35 statements); whereas the last program is quite efficient and involves only 9 statements. On the other hand, the last program causes some 133 separate instruction executions; whereas the first program is complete after only 35 instruction executions. Thus, the first program, although inefficient in the amount of space required, is very much faster in accomplishing the purpose of the program. It is almost invariably true that the more compact programming is more wasteful of machine time. Thus, the actual shape of a program may depend on what is more crucial, *space* or *time*.

BIBLIOGRAPHY

1. Wegner, P.: "Programming Languages, Information Structures, and Machine Organization," Chaps. 1 and 2, McGraw-Hill, New York, 1968.
2. Knuth, D. E.: "Art of Computer Programming," vol. I, chap. 1, Addison-Wesley, Reading, Mass., 1968.
3. Gear, C. W.: "Computer Organization and Programming," chaps. 2 and 3, McGraw-Hill, New York, 1969.
4. "A Pocket Guide to Hewlett-Packard Computers," Hewlett-Packard, Cupertino, Calif.
5. "Introduction to Programming," Digital Equipment Corp., Maynard, Mass., 1972.
6. "How to Use the NOVA Computers," Data General Corp., Southboro, Mass., 1971.
7. "Varian 620 Computer Handbook," Varian Data Machines, Irvine, Calif., 1971.

3
Simple Arithmetic Programming Algorithms

The ninth-century Arabian arithmetician, al-Khuwārizmi, reportedly was the first to demonstrate that arithmetic could be done by symbol manipulation, as opposed to counting fingers, toes, pebbles, etc. Specifically, he is credited with developing symbolic decimal arithmetic using nine figures and a zero. Today, we refer to any logical sequence involving manipulation of quantities representing numerical values to accomplish a specific arithmetic objective as an *algorithm* or *algorism*, after al-Khuwārizmi. Thus, the stepwise sequences of computer operations described in this section for accomplishing simple arithmetic operations can be referred to as algorithms. These can usually be illustrated in the format of a *flowchart*.

3-1 FLOWCHARTS

The step-by-step logical sequencing of a program or algorithm can best be conveyed by means of a flowchart. A flowchart is simply a sequential listing of logical and/or arithmetic steps in the program, with provision for branching and looping. Simple and complex steps alike are reduced to brief descriptive "blocks" in the overall flowchart. Thus, the wise programmer precedes the actual writing of a program by a flowchart of the desired program. The program can then be written with a clear understanding of how all the detailed parts hang together. That is, the possibility of the programmer losing sight of the forest for the trees is minimized when he follows a flowchart. In addition, the flowchart will allow the programmer to return to a program many months after having written it and be able to reacquaint himself quickly with the intricacies of program execution. (Along these same lines, frequently spaced comments within the program itself are invaluable information to a programmer wishing to reacquaint himself with a program he has written previously or to another programmer wishing to use or modify the program.)

To provide an example of a simple flowcharting problem, consider again the Clear Core programming discussion of the previous chapter. A flowchart for the first direct approach to solving the problem (Example 2-3) is written as in Fig. 3-1. The flowchart corresponding to the program presented in Example 2-6 is shown in Fig. 3-2.

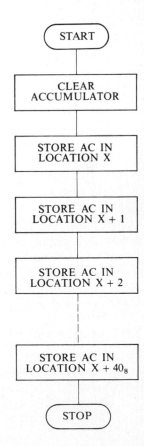

Fig. 3-1 Flowchart for direct approach to Clear Core program of Example 2-3.

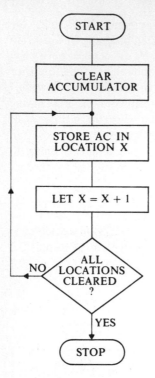

Fig. 3-2 Flowchart for Clear Core program of Example 2-6.

Note that the flowchart does not generally specify how a given task is to be accomplished programmatically. (In fact, the programs of *both* Examples 2-5 and 2-6 fit the flowchart of Fig. 3-2.) It is the programmer's responsibility to be aware of whether or not the logical or arithmetic step(s) indicated can be implemented with the appropriate combination of program instructions.

The particular geometric shapes of blocks used in flowcharting have been standardized, with particular shapes corresponding to particular functions. For example, the rectangular blocks are normal sequential program statements usually representing only a single logical or arithmetic function; whereas the diamond-shaped blocks are used to represent *decision* steps. The rounded-end rectangles are used as *terminal* indicators. These three are the most common blocks to be used here. Other symbols will be defined as they are used.

3-2 ILLUSTRATIVE PROGRAMS

The following group of programs, flowcharts, and discussions is represented here to provide an introduction to various fundamental programming concepts not yet introduced.

The arithmetic operations indicated in the flowchart of Fig. 3-3 include subtraction, for which no instruction has yet been given. In fact, most small

Program: Result = A + B − C
If result G.T. 100_{10}, go to STOPA;
otherwise go to STOPB

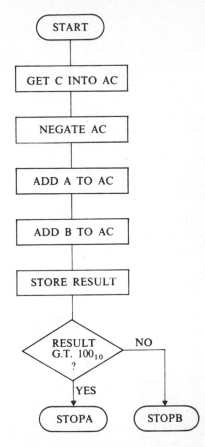

Fig. 3-3 Flowchart for Simple Arithmetic and Test program.

computers do not have a subtract instruction. What the programmer must do is negate and add to accomplish subtraction. Moreover, using 2's-complement arithmetic, negation involves complement and increment operations. These two instructions are invariably included in an instruction set. For our computer, the contents of the accumulator (AC) will be complemented with a CMA instruction and incremented with an INA instruction. Thus, the arithmetic operation A + B − C could be written as

```
LDA  C        /GET  C
CMA           /NEGATE  C
INA
ADA  A        /ADD  A
ADA  B        /ADD  B--(A + B − C NOW IN AC)
```

Since both instructions CMA and INA operate on the AC only and no reference to memory is made, both operations may be executed rapidly within one machine cycle. Thus, it is possible to combine the two instructions into one statement—CMA,INA.

The program of Fig. 3-3 must also include a decision-making sequence based on the magnitude of the arithmetic result. That is, if the result is greater than 100_{10} the program branches in one direction; if not, it takes an alternate course. Instructions which allow the selection of alternate paths based on arithmetic tests are called *branch, skip,* or *test* instructions. One example of this category is the ISZ instruction already discussed. Other instructions are available which can test the contents of the AC for certain conditions. The computer's instruction set includes several such instructions, some of which are listed here:

Instruction[a]	Operation
SZA	Skip next instruction if AC = 0
SZA,RSS	Skip next instruction if AC ≠ 0
SSA	Skip next instruction if AC ⩾ 0
SSA,RSS	Skip next instruction if AC < 0
RSS	Unconditional skip of next instruction

[a] A complete listing of test instructions is given in Appendix A. Any logical combination of the test instructions can be incorporated into a single machine instruction.

Now the complete assembly-language program corresponding to Fig. 3-3 can be written, as shown in Example 3-1.

Program to sum a block of data   A simple, but quite useful, arithmetic routine is one to add together all values stored in a block of core memory. The utility of such a program can be appreciated when one considers that mathematical integration of a curve can be obtained by the summation of all points on that curve when the distance between points approaches 0. Thus, digital integration of peak areas in gas chromatography and other chemical experiments is accomplished basically by calculating the sum of data points on individual peaks.

We will discuss digital integration of experimental data in some detail later. For now, let us consider the simple problem of summing a block of data stored

Example 3-1   Program for flowchart of Fig. 3-3

```
          ORG   2000B
          LDA   C          /GET C
          CMA,INA          /NEGATE C
          ADA   A          /ADD A
          ADA   B          /ADD B
          STA   RSULT      /SAVE RESULT
          ADA   M100       /SUBTRACT 100
          SSA,RSS          /REMAINDER = NEG?
          SZA,RSS          /NO–REMNDR = ZERO?
          JMP   STOPB      /RMNDR = NEG OR ZERO, GO TO STOPB
          JMP   STOPA      /RMNDR = POS. RSULT G.T. 100. GO TO STOPA
M100      DEC   –100
A         DEC   – – –
B         DEC   – – –
C         DEC   – – –
RSULT     OCT   0
          .
          .
          .
STOPA     –
          –
          –
          .
          .
          .
STOPB     –
          –
          –
          .
          .
          .
          END
```

in core memory. A program could be written to add all values in a block of core memory by following the direct approach similar to that illustrated in the Clear Core program (Example 2-3). That is, a series of instructions could be written which sequentially add memory contents to the AC, with one memory reference instruction for each location referenced. As already pointed out, this type of programming is very wasteful of memory space; although it may provide for very rapid execution time. A more acceptable algorithm for adding a block of data is described by the flowchart of Fig. 3-4.

The program corresponding to the flowchart of Fig. 3-4 obviously assumes that the data are stored sequentially in core memory. The locations are accessed sequentially by a program loop which uses indirect addressing and address modification. The Block Sum program could be written simply, as shown in Example 3-2.

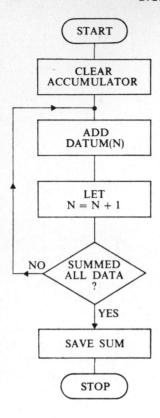

Fig. 3-4　Flowchart for Block Sum program.

Example 3-2　Block Sum program

```
            ORG  1000B
            CLA
            ADA  PDATA,I   /ADD DATUM(N) TO AC
            ISZ  PDATA     /INCRMNT DATA BUFFR POINTR
            ISZ  DCNTR     /ALL DATA SUMMED?
            JMP  *-3       /NO-ADD IN NEXT DATUM
            STA  SUM       /YES-SAVE SUM
            HLT
PDATA       DEF  DATA      /DATA BUFFR POINTR
DCNTR       DEC  -N        /SET DCNTR = -MINUS NUMBR DATA
SUM         OCT  0         /LOC'N SET ASIDE FOR SUM
            ORG  2000B
DATA        OCT  0         /FIRST ADDRSS OF DATA BLOCK
            END
```

The program of Example 3-2 introduces two new points. First, note that the initial data point could have been taken into the AC with an LDA instruction. However, subsequent additions would have required ADA instructions. Thus, to allow more efficient programming, the AC was cleared first with a CLA instruction, allowing the first data point to be loaded in with an ADA instruction. The second innovation was the use of *relative addressing*. The JMP   *−3 instruction obviously means that the computer should go back three locations in memory to obtain the next instruction. One can also refer to other *unnamed* locations by modifying a *named* location in a memory reference instruction (e.g., LDA   DATA + 1, STA   SUM−6, etc.). This type of memory reference option is possible because the Assembler can translate the address portion into the appropriate numerical value by simple addition or subtraction. However, the programmer must be careful not to let the resultant memory reference cross a page boundary because if the Assembler does not detect the error and give a diagnostic message, the assembled program will give an erroneous memory reference instruction.

Double-precision addition   It should be obvious that the results of the repetitive addition of 16-bit numbers might quickly exceed the bounds of a 16-bit AC. Thus, for this operation and for many others, it is necessary to accommodate the result in *double precision*, i.e., with two 16-bit words. One of the hardware features of the computer which allows this to be done conveniently is the incorporation of a 1-bit *Extend* register (sometimes called a *Link* or *Carry* register). The Extend register (E register) is set to a one whenever a binary 1 is carried beyond the 16-bit limit of the AC as a result of binary addition. Because of this fact, it is possible to keep track of arithmetic carries by checking the status of the E register. Appropriate instructions pertaining to the E register are CLE, clear the E register (set to 0); CCE, set the E register to a binary 1; CME, complement the E register; SEZ, skip on zero E register; SEZ,RSS, skip on nonzero E register.

The algorithm for carrying out double-precision addition is illustrated in Fig. 3-5. It involves, first of all, defining two locations in core memory to contain the 32-bit sum. These are labeled MSH and LSH for most significant half and

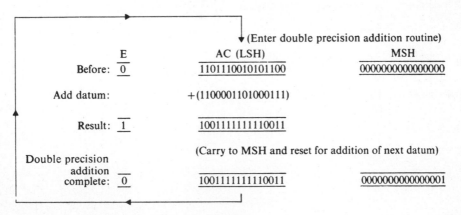

Fig. 3-5   Algorithm for double-precision addition.

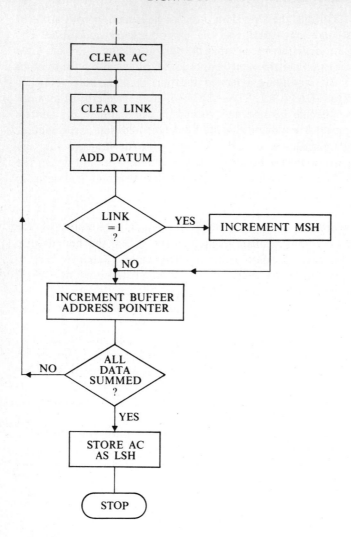

Fig. 3-6   Flowchart for double-precision addition of positive numbers.

least significant half. The status of the E register is checked after each addition of 16-bit data words into the 16-bit AC. Whenever a carry is detected, the 1 is added to the contents of MSH. The E register must be cleared before each addition. A flowchart for double-precision addition of positive numbers is given in Fig. 3-6, and the corresponding program is given in Example 3-3.

Example 3-3.   Program for double-precision addition of positive numbers

```
          ORG   2000B
          CLA
          STA   MSH        /CLEAR MSH AND "E".
DLUPE     CLE
          ADA   PDATA,I    /ADD DATUM TO AC
          SEZ              /CHECK LINK FOR CARRY?
          ISZ   MSH        /YES–INCRMNT MSH AND CONTINUE
          ISZ   PDATA      /NO--INCRMNT DATA POINTR AND
                             CONTINUE
          ISZ   CNTR       /SUMMED ALL DATA?
          JMP   DLUPE      /NO–RE-LOOP
          STA   LSH        /YES--STORE AC IN LSH
          HLT
MSH       OCT   0
LSH       OCT   0
PDATA     DEF   DATA
CNTR      DEC   −N
          .
          .
          .
          ORG   4000B
DATA      OCT   0
          .
          .
          .
          END
```

3-3.   INITIALIZATION

One fundamental problem with all the programs illustrated so far is that there has been no consideration of how to provide for reexecution of the program. For example, the Block Sum program of Example 3-2 might be utilized several times during the course of processing several different blocks of data. However, after the first execution, the values of certain parameters critical to the proper execution of the program will have been altered (i.e., DCNTR = 0; PDATA = DATA+N). As written the program can be reset (*initialized*) only by reloading the entire binary program tape.

A more practical approach to allowing the reexecution of a particular program segment is to provide an initialization procedure within the program itself. For example, the Block Sum program, rewritten to include an initialization segment, might be given as in Example 3-4.

Example 3-4.   Block Sum with initialization

```
              ORG    1000B
     STRT     LDA    PDTST       /GET INITIAL VALUE FOR DATA BLOCK
                                   POINTR
              STA    PDATA       /INITIALIZE PDATA
              LDA    DCTST       /GET DATA CNTR INITLZTN PARAMATR
              STA    DCNTR       /INITIALIZE DCNTR
              CLA
              ADA    PDATA,I
              .
              .
              .
              HLT
     PDTST    DEF    DATA
     PDATA    OCT    0
     DCTST    DEC    −N
     DCNTR    OCT    0
     SUM      OCT    0
              .
              .
              .
              END
```

Note: The parameters PDATA and DCNTR are set to 0 initially since their initial values do not matter. They are initialized before the main part of the program can be executed.

Often the console switch register (SR) is used to allow communication between the operator and the computer to initialize or modify a program. This is accomplished by including in the program an instruction to load the binary number set on the SR into the AC. This value can then be used to set some parameter within the program. The instruction to load from the SR is LIA   1. (The operand 1 refers to the fact that the SR is input channel 1.)

Consider, for example, that you wanted to use the Block Sum program repetitively, loading new sets of data from data tapes into the reserved core memory block after each summation is completed. If the number of data in each set varied, it would be necessary to change the value for initializing DCNTR for each data set. This could be accomplished in a number of ways, one of which is to set the SR to represent the total number of data points in the set to be processed and to use the LIA   1 instruction to bring that information in for initialization. The flowchart of Fig. 3-7 and the program of Example 3-5 illustrate this application.

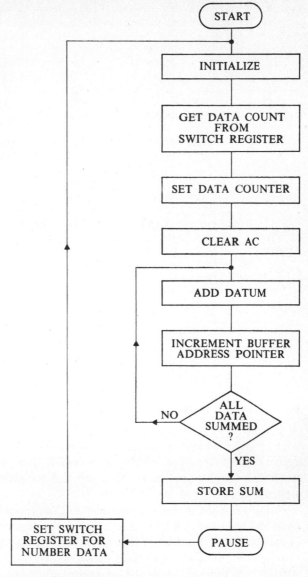

Fig. 3-7 Flowchart for Block Sum program with initialization for restart and selection of the number of data points through console switch register.

Example 3-5   Block Sum program with (1) initialization for restart and (2) specification of the number of data points to be summed made through the SR

```
                ORG    1000B
    START       LDA    PDTST      /INITIALIZE DATA POINTR
                STA    PDATA
                LIA    1          /GET DATA COUNT FROM SW. REG.
                CMA,INA           /NEGATE FOR DCNTR
                STA    DCNTR      /INITIALIZE DCNTR
                CLA
                ADA    PDATA,I    /ADD DATUM(N)
                ISZ    PDATA      /INCREMNT DATA BUFFR POINTR
                ISZ    DCNTR      /ALL DATA SUMMD?
                JMP    *-3        /NO
                STA    SUM        /YES
                HLT               /PAUSE–ENTER NEW VALUE IN SW. REG.
                JMP    START      /RE-START PROGRAM
    PDTST       DEF    DATA
    PDATA       OCT    0
    DCNTR       OCT    0
    SUM         OCT    0
                ORG    4000B
    DATA        OCT    0
                END
```

Note: In Example 3-5 the AC was not cleared before the LIA   1 instruction was executed since the resultant operation is a *load* operation replacing the contents of the AC with the SR setting.

3-4   SUBROUTINES

Often a given program segment becomes useful in several different parts of the total program. Whenever this happens, it is more efficient to write that program segment as a subroutine, which can be called from any part of the program, rather than duplicating the program segment for every time it is used in the program.

For example, it may be necessary to clear a block of core memory many times and at many different places during the execution of a given program. Thus, it would be worthwhile to rewrite the Clear Core routine of Example 2-6 as a subroutine. It is also necessary to define a new type of memory reference instruction which will allow the convenient entry and exit of subroutines. The new instruction is JSB (jump subroutine) which causes the following events to occur: The address of the next sequential location (after the JSB instruction) is stored in the location whose address is referred to by the JSB instruction; then control is transferred to the second location in the subroutine. The storage of address $Z + 1$ in the first location of the subroutine gives the subroutine the needed information to get back to the main program. That is, the address information stored in the first location is the linkage between program segments.

This operation is illustrated below:

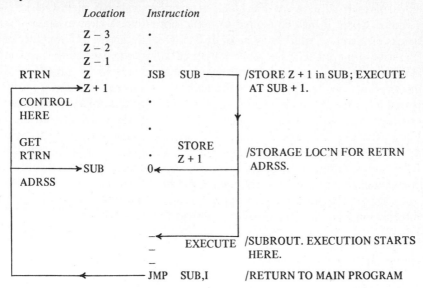

Thus, if we write the Clear Core routine of Example 2-6 as a subroutine, it will appear as in Example 3-6.

Example 3-6   Simple Clear Core subroutine

```
          ORG   3000B
CLRCR     OCT   0
          LDA   CTST        /INITIALIZE CNTR, PBLCK.
          STA   CNTR
          LDA   PBLST
          STA   PBLCK
          CLA               /CLEAR AC.
LOOP      STA   PBLCK,I      /CLEAR BUFFR LOCN N.
          ISZ   PBLCK        /LET N = N + 1.
          ISZ   CNTR         /ALL LOCNS CLRD?
          JMP   LOOP         /NO–RELOOP.
          JMP   CLRCR,I      /YES–EXIT–RETRN MAIN PRGRM THRU
                              CLRCR.
CTST      DEC   –Z
CNTR      OCT   0
PBLST     DEF   BFFR
PBLCK     OCT   0
          END
```

Note: The Clear Core subroutine of Example 3-6 required initialization at the beginning since the parameters CNTR and PBLCK are modified each time the subroutine is used. Also note that the first location of the subroutine does not contain an instruction or other useful information because it is altered whenever the subroutine is entered.

As written in Example 3-6, the Clear Core subroutine is not generally useful since it clears a specific block of memory and that is all. Consider the situation where it may be desired to use a subroutine to clear several different blocks of memory, which may be of different lengths and at different locations. The subroutine would have to be initialized *differently* each time it was used. How is the subroutine to know in which manner it is to be initialized? This information can be provided to the subroutine through the *linkage* address stored in its first location when it is called. Through this linkage the subroutine can reach back to the main program from which it was called and retrieve vital initialization information. This not only allows the same subroutine to be used differently within the same program but also allows the subroutine to be called from several different programs, provided it is well documented with regard to what initialization information is required and what the proper call sequence is.

To see the use of the subroutine linkage location to extract initialization information from the main program, consider Example 3-7.

Example 3-7   General-purpose Clear Core subroutine

```
             .
             .
             .
CALL     JSB   CLRCR          /CALL SEQUENCE: JSB; BUFFR ADDRSS;
                                 BUFFR LENGTH.
         DEF   BFFR
         DEC   N
             .
             .
             .
CLRCR    OCT   0
         LDA   CLRCR,I         /GET BUFFR ADRESS FROM MAIN PRGRM.
         STA   PBLCK           /INITIALIZE PBLCK.
         ISZ   CLRCR           /INCRMNT LINKAGE TO MAIN PRGRM.
         LDA   CLRCR,I         /GET BUFFR LENGTH FROM MAIN PRGRM.
         CMA,INA               /NEGATE LENGTH FOR CNTR.
         STA   CNTR            /INITIALIZE CNTR.
         CLA
LOOP     STA   PBLCK,I
         ISZ   PBLCK
         ISZ   CNTR
         JMP   LOOP
         ISZ   CLRCR           /INCRMNT LINKAGE TO RETRN TO MAIN
                                 PRGRM.
         JMP   CLRCR,I         /RTRN TO MAIN PRGRM AT CALL+3.
PBLCK    OCT   0
CNTR     OCT   0
             .
```

The general-purpose Clear Core subroutine (Example 3-7) should be documented with the fact that the call sequence should include three sequential statements: the JSB   CLRCR instruction, the address of the data block to be cleared, and the positive length of the buffer (the total number of locations to be cleared). The subroutine, by successive modification of the linkage address, is able to reach back into the main program (using indirect addressing) and obtain the information needed for proper execution.

EXERCISES

3-1. Write a program for double-precision addition of positive and negative (16-bit) numbers.

3-2. Write a program for triple-precision addition of positive numbers.

3-3. Write a subroutine for integration involving double-precision addition. Provide a call sequence allowing variable length and location of data.

3-4. Write a program to find the maximum value in a block of data.

3-5. Write the program in Exercise 3-4 as a subroutine.

3-6. Carry out the assignments of Exercises 3-4 and 3-5 for the minimum value in a block of data.

3-7. Write a program which will count to 1 million and halt. Provide initialization steps.

3-8. Write a program which will add all positive, nonzero numbers in a block of data. Provide initialization steps.

3-9. Write a program which will add the first n integers $(1 + 2 + 3 + \cdots + n)$, where n is entered from the SR. Provide initialization steps.

3-10. Write a program which will add the first n digits with alternating signs $(1 - 2 + 3 - 4 + \cdots n)$. Provide for entering n through the SR and for initialization steps.

3-11. Write a subroutine which will count all the *negative values* stored in a block of memory. Provide a call sequence which can be applied to a variable data block length and location.

3-12. Write a subroutine which will add all values less than 50_{10} in a block of data. Provide a call sequence allowing variable length and location of data.

4

Introduction to Chemical Data Handling with Assembly-Language Programming

The previous chapters have introduced the fundamental principles and techniques for assembly-language programming. This chapter will serve to introduce the reader to some of the techniques applied to typical chemical data-handling problems, utilizing assembly-language programming principles. The specific example used is the handling of gas chromatographic (GC) data since these data are quite often computer-processed and since GC output signals are typical of outputs from other laboratory measurement devices in chemistry and other sciences.

4-1 DATA INPUT AND FORMAT

For this discussion, we will consider that the GC data input and formatting will be set up as described in Fig. 4-1. That is, the analog output from a GC experiment is digitized using an analog-to-digital converter (ADC) and a data-acquisition rate of 1 point per second. The digitized data are stored sequentially in a block of core memory. Data points are taken and stored continuously during the entire experiment; thus, for a 500-sec chromatogram,

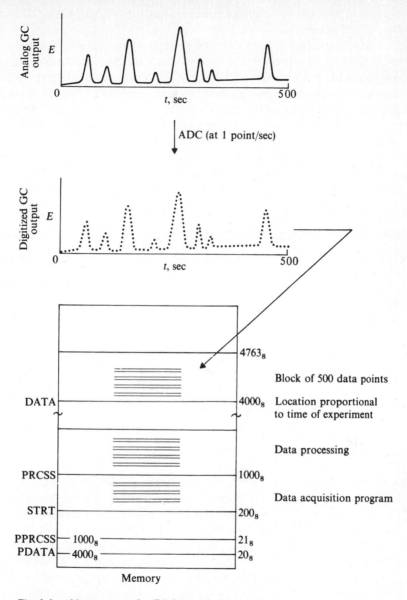

Fig. 4-1 Memory map for GC data and programming.

500 data points would be taken, and 500 memory locations would be required. For a chromatogram requiring a more typical 10 point per second data rate, the number of memory locations required would go up to 5,000! This could be a severe strain on the computer facility, particularly if very little bulk data storage

capacity is available. (We will consider methods for circumventing this problem later in this chapter.)

It is important to note that the position of storage of a data word in the memory block is related to the time at which the data point was taken during the experiment. Thus, the address of a given data word (relative to the first address of the memory block) provides information to the program regarding the experimental time to which that data word corresponds. Access to each data point stored in memory for data processing can be accomplished easily by using indirect addressing with sequential address modification. For example, in Fig. 4-1, the symbolic address DATA is assigned to the first location of the data block. The location PDATA—placed on the zero page—is the linkage or pointer to the data block and contains initially the address of location DATA.

4-2   PROCESSING EXPERIMENTAL DATA

The analytical information normally extracted from GC data includes peak heights, peak areas, and retention times (i.e., the time corresponding to the peak maximum, relative to the initiation of the experiment). The following discussions will describe algorithms for obtaining this information based on the data input and format described above. One inherent assumption is that the base line remains at 0 throughout.

Peak integration   The simplest processing of GC data would involve only measuring the areas of individual peaks. The flowchart for such a program is given in Fig. 4-2. Figure 4-3 illustrates the actual area measurements taken from a chromatogram. The algorithm is based simply on the establishment of a signal threshold, with the start and stop of integration determined by whether the signal is above or below the threshold. An assembly-language program based on the flowchart of Fig. 4-2 is given in Example 4-1.

Several aspects of this program should be discussed: First of all, the integration segment of the program simply involves the implementation of a double-precision addition routine, similar to that discussed earlier (Chap. 3). The resultant sum is a 32-bit integer value proportional to the peak area.

A second aspect of the program to be considered is the concept of a software *flag*. This simply involves the use of the contents of a memory location or working register to indicate whether or not a particular event has occurred. For example, as the computer is processing GC data, it can record the fact that it has recognized the beginning of a peak and that integration should start by setting an integration flag INTFLG. That is, the contents of some memory location called INTFLG will be set to a 1 state. Thus, as sequential data points are taken, the computer can easily decide whether these should be incorporated into a peak integral by checking the status of INTFLG. If it is a 1, continue to

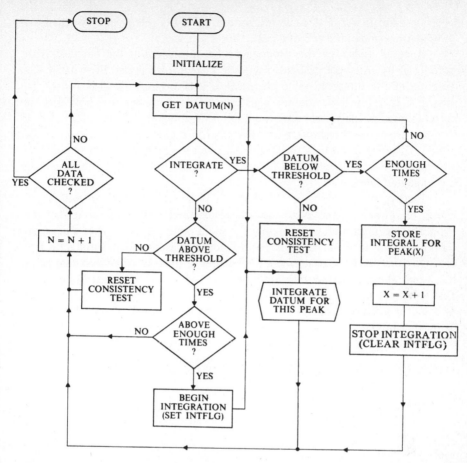

Fig. 4-2 Flowchart for program to integrate peak areas for GC data.

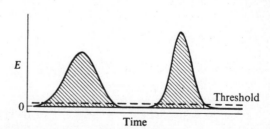

Fig. 4-3 Peak-area measurements using simple thresholding.

integrate data; if it is a 0, cease integration. INTFLG gets reset to 0 when the computer recognizes the end of a peak.

Another aspect of the program to be considered is the inherent error introduced by using a threshold diagnostic to recognize peaks. First of all, the selection of the threshold value is critical. If it is too high, small peaks will be missed; if it is too low, random signal excursions above the threshold may be misinterpreted as analytical peaks. Moreover, since integration is not initiated until the data are above threshold, a negative error in the computed peak area must result. (This may be serious for small peaks, but the problem can be handled by arbitrarily including several data points prior to and following the thresholds in the computed integral.)

Finally, it should be noted that the program is written to nullify the effects of spurious signal excursions above the threshold by requiring that the signal be *consistently* above or below threshold before deciding that a peak has commenced or terminated. This consistency test is provided simply by two program counters, which determine the number of times in a row that data point excursions above or below threshold are observed. Thus, the Reset Consistency Test statement in the flowchart of Fig. 4-2 requires that these counters be reset whenever a sufficient number of data points in a row is *not* observed above or below threshold.

Example 4-1   Simple program for peak integration in gas chromatography

```
          ORG   2000B
START     CLA                     /INITIALIZE ALL PROGRAM
          STA   INTFG             PARAMETERS
          STA   MSH
          STA   LSH
          LDA   ACTST
          STA   ACNTR
          LDA   BCTST
          STA   BCNTR
          LDA   DCTST
          STA   DCNTR
          LDA   PDTST
          STA   PDATA
          LDA   PLSST
          STA   PLSH
          LDA   PMSST
          STA   PMSH
PRCSS     LDA   INTFG
          SZA                     /INTFG SET?
          JMP   BCHK              /YES–CHK IF BELOW THRSHLD TO STOP
                                   INTGTN
          LDA   PDATA,I           /NO–GET DATUM(N)
          ADA   TRSLD             /SUBTRACT THRSHLD
          SSA                     /DATUM G.T. THRSHLD?
```

Example 4-1   (*continued*)

```
                JMP   RSETA        /NO--GO TO CNSISTNCY TEST RESET
                ISZ   ACNTR        /YES--ABOVE ENUFF TIMES IN ROW?
                JMP   NEXT         /NO--GET NXT DATUM
                ISZ   INTFG        /YES--SET INTFG
                LDA   ACTST        /RESET ACNTR
                STA   ACNTR
       INTGL    LDA   PDATA,I      /GET DATUM(N)
                CLE
                ADA   LSH          /ADD CURRENT LSH
                SEZ                /CARRY?
                ISZ   MSH          /YES
                STA   LSH          /NO
       NEXT     ISZ   PDATA        /GENERATE ADDRESS FOR NXT DATUM
                ISZ   DCNTR        /ALL DATA CHKD?
                JMP   PRCSS        /NO--STRT PRCSSNG NXT DATUM
                HLT   77B          /YES--STOP
       BCHK     LDA   PDATA,I      /GET DATUM(N)
                ADA   TRSLD        /SUBTRACT THRSHLD
                SSA                /BELOW THRSHLD?
                JMP   TEST         /YES--CHK # TIMES
                LDA   BCTST        /NO--RESET CNSISTNCY TEST AND
                                    INTGRATE

                STA   BCNTR
                JMP   INTGL
       RSETA    LDA   ACTST
                STA   ACNTR
                JMP   NEXT
       TEST     ISZ   BCNTR        /BELOW ENUFF TIMES
                JMP   INTGL        /NO--INTGRATE
                LDA   BCTST        /YES--RESET BCNTR
                STA   BCNTR
                LDA   LSH          /STORE PK. INTGRL IN D.P. BFFR
                STA   PLSH,I
                LDA   MSH
                STA   PMSH,I
                ISZ   PLSH         /INCRMNT D.P. BFFR POINTRS
                ISZ   PMSH
                CLA
                STA   MSH          /RE-INITIALIZE MSH, LSH
                STA   LSH
                STA   INTFG        /CLR INTFG
                JMP   NEXT
       INTFG    OCT   0
       TRSLD    DEC   -Z
       ACTST    DEC   -M
       ACNTR    OCT   0
       BCTST    DEC   -M
       BCNTR    OCT   0
       DCTST    DEC   -Q
       DCNTR    OCT   0
```

Example 4-1   *(continued)*

```
PDTST      DEF   DATA
PDATA      OCT   0
PLSST      DEF   LSBUF
PLSH       OCT   0
PMSST      DEF   MSBUF
PMSH       OCT   0
MSH        OCT   0
LSH        OCT   0

           ORG   5000B
MSBUF      OCT   0

           ORG   5100B
LSBUF      OCT   0

           ORG   4000B
DATA       OCT   0
           END
```

Peak-height and retention-time determination   Generally, the analyst does not require peak-height data in gas chromatography. However, because this is a common measurement for other kinds of experimental data, it will be discussed here. Also, it is important to determine the time at which the peak appears (the retention time). The retention time is conventionally assigned to the time at which the peak maximum is observed. Thus, the program should provide for recognition of the peak maximum and the precise calculation of the retention time.

The flowchart presented in Fig. 4-4 is designed to provide for the recognition of bona fide peaks in the output data. This program is presented for GC data, but it is generally applicable to any data where symmetrical peaks are normally observed. The program assumes smooth data, but it also can accommodate some amount of random noise.

The algorithm represented by the flowchart of Fig. 4-4 is illustrated in Fig. 4-5. The program is simply looking for a portion of the data where a sharply increasing region is followed by a sharply decreasing region. The point in between, where the first derivative goes through 0, is the peak maximum. Rigorous mathematical differentiation is not necessary, but rather sequential *delta* values can be obtained. These differentials provide discontinuous values proportional to the instantaneous first-derivative value involved. Division by Δt is not necessary, provided that the data-acquisition rate is constant since Δt will then be a constant divisor for all derivatives obtained.

The program applies two tests to the data in order to recognize a bona fide peak maximum: (1) Are the data increasing (or decreasing) *sharply*? (2) Are the sharp increases (or decreases) *consistent*? The consistency test is used to

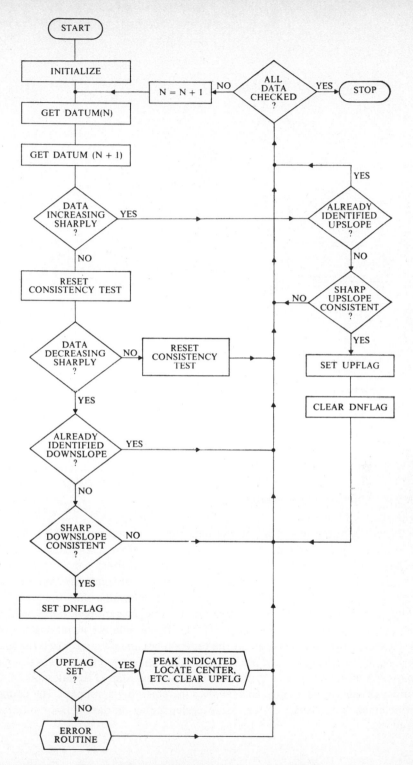

Fig. 4-4 Flowchart for peak-identification routine for GC data.

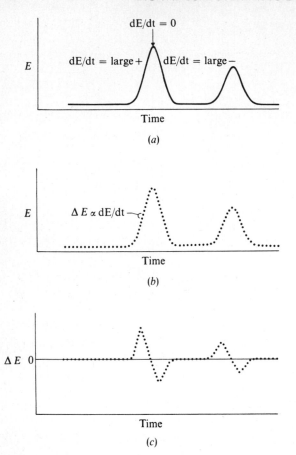

Fig. 4-5 Characteristics of digitized GC data for processing.

distinguish between real trends in the data and any random noise. This test can be applied in the same manner as in Example 4-1, where a consistency test was applied to determine if data were truly above or below threshold.

  If one goes through the flowchart step by step, one sees that the first operation is to get the first two data points and determine the differential. By comparison to some arbitrary standard value the program determines whether the data are increasing or decreasing sharply. If the data are increasing sharply, the next question asked is whether the program has already recognized the sharp upslope region of the current peak. Since it is considering only the first two data points initially, the answer should be no. The next step is to determine if the sharp upslope is consistent. This involves incrementing a counter. The counter might be set initially to -3 or -5, etc., depending on data density and noise problems.

After the no answer on the consistency test (i.e., the counter was not incremented to 0) the program asks whether all data have been processed. If not, the next data point is fetched and compared to the previous one in the same manner as described above. If the differential does not indicate a sharp increase, the upslope consistency test is reset. (That is, the upslope counter is reset to its initial value.) This feature ensures that the required number of sharply increasing measurements must be made *in a row* in order to recognize an upslope of a peak.

Continuing, when a sharp increase is not observed for these two data points, the next test applied is whether the data represent a sharp decrease. If they do, then a pathway similar to that described above after the first pair of sharply increasing data points were observed is followed.

The program continues to sequentially fetch data points until a consistent sharp increase is observed (the upslope counter incremented to 0), and then the UPFLAG is set to indicate to the program that the sharp upslope region of the peak has been recognized. Next, the DNFLAG is cleared (memory word set to 0) to be sure that it is in the initialized state because now the program will be looking for the downslope side of the peak. A set DNFLAG would indicate that it had been observed; just as a set UPFLAG indicates that the upslope side of the peak has been recognized.

As the program continues to process successive data points, it may observe several more sharply increasing data. However, when the question is asked ALREADY IDENTIFIED UPSLOPE?, the answer will be determined (as always) by checking the status of the memory location called UPFLAG. Now, the UPFLAG location will be set to 1, indicating that the upslope has been recognized; a YES answer is indicated, and the program continues on checking successive data points until sharply decreasing data are observed.

When consistently sharply decreasing data are observed, the DNFLAG is set, and the question is asked UPFLAG SET?. If the UPFLAG is set, then the criterion for recognizing a peak is met—a sharply increasing region followed by a sharply decreasing region. Thus, the program shifts to a routine designed to locate the peak maximum, determine the peak height and the retention time, store this information, and finally clear the UPFLAG before continuing on to process the rest of the data.

If the UPFLAG had not already been set when the sharp downslope was recognized, this would have indicated something abnormal about the data, and the program would have shifted into an error routine, which would at least have recorded the abnormality in memory for later readout before resetting the program parameters and continuing.

That part of the program which determines the peak maximum and the retention time has not been included in the flowchart. (Nor has any provision for peak integration been included. However, an appropriate combination of the flowcharts in Fig. 4-2 and 4-4 would provide this feature, and this is assigned at the end of the chapter.) There are several possibilities for determining the peak maximum, some of which would impose other steps on the other parts of the program. Some possibilities are considered below.

Locating the peak maximum   First of all, the simplest approach for locating the peak maximum would be to find the maximum value in the data points falling between the points at which the sharp upslope and downslope, respectively, were recognized. This approach would necessarily have to assume smooth data. The memory location at which it is found can be directly related to the retention time, knowing what the data-acquisition rate is.

Mathematically rigorous peak-location algorithms, like differentiation, are feasible when data smoothing is applied first. (See later discussion, this chapter.) The point at which the first derivative of the data goes through 0 indicates the position of the peak maximum.

A third approach might involve simply taking the midpoint between the regions where the sharp upslope and downslope were recognized. However, the peak-recognition part of the program would have to be altered somewhat from its present state. That is, currently the sharp upslope might be recognized near the foot of the peak; while the sharp downslope might be recognized shortly after passing through the maximum. Thus, the program would have to be rewritten to provide for a more symmetrical recognition of sharp upslope and downslope regions. (For example, the program could be written to record the *last* set of data points consistently showing a sharp upslope. Then, assuming a symmetrical peak, these points should be symmetrical with the first set of points showing a consistent, sharp downslope; the midpoint ought to correspond to the peak maximum.)

The advantage of this latter approach is that it is less susceptible to random-noise problems than the two alternatives mentioned. However, it does appear limited to symmetrical peaks; although this limitation becomes less serious when the data density is high and the sharp upslope and downslope regions can be recognized shortly before and after the peak maximum.

Effect of data density   One point that should be considered here is that the successful operation of any of these peak-location algorithms depends critically on the data density. For example, if the data density is low, perhaps only a few data points will be taken during a peak. There might not even be enough data to make the arithmetic and logical tests necessary to decide that a peak had been observed. Moreover, if the peak is recognized, there will probably not be a data point corresponding to the true peak maximum. On the other hand, a low data density tends to nullify the effects of small-amplitude random noise.

Taking the other point of view, if the data density is very high, all the minute detail of the peak will be seen. However, some complications might arise. For example, any small-amplitude random noise will be seen and could lead to the erroneous identification of peaks. Also, it will be necessary to test slopes over a span of several data points in order to detect significant delta values.

Thus, the utility of the very high data density may be nonexistent, depending on the peak-identification algorithm used. Algorithms involving smoothing techniques are enhanced by the very high data density. However, the simple algorithm proposed in the flowchart of Fig. 4-4 works poorly with a very high data density.

Simple-minded algorithms   The peak-characterization algorithm suggested by the flowchart of Fig. 4-4 might be called a *simple-minded* algorithm. It does not involve any sophisticated arithmetic operations at all. Consequently it can be written quickly, briefly, and in a form which requires a minimum of computer execution time. The first two considerations—requiring a minimum of programming effort and time and occupying a minimum of core memory space—are important to the small-computer user, who is, first of all, more interested in applying the computer to his chemical problems than he is in sophisticated program development, and, secondly, who usually must operate with relatively small core memory sizes (4K to 8K). The third consideration, rapid execution time, may or may not be important as the computer can generally keep way ahead of the human operator. However, for some systems execution time can be very critical. One example of this situation is given below in the discussion of the real-time GC data processing algorithm. Many other examples will be given throughout the text.

The important point to be recognized here, however, is that simple data processing algorithms may be successfully applied only when the programmer is intimately aware of the general characteristics of the data which will be processed. For example, if the data are generally going to involve symmetrical peaks which have some small-amplitude high-frequency noise associated with them, the programmer may choose to use a relatively low data density and locate peaks by the midpoint approach outlined above. On the other hand, if the data will normally include asymmetrical, smooth peaks, the programmer may choose to use a higher data density and locate peaks by searching for the maximum value in the peak region. Neither of these algorithms is appropriate for data which include peaks with high-amplitude low-frequency noise superimposed.

If the programmer is not aware of the type of data to which his program is to be applied, he cannot get by with a simple-minded algorithm. The program will have to be written to accommodate all types of peaks. Very likely the programming will become quite involved—meaning that it will probably be long, taking up lots of core memory, and slow, taking up considerable execution time.

4-3   SOME ALTERNATIVE ALGORITHMS FOR PROCESSING GC DATA

Real-time thresholding   One of the serious drawbacks of the GC data-acquisition and data processing schemes described above is that *all* data are taken and stored during an experimental run. Since we are assuming that the base line remains at 0 throughout the run, the base-line data between peaks are nonessential and do not have to be stored. Thus, one approach would be to take a preliminary look at data as they are being acquired and determine whether they are above threshold. Only the data above threshold would be stored for later processing. This would greatly reduce the amount of core memory storage needed for a chromatogram.

The processing of data as they are being acquired is referred to as a *real-time* operation. In this example, the processing is quite simple—comparing each data point to an arbitrary threshold to decide whether data should be stored or not. However, such steps take up execution time and must be completed before the next data point is ready to be taken into the computer. Thus, the execution time associated with real-time processing can be very critical and places an upper limit on the data-acquisition rate. (These problems are discussed in more detail in Chap. 9.)

For this particular case, where all that is required is real-time thresholding, the programming will be very short and simple. Execution time will probably require much less than 100 μsec. This includes determining and storing the time associated with the beginning of each peak as well as generating the linkage steps to the real-time subroutine. For GC experiments, maximum data-acquisition rates are the order of 10 to 50 points per second. Therefore, real-time thresholding will not impose any serious data rate limitations.

Complete real-time GC data processing   Because the normal GC data-acquisition rate is relatively slow compared to the computational speed of the digital computer, it is possible to carry out the fundamental data processing operations—peak integration, peak-maximum determination, and retention-time calculation—in real time. One might ask, "Why complicate the programming effort with real-time functions, when the computer can complete the appropriate calculations in a fraction of a second as soon as the chromatogram is complete?" There are several answers to this question: First, even with real-time thresholding, where only data taken on peaks are stored, a considerable amount of memory will be required to store peak information. (For example, a chromatogram with 20 peaks and 50 points per peak will require 1,000 core memory locations for storage.) With real-time processing, only enough storage space will be required to accommodate the data for one peak at a time, drastically reducing the storage requirements. The program will use the same storage area over and over again, with the data for previous peaks deleted since they are no longer needed once processed. (One disadvantage here, of course, is that it is not possible to maintain a record of the complete digitized chromatogram, which is sometimes useful.)

A second reason for complete real-time data processing is that the computer may be able to exercise some control over experimental conditions or measurement parameters as a result of evaluation of incoming data. For example, a critical GC separation parameter—such as column temperature—might be modified under computer control during the experiment on the basis of a continuous monitoring of the peak resolution. Also, the amplification factor of GC output electronics might be computer-controlled and adjusted on the basis of a continuous evaluation of incoming data.

A third reason for real-time data processing is that one might desire a printout of peak characteristics as each peak is observed, rather than waiting until the entire chromatogram is completed. The computer can be programmed to be printing out this information simultaneously with its data-acquisition and data processing operations.

Data smoothing It should be obvious from preceding discussions in this chapter that noisy data can raise havoc with data processing algorithms. Also, it should be recognized that it is impossible to completely eliminate noise from the output of any experimental device. The particular measurement made may be insensitive to the noise (as discussed above), but the noise will be there. Whenever measurements are made which are particularly demanding, sensitive, or precise, the noise level will be a factor. Thus, it is useful to consider some computer-oriented mathematical methods for smoothing out the random data fluctuations due to superimposed noise.

Mathematical data-smoothing techniques may involve a simple average, least-squares curve fitting, or more complex numerical approaches [1 to 8]. Smoothing can be done in real time or on stored data, depending, of course, on the data rate limitations and need for real-time processing. A detailed discussion of data-smoothing techniques is beyond the scope of this chapter, but some important considerations will be presented here.

First of all, mathematical data-smoothing procedures can very easily and insidiously cause a distortion of the signal of interest [4]. The utility of data-smoothing techniques is directly related to the programmer's awareness of the data characteristics and his willingness to exercise caution and restraint in their application. In short, indiscriminate application of mathematical smoothing techniques can be disastrous!

For example, when data are obtained where the signal of interest is superimposed on a noise background of relatively small amplitude and relatively high frequency, almost any reasonable smoothing approach can be applied with success. However, if the noise frequency approaches the frequency of the signal of interest, normal mathematical smoothing techniques become useless, and such attempts to minimize the noise contribution will cause a distortion of the desired signal.

Handling overlapping peaks and nonzero base lines GC (and other) experimental data do not always follow the idealized characteristics suggested by Fig. 4-1. One not uncommon observation is that peaks are asymmetrical. This is particularly true in gas-solid chromatography where very badly tailing peaks are observed. This is not a serious problem if the programmer is aware of such data characteristics ahead of time, as was discussed above.

More difficult programming problems are presented when GC data involve overlapping (incompletely resolved) peaks or nonzero base lines as shown in Fig. 4-6. There is no question but that such problems can be handled. A detailed discussion of the algorithms involved will not be presented here, and the reader is referred to the original literature [9-12]. However, the problems involved should be mentioned briefly at this point.

In curve I of Fig. 4-6, three different situations of peak overlap are demonstrated. One is the case where both peaks are nearly the same size and shape, and they can readily be distinguished although not completely resolved (peaks A and B). The second case is where there is a large discrepancy between the sizes of the overlapping peaks, and the smaller peak (peak D) appears as a "shoulder" on the larger (peak C). The third case is where the two peaks overlap so closely that to all appearances only a single broad peak is observed. (There are other possibilities, of course, such as the overlap of asymmetrical peaks, or more than two peaks.)

The programmer could extract quantitative data from overlapping peaks A and B by mathematically dividing the two peaks arbitrarily by the perpendicular (line a) which would be placed at the minimum between the two peaks. For these particular peaks, the resultant, arbitrary peak-area measurements would

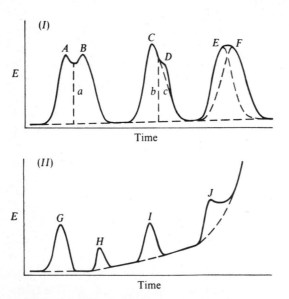

Fig. 4-6   GC data with overlapping peaks and non-zero base lines.

probably not be far off, particularly if the peaks are identical in shape and size. However, if the same procedure is applied to peaks C and D (using line b), a large error will probably result. The more realistic dividing line is line c. Thus, in order to be effective in resolving overlapping distinguishable peaks, the processing program must be able to recognize and handle these two distinct cases described by peaks A, B, C, and D.

The combined peak contributed to by peaks E and F presents a more complex problem. It is difficult to recognize that peak overlap occurs here, first of all, and it is even more difficult to mathematically resolve the contributing peaks. What is required is some foreknowledge of the expected peak shapes. Thus, the program can test each observed peak to determine if it deviates sufficiently from the predicted norm such that peak overlap is suspected. Once the decision is made that a given peak is really a composite of more than one peak, the problem of resolving the contributing components becomes a complex mathematical deconvoluting problem which relies critically on how accurately expected shapes of the contributing peaks can be predicted. It can be done, and specific techniques have been discussed [10, 12], but it should be realized that the restrictions are severe.

Curve II of Fig. 4-6 represents a case where the base line is nonzero and nonlinear. In gas chromatography this sort of base line is observed under conditions where *column bleed* has occurred and the column temperature was increased continuously during the experiment.

There are at least three approaches to handling the base-line problems typified by curve II of Figure 4-6. One of these is obviously to modify experimental conditions, if possible, so that a zero or constant base line might be obtained. If this is not possible, an alternative approach would be to measure the base line in the absence of a sample, i.e., obtain a *blank*. In many cases, however, blanks are difficult or impossible to obtain, such as the situation where the very presence of a sample influences the base line.

The third alternative is simply to mathematically connect the ascending and descending "feet" of each peak by a line tangent to the base-line portion on either side of the peak. This programming is not difficult to do if the programmer assumes that the projected base line should be linear. Although such an assumption is adequate for peaks G, H, and I, it will obviously lead to an erroneous measurement for peak J.

Effect of the dynamic range of data   One very subtle, but very important, factor which determines the characteristics of a data processing program is the *dynamic range* of the data. The dynamic range can be defined as the ratio of the maximum data value to the minimum over the largest linear (or otherwise well-defined) data scale. The minimum value is assumed to be twice the background noise level unless the scale is nonlinear at that point. Unfortunately, the dynamic range of the data seen by the computer does not necessarily correspond to the dynamic range of the data output by the detector/transducer combination of a gas chromatograph. The picture is complicated by the other parts in the data-acquisition system, such as the dynamic range of the electronic amplifier and the ADC.

Thus, the dynamic-range characteristics of a given data-acquisition system are a composite of several contributing factors. At this point, we will consider only how the dynamic range of the resultant data should be considered in the development of a data processing program. For example, assume the range is limited by a 10-bit ADC used in the data-acquisition system, and that the noise level is ±2 units on the ADC scale. Then the dynamic range of the data is 128. If the peak-analysis algorithm requires a minimum signal-to-noise ratio of 10:1, the dynamic range of peaks that may be handled is about 25.

It is possible that the same processing criteria used for recognizing and processing the largest peaks might be applicable to the smallest peaks over a dynamic range of 50. On the other hand, should a much larger dynamic range (>500) be inherent in the stored data, a much greater programming effort will be required to allow the analogous identification and characterization of small and large peaks. Some sort of scaling factor will have to be included in the program. Such techniques are discussed in detail in later chapters.

EXERCISES

4-1. Write a program corresponding to the flowchart of Fig. 4-4.

4-2. Prepare a flowchart and corresponding program which processes stored GC data. Include in your flowchart and/or program the capabilities for peak recognition, peak integration, peak-height determination, and retention-time evaluation. Assume the GC peaks are base-line-resolved with a dynamic range of 50 and a maximum value of 1,023.

4-3. What will be the dynamic range of digitized GC data where the inherent dynamic range of the detector/transducer is 10^4, that of the electronic amplifier is 10^3, the ADC is a 14-bit converter, and the background noise level amounts to ±1 unit on the ADC scale?

4-4. Carry out the assignment of Exercise 4-2 for asymmetrical GC peaks, where you have prior knowledge that all peaks will be severely tailed.

4-5. Prepare a flowchart and corresponding program for a real-time thresholding subroutine, assuming that the routine is entered each time a digitized data point is taken into the AC. Provide for storage of the appropriate peak data and times.

4-6. Carry out the assignment of Exercise 4-5, except include all data processing in the real-time subroutine. If we assume that each statement execution requires an average of 3 μsec, what is the maximum data-acquisition rate allowed by this routine?

4-7. Prepare a flowchart and corresponding program which will process GC peaks as shown for peaks A and B in Fig. 4-6.

4-8. After reading Chap. 5, do Exercise 4-7 for peaks C and D in Fig. 4-6.

BIBLIOGRAPHY

1. Mertz, P.: in E. M. Grabbe, S. Ramo, D. E. Wooldridge (eds.), "Handbook of Automation, Computation, and Control," vol. 1. chap. 17, Wiley, New York, 1958.
2. Whitaker, S., and R. L. Pigford: *Ind. Eng. Chem.*, 52:185 (1960).
3. Guest, P. G.: "Numerical Methods of Curve Fitting," p. 349 ff. Cambridge, London, 1961.
4. Savitzky, A., and M. J. E. Golay: *Anal. Chem.*, 36:1627 (1964).
5. Mandel, J.: "The Statistical Analysis of Experimental Data," Interscience, New York, 1964.
6. Ralston, A.: "A First Course in Numerical Analysis," McGraw-Hill, New York, 1964.
7. Groves, W. E.: "Brief Numerical Methods," Prentice-Hall, Englewood Cliffs, N.J., 1966.
8. Bevington, P. R.: "Data Reduction and Error Analysis for the Physical Sciences," McGraw-Hill, New York, 1969.
9. McCullough, R. D.: *J. Gas Chromatogr.*, 5:635 (1967).
10. Westerberg, A. W.: *Anal. Chem.*, 41:1770 (1969).
11. Baumann, F., A. C. Brown, and M. B. Mitchell: *J. Gas Chromatogr.*, 5:635 (1967).
12. Hancock, H. A., Jr., L. A. Dahm, and J. F. Muldoon: *J. Chromatogr. Sci.*, 8:57 (1970).

5
Integer, Floating-Point, and Abbreviated Arithmetic Algorithms

Up to this point, we have considered assembly-language programming from the point of view strictly of integer operations. Nothing more elegant than an addition or subtraction operation has been considered. However, much more complex mathematical operations can be carried out with the digital computer. The question here is how to go about programming arithmetic operations with the digital computer whose *only* arithmetic instruction is an ADD instruction. The answer is that one can take advantage of other logical operations, and by combining the simple, elementary machine-language instructions, the extremely complex mathematical operations of which the computer is capable can be generated.

5-1   INTEGER-MULTIPLY ALGORITHM

Consider the binary multiplication example shown below:

$$
\begin{array}{ll}
1011 & \text{(multiplicand)} \\
\underline{1001} & \text{(multiplier)} \\
1011 \\
00000 \\
000000 & \text{(partial products)} \\
\underline{1011000} \\
1100011 & \text{(product)}
\end{array}
$$

The operations are exactly analogous to familiar decimal multiplication. To develop a computer program to reproduce these operations, analyze the individual steps involved: Starting with the least significant bit of the multiplier, each succeeding bit is checked to see if it is a 1 or 0. If it is a 1, the multiplicand is multiplied by the appropriate power of 2 designated by the location of the 1 bit in the multiplier. (For example, the fourth bit to the left represents 2^3.) Each partial product is then added to the last until all multiplier bits have been checked. The sum of partial products is the final product.

An algorithm to carry out binary integer multiplication of positive values is diagramed in Fig. 5-1. It is based on the use of two 16-bit arithmetic registers, labeled A and B. The multiplier is originally placed in the A register, while the multiplicand is stored in memory. The B register is cleared initially. The least significant bit of A is checked with an SLA instruction (skip on the least significant bit of A). When it is a 1, the multiplicand is added to the B register. Successively higher-order multiplier bits are checked sequentially by shifting A right. With each shift, the accumulated partial products in B are shifted right through the 1-bit E register into the upper bits of A. Thus, successive additions of the multiplicand to the B register are *effectively* shifted *left* with respect to the earlier partial product additions. After all 16 bits of the A register have been checked, the final 32-bit product is found in the A and B registers. The most significant half (MSH) is in B; the least significant half (LSH) is in A.

Example 5-1 illustrates the program defined by the algorithm of Fig. 5-1. It is the integer-multiply subroutine supplied with standard Hewlett-Packard software.

Two aspects of the multiply routine of Example 5-1 should be noted: (1) The first few instructions are used to determine the signs of the multiplier and multiplicand. The values are negated, if necessary, to make both positive. The sign of the final product is determined, and the value of SIGN is set accordingly. (2) The 16-bit shift program includes 15 successive identical 3-word instructions. Why was this not written as a loop? Much fewer instructions would be required; however, the execution time would be increased significantly. This is because any program loop must have some amount of bookkeeping built into it. The

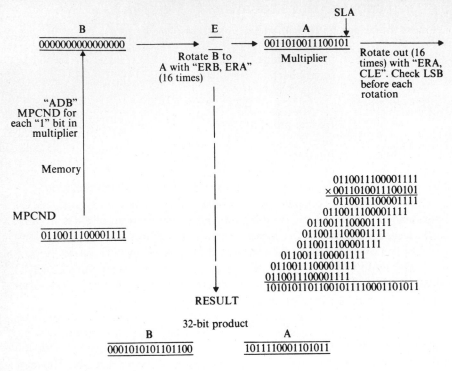

Fig. 5-1   Integer-multiply algorithm.

Example 5-1   Integer-multiply subroutine

	Call sequence:	/ENTER W/MLTPLR IN "A"
	JSB MPY	
	DEF MPCND	
	.	
	.	
	.	
	NAM MPY	
	ENT MPY	
MPY	NOP	
	LDB MIN2	/SET SIGN TO−2 ASSUMING POSITIVE
	STB SIGN	
	LDB MPY,I	/GET ADRSS OF MPCND
	LDB 1,I	/GET MPCND
	CLE,SSA	/IS MLTPLR NEG?
	CMA,CME,INA	/YES, MAKE POS. & COMPLMNT "E"

Example 5-1   (*continued*)

```
              SSB                    /SAME TEST FOR MPCND
              CMB,CME,INB
              SEZ                    /IF RESLT TO BE NEG, INCRMNT
                                     SIGN TO −1
              ISZ   SIGN
              ISZ   MPY              /GENERATE RETRN ADRSS
              STB   B                /SAVE MPCND
              CLB                    /INITIALIZE PRODUCT TO ZERO
              CLE,SLA                /SHIFT, TEST & ADD SEQUENCE
              ADB   B                /ADD MPCND TO "B" IF LSB "A"
                                     = "1"
              ERB                    /ROTATE INTO "A," FOR BIT 0 OF
                                     "A"
              ERA,CLE,SLA
              ADB   B                /REPEAT
              ERB                    /BIT 1
              ERA,CLE,SLA
              ADB   B
              ERB                    /BIT 2
              ERA,CLE,SLA
              ADB   B
              ERB                    /BIT 3
              ERA,CLE,SLA
              ERB                    /BIT 4
                .
                .
                .
              ADB   B
              ERB                    /BIT 15
              ERA,CLE
              ISZ   SIGN             /TEST FOR RESLT NEGATIVE
              JMP   *+4              /NO
              CMB                    /DOUBLE LNGTH COMPLEMENT
              CMA,INA,SZA,RSS
              INB
              CLO
              JMP   MPY,I            /EXIT
MIN2          DEC   −2
SIGN          BSS   1
B             BSS   1
              END
```

number of times through must be checked each time, and a JMP to the starting point must be provided. In fact, the execution time of the multiply subroutine would nearly *double* if the multiply algorithm were written as a loop. Thus, the choice was made to trade memory space for execution time in this case.

5-2 INTEGER-DIVIDE ALGORITHM

Consider the following binary division problem:

$$
\begin{array}{r}
0001001 \quad \text{(quotient)} \\
\text{(Divisor)} \quad 1001\,\overline{)\,1010011} \quad \text{(dividend)} \\
\underline{1001000} \\
1011 \\
\underline{1001} \\
10 \quad \text{(remainder)}
\end{array}
$$

The individual steps in the divide algorithm involve repetitive subtraction of the largest possible power-of-2 multiples of the divisor from the dividend and subsequent remainders. For every possible subtraction, a 1 bit is recorded in the quotient in the column corresponding to the appropriate power-of-2 multiple of the divisor which has been subtracted. The 0s are recorded for multiples which cannot be subtracted from the dividend or intermediate remainders.

The integer-divide algorithm diagramed in Fig. 5-2 is based on the divide subroutine included in Hewlett-Packard standard software. It assumes positive values or converts to positive values, with the sign of the quotient being assigned

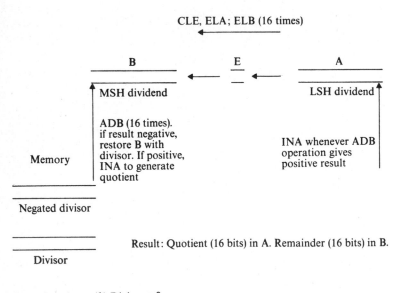

Fig. 5-2 Integer-divide algorithm. Enter with 32-bit dividend and 16-bit divisor.

at the end, just as for the multiply subroutine. The dividend is represented by a 32-bit word—MSH in B, LSH in A initially. The divisor and the *negated* divisor are both stored in memory. The *negative* divisor is added to the B register (MSH), and the intermediate remainder is checked to see if it is positive or negative. If it is negative, the divisor was too *large* to be subtracted; the original dividend is restored by adding the positive divisor to the B register, and a 0 bit is recorded in the quotient. If the intermediate remainder is positive, then the divisor was *not* too large to be subtracted, and a 1 bit is recorded in the quotient. The remainder from the successful subtraction becomes the new dividend. The quotient is generated by shifting A left and then placing a 1 or 0 in the least significant bit of A. The A register is shifted left (through E) into B, and B is shifted left after each attempted divisor subtraction. This procedure is repeated 16 times with the divisor being subtracted from successively smaller values, each shift dividing the dividend by a factor of 2. The final 16-bit quotient is then contained in the A register, with the final 16-bit remainder in the B register. The subroutine then assigns the proper sign to the quotient.

There are two important error situations which must be detected by the subroutine. One of these is the case where the divisor is 0. The other is when the quotient will be greater than a 16-bit value (15 bits plus a sign bit in 2's-complement notation). The former situation is easily checked when the divisor is picked up by the subroutine. The latter situation is detectable when the *first* subtraction of the divisor from MSH of the dividend is attempted. If this initial result is *positive*, the eventual quotient will be too large. In either of these error cases, the subroutine is exited immediately with an error condition indicated.

The execution time of this routine is somewhat longer than the multiply routine. Part of the reason is because it is written with a programmed loop for the repetitive subtraction. This alternative was chosen because the sequential coding approach would have resulted in an unacceptably long (memory space required) subroutine in this case.

It should be obvious to the reader by now that a more thorough familiarity with those machine instructions which execute logical and test operations on the arithmetic registers is required to master the material in this chapter. A detailed list of the alter-skip and shift-rotate instructions is provided in Sec. A-1. These should be reviewed before proceeding with the rest of this chapter.

It should be pointed out that the arithmetic algorithms described here are not restricted to computers with two arithmetic registers. They can be applied to a computer with one working register and an Extend 1-bit register, using a core memory location as the second register. The actual programming may be more awkward, however, unless one can add to or execute logical operations·on the contents of memory locations.

5-3   FLOATING-POINT ARITHMETIC

Obviously, if only integer arithmetic operations are used, the resolution is limited to that of a single computer word. In many cases, that resolution may not be adequate. For example, for a 16-bit word length, single-precision range is ±32,768. Double-precision range may be required to handle the numbers in a particular arithmetic operation, and programs can be written that handle all the arithmetic operations in double precision. For example, one could multiply 32-bit words by 16-bit words, divide 48-bit words by 16- or 32-bit words, etc. The range of integer values for a 32-bit word is $\pm 2 \times 10^9$. However, in many cases even this range may be inadequate.

Another problem with integer arithmetic is that truncation errors occur. For example, if one divides 5 by 2, the answer is 2. There is no fractional part for the quotient. The integer representation does not provide for values between 0 and 1. Obviously, serious mathematical errors can result when truncation occurs in integer arithmetic operations.

To provide extended range and to avoid truncation errors, *floating-point* arithmetic operations can be used. The machine format for floating-point arithmetic is simply this: Two words (or parts of words) are used to store each numerical value. The significant digits of each value are stored separately from an exponential term. Thus, there are two parts of each floating-point word saved in memory. One part is the *mantissa*, containing the significant digits; the other is the *exponent*, containing the power-of-2 multiplier to obtain the proper magnitude of the stored value. For example, consider Fig. 5-3 and the floating-point value 1011.11011. If this binary number were evaluated, it would be 11.85 (decimal). But this is a *binal* number, i.e., a binary number where both integer and fractional parts are represented. It is exactly analogous to the decimal system. Now, to store this floating-point number in memory, the computer must shift the binal point four places to the left to obtain the mantissa. The exponential term should then be set equal to 4 to retain the correct magnitude.

For the Hewlett-Packard software, the resultant floating-point value will be stored, as shown in Fig. 5-3, in 2 words. The mantissa is in 24-bit 2's-complement notation filling 1½ computer words. The lower 8 bits of the second word include a 7-bit exponent and the sign. The sign bit of the exponent is the least significant bit of the second word; the sign bit of the mantissa is the most significant bit of the first word.

It should be emphasized that the specific format described here for representing floating-point values in two 16-bit words is peculiar to the particular computer system considered. Other computers may have a different format. For example, the Digital Equipment Corp. (DEC) PDP-8 uses three 12-bit words to represent a floating-point value. A 24-bit mantissa and a 12-bit exponential term give quite a large range, but 3 words are required to represent each number. The range of values for the Hewlett-Packard format is $\pm 10^{\pm 38}$, with seven

Use two data words to store each floating-point datum;
e.g., consider the binary floating-point number:

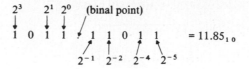

To express this floating-point number in memory:

(1)   Shift binal point left 4 bits, and multiply by 2^4.

$$1011.11011 = 0.101111011 \times 2^4$$

(2)   Store shifted binary number (plus sign) in one word
plus 8 bits of second word. Store power of 2 (plus
sign) in lower 8 bits of second word.

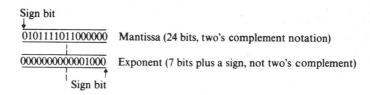

Sign bit

0101111011000000   Mantissa (24 bits, two's complement notation)

0000000000001000   Exponent (7 bits plus a sign, not two's complement)

Sign bit

Fig. 5-3   Machine format for floating-point arithmetic.

significant figures. The range for the DEC PDP-8 format is $\pm 10^{\pm 614}$ with seven
significant figures.

5-4   EXAMPLES OF FLOATING-POINT OPERATIONS

Example 5-2 demonstrates some simple conversions of binal numbers to the
machine-storage format described in Fig. 5-3. The examples are self-explanatory.
It should be noted that a value requiring more than 24 bits for storage is
truncated, with the least significant bits being dropped.

Multiplication and division of floating-point values are fairly straightforward. The mantissas are multiplied (or divided) like integer values; the
exponents are added (or subtracted); the binal point of the resultant mantissa
value is adjusted to retain only the significant digits, and the exponent value is
adjusted accordingly. This operation is illustrated for multiplication in Example
5-3. Note that the mantissa multiplication is shown as a 16-bit integer-multiply
operation. The important point is that an *integer*-multiply routine can be used.
However, a 24-bit integer-multiply routine should be used to ensure retention of
the same significance in the product as in the multiplied values.

Example 5-2   Conversion of binal numbers to floating-point format

a. $1011110110111. = .1011110110111 \times 2^{13}$

$$\underline{0101111011011100}$$

$$\underline{0000000000011010}$$

b. $-0.0001011110110111 = -0.1011110110111 \times 2^{-3}$

$$\underline{1010000100100100}$$

(*Note:* 2's complement of mantissa.)

$$\underline{0000000000000111}$$

Note: In this format, if the mantissa required more than 24 bits, the least significant bits would be dropped.

c. $11001.001101111011110100011 \times 2^{13} = .110010011011110111110100 \times 2^{18}$
(Truncated)

$$\underline{0110010011011110}$$

$$\underline{1111010000100100}$$

Example 5-3   Floating-point multiplication

$$1011.11 \times 1.01 = (0.101111 \times 2^4) \times (0.101 \times 2^1)$$

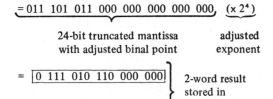

$$
\begin{array}{ll}
0101\ 111\ 000\ 000\ 000 & (\times 2^4) \\
\times\, 0101\ 000\ 000\ 000\ 000 & (\times 2^1) \\
\hline
0011\ 101\ 011\ 000\ 000\ \ 000\ 000\ 000\ 000\ 000 & (\times 2^5)
\end{array}
$$

32-bit product

$$= 011\ 101\ 011\ 000\ 000\ 000\ 000\ 000, \quad (\times 2^4)$$

24-bit truncated mantissa   adjusted
with adjusted binal point   exponent

$$= \boxed{0\ 111\ 010\ 110\ 000\ 000}$$ 2-word result
          stored in
$$\boxed{0\ 000\ 000\ 000\ 001\ 000}$$ memory

Addition and subtraction of floating-point values are considerably more difficult than integer-add and integer-subtract operations. The main problem is that the values added together must have the same exponential terms. Thus, in order to carry out add and subtract operations, the procedures illustrated in

Example 5-4 are involved. For values having largely different exponential terms, much time is spent in shifting the mantissa to allow the add operation.

Example 5-4   Floating-point addition

$$100.1001 + 1.0110 = (0.101001 \times 2^3) + (0.10110 \times 2^1)$$

$$\begin{array}{r} 0.101001 \times 2^3 \\ +\ 0.001011 \times 2^3 \\ \hline 0.110100 \times 2^3 \end{array}$$

$$\left. \begin{array}{l} \boxed{0110\ 100\ 000\ 000\ 000} \\[2em] \boxed{0000\ 000\ 000\ 000\ 110} \end{array} \right\} \begin{array}{l} \text{2-word result} \\ \text{stored in} \\ \text{memory} \end{array}$$

Table 5-1 summarizes the time and space requirements of some integer and floating-point mathematical subroutines for the H.P. 2116 computer with a 1.6-μsec memory cycle time. Note that the core memory required for each subroutine does not include space needed for other subroutines called from the subroutine.

5-5   ABBREVIATED MATHEMATICAL TECHNIQUES

The types of mathematical subroutines described above are typical of *general* routines. You need know very little about the data operated upon in order to use the subroutine. This advantage can prove costly in both execution time and memory space. A simple example of the cost can be seen in contrasting the

Table 5-1   Comparison of time and space required for integer and floating-point operations (H.P. 2116)

Operation	Integer	Floating-point
Time required		
ADD	3.2 μsec	300-900 μsec
MPY	90-150 μsec	640-750 μsec
DIV	300 μsec	1,200-1,500 μsec
Core memory required		
ADD	2	94_{10}
MPY	73_{10}	51_{10}
DIV	49_{10}	70_{10}

time/space requirements of floating-point-add and integer-add operations. If the programmer can be sure that his sums will never exceed single-precision integer capabilities, he can save much program time and space by *not* using floating-point-add operations.

Considerations of computational time/space requirements are particularly critical when the digital computer is to be used for real-time experimental control. That is, where the computer must be executing mathematical computations on experimental data—during the data-acquisition process—the mathematical execution time may be very critical. (See Chap. 9 for detailed discussions.) In such a situation, the programmer may be forced to use abbreviated mathematical operations, some of which will be described here.

One obvious situation to avoid when programming for real-time calculations is that where floating-point operations are required. One alternative is to use integer arithmetic. If the single-precision range of integer calculations is inadequate, one may be able to use double-precision integer routines. Generally, even double-precision integer mathematical routines can be executed more rapidly than floating-point.

Of course, the problem of speeding up required real-time calculations can be solved by including a hardware arithmetic device in the digital computer. For example, integer-divide and integer-multiply operations can be carried out in 16.7 and 10.7 μsec in an H.P. 2100 computer with integer arithmetic hardware. Similar savings in time can be gained by incorporating floating-point arithmetic hardware, an option which is becoming more available for small digital computers. (For example, maximum floating-point divide and multiply times are 55.9 and 41.1 μsec for an H.P. 2100 computer with floating-point hardware.) Moreover, the memory space usually allocated to the mathematical routines becomes available when a hardware arithmetic device is used.

In the absence of a hardware arithmetic unit, there are many shortcut approaches that can be taken to provide fast, adequate mathematical operations for real-time programming. Generally, these shortcuts require that the programmer have some advance information about the magnitude of the numbers which will be processed. For example, if one knows that the multiplier for a required real-time multiplication will always be an integer value between 1 and 10 and that the anticipated product will never exceed the single-precision word size of the computer, multiplication can be carried out very efficiently with a repetitive addition loop. The loop will add the multiplicand to itself n times, where n is the multiplier.

Analogous considerations allow a programmer to generate an abbreviated real-time divide routine involving repetitive subtraction. In either case, the approach is only advantageous for multiplication or division by small integer values.

Arithmetic program execution time can be diminished even when the numbers involved are large if it is possible always to multiply or divide by integral powers of 2. Thus, multiplication by 512 will involve only shifting the multiplicand left nine places. Division by 128 will involve a seven-place right shift. Of course, the additional limitation of being able to represent initial and final values within the single-precision word size must be recognized. Also, in the particular case of division it should be noted that significant bits may be lost with a right shift. Some illustrations are given in Example 5-5. (Note that the result of the division is observed as 709. However, because the least significant bits of the dividend were shifted out, some precision was lost.)

Example 5-5   Multiplication and division by powers of 2

   a. Multiplication by powers of 2

   $100_{10} \times 128_{10}$ $= 0$ 000 000 001 100 100 $(\times 2^7)$
   $= 0$ 011 001 000 000 000
   $= 12,800_{10}$
   (Product obtained with seven-place left shift)

   b. Division by powers of 2

   $22,691_{10} \div 32_{10}$ $= 0$ 101 100 010 100 011 $(\div 2^5)$
   $= 0$ 000 001 011 000 101
   $= 709_{10}$

   (Quotient obtained with five-place right shift)

   c. Quick multiplication by nonintegral power of 2

   $612_{10} \times 10_{10}$ $= 0$ 000 001 001 100 100 $[\times (2^3 + 2^1)]$

   0 000 010 011 001 000 (shift left one)
   $+ 0$ 001 001 100 100 000 (shift left three)
   0 001 011 111 101 000

   $= 6120_{10}$

It is also possible to provide quick multiply and divide routines for certain integer values which are not integral powers of 2. For example, multiplication by 10 involves multiplication by $(2^3 + 2^1)$. Thus, the multiplicand can be shifted left one place, this result $(\times 2^1)$ saved, and then shifted left two more places $(\times 2^3)$. The saved value added to the latter $(\times 2^3)$ value gives the desired $(\times 10)$ product. This is shown also in Example 5-5.

5-6   SCALING TECHNIQUES

As indicated, the abbreviated mathematical techniques described above will fail whenever the range of initial or final numerical values exceeds the integer range available. The calculations may require floating-point arithmetic, and the computational time may prove excessive and incompatible with the real-time programming objectives. A general approach that may allow the use of integer arithmetic and still provide for a wide, effective numerical range involves the technique of *scaling* data. That is, any number, integer or fraction, can be scaled so that it can be represented within a single computer word. Also, by appropriate selection of scaling factors and sequencing of mathematical operations, intermediate and final results can be restricted to single- or double-precision integer values. Thus, standard integer arithmetic subroutines (or hardware) can be used; while the large-range capability of floating-point mathematics is still provided for (within limits).

Some examples of scaling values for integer calculations are shown in Example 5-6. Note that in each case scaling factors were chosen to keep each individual operation within integer mathematical limits. Also note that numbers were not *rounded* but *truncated* when the least significant bits could not be included in the scaled value. The scale factors were selected for 16-bit integer precision. In Example 5-6b and c, scaling was arbitrarily adjusted to allow only

Example 5-6   Scaling examples

a. $\underline{135.7} \times \underline{0.00013}$

Scale by   Scale by
10^1 10^5

$\Downarrow$

$\underline{(1357 \times 13)} \times 10^{-6}$

Integer
multiply
$\Downarrow$
$\underline{17,641} \times \underline{(10^{-6})}$

Integer   Total
product   scale
              factor

b. $\underline{317,542} \times \underline{1.016}$

Scale by   Scale by
10^{-3} 10^2

$\Downarrow$

$(317 \times 101) \times 10^1 = \underline{32,017} \times \underline{10^1}$

Integer   Total scale
product   factor

c. $\dfrac{3.217 \times 1.106}{2.571} = \dfrac{32,170 \times 10^{-4}}{257 \times 10^{-2}} \times (110 \times 10^{-2})$

$= (125 \times 10^{-2}) \times (110 \times 10^{-2})$

$= 13,750 \times 10^{-4}$

about 1 percent precision in the numbers used in order to avoid exceeding 16-bit precision in the final products. Each of these calculations could have been carried out using standard single-precision integer-multiply and integer-divide routines. However, the range of values used would have had to be known in advance to select scale factors. Moreover, the total scale factor (exponent) would have to be saved in memory in order to allow appropriate readout or other mathematical usage of the final data.

The various considerations for handling a typical computational problem by scaling techniques are illustrated in Example 5-7. Several aspects of this example should be pointed out. Most importantly, note that C had to be scaled by 10^4 because it was to be added to the scaled product of $A \times B$, which had a combined scale factor of 10^4. Also, note that $A_S \times B_S$ and C_S could exceed 16 bits because they form the dividend for a subsequent division step; the integer-divide routine assumes a 32-bit dividend. However, the addition step must be a double-precision addition.

Example 5-7   Scaling a general computation formula

$$Q = \frac{A \times B + C}{D}$$

Known:

Variables	Range of variables	Suggested scaling factors
A	10.0-30.0	$A_S = A \times 10^1$
B	0.100-0.900	$B_S = B \times 10^3$
C	1.000-10.00	$C_S = C \times 10^4$
D	5.00-7.00	$D_S = D \times 10^2$
		$\therefore Q = Q_S \times 10^{-2}$

Note: The scale factor for C must equal that for $A \times B$ since they are added.

Problem:
Will any of the *scaled* values exceed a 16-bit word? (This is OK for the dividend since integer division assumes a 32-bit dividend, but the divisor cannot exceed 16 bits, nor can the quotient.)

Check out:

$$A_S(\text{max}) \times B_S(\text{max}) = 27 \times 10^4 \quad > 16 \text{ bits} < 32 \text{ bits} = \text{OK}$$
$$A_S(\text{max}) \times B_S(\text{max}) + C_S(\text{max}) = 37 \times 10^4 = \quad \text{OK}$$
$$D_S(\text{min}) = 5 \times 10^2$$
$$\therefore Q_S(\text{max}) = 7.4 \times 10^2 \quad < 16 \text{ bits} = \text{OK}$$

In setting up the scale factors, the programmer must determine if any operations will exceed the integer arithmetic capabilities. The simple check-out steps are listed in Example 5-7. If one knows the expected range of each variable in Example 5-7, A_s(max), B_s(max), and C_s(max) can be calculated and found to be within double-precision limits. Also, Q_s(max) can be computed if one knows D_s(min), and it is found to be within 16-bit integer range.

Although there is a considerable effort involved in setting up a scaled computational formula, the benefits can be substantial when that calculation must be performed in real time. Quite often most of the variables are known before experimental data acquisition starts, and only scaling of experimental data may be required in real time. If real-time scaling is necessary, powers-of-2 scale factors should be used to save time.

5-7 PERSPECTIVE

The discussions presented in this chapter are not intended to encourage computer users to write their own mathematical subroutines. The mathematical subroutines available through manufacturers' standard software libraries should be used under most computational circumstances. However, the discussions here should provide some insight into the nature of simple mathematical programming. Moreover, an appreciation of the machine format requirements for integer and, particularly, for floating-point numbers should be obtained. Most importantly, this chapter considers the limitations of standard mathematical programming for high-speed computations. It is hoped that the elementary discussions of abbreviated mathematical techniques will be of some assistance to the experimentalist using the small digital computer in a real-time environment.

EXERCISES

5-1. Write a subroutine which will apply any given scaling factor to a floating-point value (stored in Hewlett-Packard 2-word format) and generate the appropriate integer value in the A or B register before exit.

5-2. Write a subroutine corresponding to the integer-divide algorithm of Fig. 5-2.

5-3. Define an algorithm, and write a subroutine for integer multiplication where only the A and E registers are used.

5-4. Determine the appropriate scale factors to utilize single-precision integer mathematical routines to carry out calculations with the following formula:

$$Q = \frac{A \times B + C^2}{D \times (E+F)^2}$$

where

Variable	Range
A	1.00 -10.0
B	0.200-0.400
C	2.000-3.000
D	10.00-100.0
E	0.500-1.000
F	0.100-10.0

In arriving at your conclusions, attempt to maintain the greatest precision you can. Also, consider alternative groupings of operations to optimize precision. (Assume that the squaring operations are executed with the normal multiplication subroutine.)

5-5. Devise an algorithm for double-precision multiplication which uses the normal single-precision-multiply subroutine. [*Hint*: $X \times Y = (a + b) \times (c + d) = ac + ad + bc + bd$. That is, X and Y are 32-bit words, each of which can be represented by the sum of a 32-bit word and a 16-bit word. For example,

$$X = x \quad xxx \quad xxx \quad xxx \quad xxx \quad xxx \qquad 0 \quad 000 \quad 000 \quad 000 \quad 000 \quad 000 \; +$$
$$x \quad xxx \quad xxx \quad xxx \quad xxx \quad xxx.$$

Thus, each cross product (*ac*, *ad*, etc.) can be obtained by a single-precision multiplication; each cross product is scaled by 2^0, 2^{-16}, or 2^{-32}. Appropriate overlapping addition of the cross products will achieve the proper result.]

6
Input/Output
with Standard
Peripheral Devices

Up to this point, the only input/output (I/O) possibilities considered have been limited to using the console register display for output and the console switch register (SR) for input. One soon realizes that with this crude form of communication with the computer, most of the computer's time is spent waiting for the operator to respond. It certainly appears that the time has come to consider more efficient ways for conversing with the computer.

This chapter will describe the features and operations of the most basic devices used for computer I/O. These include the teleprinter (the *Teletype* will be discussed here), the high-speed punched paper-tape reader, and the high-speed paper-tape punch. These are the most common standard I/O peripherals to be found on small computer systems, and they will be discussed in detail here. Moreover, the same principles of I/O programming and interfacing are generally applicable to all types of peripheral devices.

6.1 THE TELEPRINTER

The Teletype teleprinter is representative of a large variety of available mechanical/electronic devices for on-line communication with the digital computer. The Teletype has a keyboard at which the operator can normally select over 100 different alphanumeric and symbolic characters to compose messages. Up to 256 different characters are possible since an 8-bit coding is used. [In practice, however, only the least significant 7 bits are used for coding, which limits the character set to 128. The most significant bit (MSB) is sometimes used as a *parity* bit to check for transmission errors.] (Refer to Fig. 6-1 for a graphic description of a Teletype teleprinter.) Keys struck on the keyboard give rise to printed characters. Also, keyboard character entries generate an appropriate 8-bit digital code, unique for each character, within the Teletype unit. These 8-bit *bytes* of information are transferred in a *serial* fashion to the computer. That is, the Teletype generates a string of precisely timed events, during each of which the presence or absence of a voltage pulse corresponds to a 1 or 0. An interface module can be employed to provide for *parallel* transfer to and from the computer while accommodating *bit-serial* transfer to and from the Teletype. This sort of hardware interface eliminates tedious I/O programming for the Teletype. Figure 6-2 illustrates the Teletype I/O system.

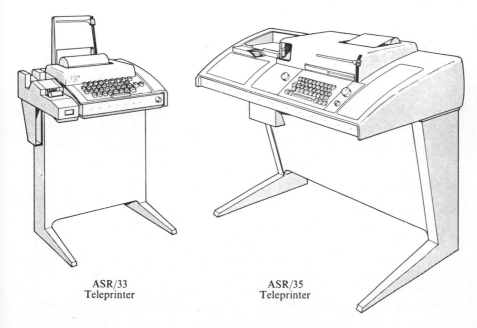

ASR/33
Teleprinter

ASR/35
Teleprinter

Fig. 6-1   Teleprinter devices. (*Teletype Corp., Skokie, Ill.*)

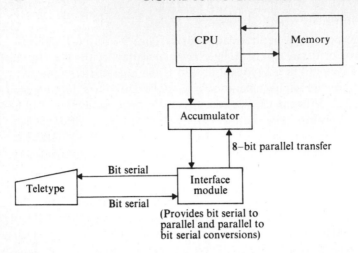

Fig. 6-2   Basic communication link between Teletype printer and computer.

Thus, a line of communication can be established between the keyboard operator and the computer, allowing the transfer of 8-bit bytes of information at a time. Conversely, the computer can output 8-bit bytes of information to the Teletype. The Teletype is capable of automatically decoding the transferred data and typing the appropriate character.

Thus, once the Teletype unit is linked to the computer by a cable and appropriate interface module which can transfer 8-bit bytes of information into or out of the computer's accumulator (AC), it would seem that the communication problem is solved. This is very far from the truth! The basic problem is simply that the computer does not know a priori how to talk to a Teletype. For example, one of the minimum requirements must be a *control* line to allow the computer to tell the Teletype unit that the CPU has transferred data which it should now print. Conversely, the Teletype should be able to tell the computer when it has completed the print operation—which requires about 0.1 sec—and is ready to print the next character. In addition, on input operations, the Teletype should be able to tell the computer that a character has been entered on the keyboard and the code is ready in the interface buffer register to be transferred to the AC. Thus, a second, *flag*, line must be provided in the complete communication link. The complete linkage is shown in Fig. 6-3. Note that two additional lines are included which allow the computer to clear the flag and control bits. These lines are necessary to provide initialization of the peripheral device. Note also that a distinct interface module is assigned to the I/O channel to which the Teletype is attached. The interface module—oversimplified in Fig. 6-3—provides device-selection logic, serial-to-parallel and parallel-to-serial conversion, and electronic buffering.

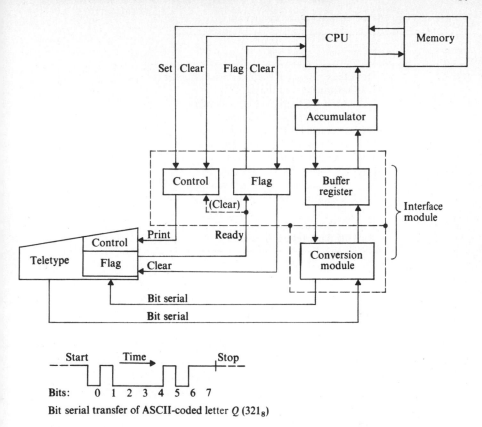

Bit serial transfer of ASCII-coded letter Q (321_8)

Fig. 6-3   Teletype printer interface and byte-transfer characteristics

The next critical step in providing for Teletype communication between the operator and the computer is to teach the computer to talk to the Teletype. In addition we must consider the I/O instructions which can be executed by the computer. These instructions are summarized in Table 6-1. As can be seen from Table 6-1, most of the I/O instructions operate on the I/O interface module for the particular channel designated. This reflects the basic configuration of the computer. That is, this computer has a built-in multichannel I/O structure. Thus, the execution of any I/O instruction will result in communication with only one selected channel. (An alternative approach is the *party-line* system where digital I/O information is transferred along common buses to or from all peripheral devices. The device to or from which information is transferred is determined by a device-selector interface module residing with the peripheral device. The party-line system is much less expensive to implement, but has the limitation that the computer does not immediately know which device has caused a set flag on the common flag line.)

Table 6-1   Input/output instruction set

Instruction	Function
LIA c	Input the contents of the interface buffer for I/O channel c into the AC
OTA c	Output the AC contents to the interface buffer register for I/O channel c
STC c	Set the control bit for I/O channel c
CLC c	Clear the control bit for I/O channel c
STF c	Set the flag bit for I/O channel c
CLF c	Clear the flag bit for I/O channel c
SFS c	Skip if the flag bit is set on I/O channel c
LIA 1	Load the AC with the contents of the SR
STF 0	Turn on the interrupt system
CLF 0	Turn off the interrupt system

Since the hardware interface buffer module is a device which really communicates with both the computer and the peripheral device, the details of how digital data transfers to and from this module occur are important. (Some readers may find it necessary to read Chaps. 7 and 8 before grasping the hardware aspects of the following discussion.) First of all, the parallel transfer of a binary word from a peripheral device to the interface buffer register can be controlled by the device ready flag. That is, the digital word from the device can be *strobed* (or *gated*) into the interface buffer register when the device flag is set, thereby causing the interface module's flag to be set. (Remember that these are flip-flops providing the flags.) The computer can then recognize the set flag on channel c and cause the data to be transferred from the buffer register into the AC by an LIA c instruction. (Note that for the Teletype peripheral the relatively uncommon bit-serial data transfer must be accommodated. Thus, the 8-bit byte is not strobed into the interface buffer, but rather is accepted 1 bit at a time into a *shift register*. The ready flag indicates when the byte transfer is complete.)

Similarly, to output a binary word to a peripheral device, the computer gates the contents of the AC into the buffer register of the interface module by an OTA c instruction. The contents of the interface buffer register are immediately available to the peripheral device by using the control bit in the interface module. That is, by executing an STC c instruction, the control bit of the interface module is set, and this can be used to tell the peripheral device that the contents of the interface buffer register are ready to be transferred and used. Thus, the STC instruction can effect an output data transfer and/or initiate operations by the peripheral device.

Output on the teleprinter Applying the I/O principles and operations discussed above, the flowchart shown in Figure 6-4 can be described for printing characters on the Teletype. The subroutine which accomplishes the printing of a single character is given in Example 6-1.

Example 6-1 Program for Teletype printer output of a single character (Exit after the printing is completed).

```
        .
        .
        .
        LDA  FCODE        /GET TTY FUNCTION CODE FOR PRINT
        OTA  TTY          /TELL TTY TO BE IN PRINT MODE
        LDA  ACHAR        /GET ASCII CODED CHARCTR IN AC
        OTA  TTY          /OUTPUT CODED CHRCTR TO TTYP
        STC  TTY          /TELL TTYP TO PRINT
        CLF  TTY          /CLEAR TTYP FLAG
        SFS  TTY          /WAIT FOR TTYP FLAG WHICH APPEARS
                             WHEN PRINT COMPLETE
        JMP  *-1
        CLF  TTY          /INITIALIZE
        CLC  TTY          /INITIALIZE
        .
        .
        .
TTY     EQU  15B          /EQUATE "TTY" TO 15B, TTYP I/O
                             CHANNL
FCODE   OCT  130000       /OCTAL CODE FOR TTY PRINT
                             FUNCTION
        .
        .
        .
```

Because the Teletype is both an input and output device, the first function of any Teletype I/O routine clearly must be to tell the device what function it is to perform. Thus, in Example 6-1 the *function code* for a print operation (130000_8) is output first to the Teletype, followed by the output of the 8-bit code for the character to be printed (ACHAR). The STC command then starts the print operation. The nature of this program is such that the computer does nothing else except output information to the Teletype. Thus, the efficiency of the computer is limited by the efficiency of the Teletype. Since the Teletype can print at a maximum rate of about 10 characters per second, the computer spends most of its time waiting for the Teletype flag, indicating that the print operation is complete. Thus, if the computer can execute one instruction every 2 μsec on the average, then it could have executed about 50,000 instructions between the time it told the Teletype to print and the time the Teletype finally completed

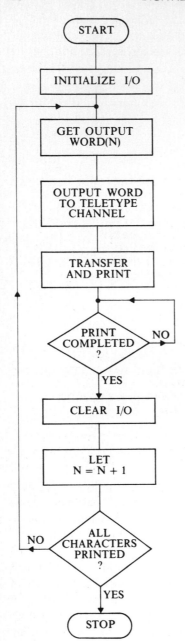

Fig. 6-4 Flowchart for typical Teletype printer print operation.

the print operation and set its ready flag. This means that the computer could really do very much useful work even while the printing is being carried out on the Teletype simultaneously. An alternative output routine which allows the computer to be executing useful programming between character prints is given in Example 6-2.

Example 6-2   Program for output of a single character on a Teletype   (Exit *before* the printing is completed.)

```
          .
          .
          .
     SFS   TTY              /TTY BUSY?
     JMP   *-1
     CLC   TTY              /INITIALIZE
     CLF   TTY              /INITIALIZE
     LDA   FCODE            /GET PRINT FUNCTION CODE
     OTA   FCODE            /TELL TTY TO BE IN PRINT MODE
     LDA   ACHAR            /GET NEW CHAR.
     OTA   TTY              /OUTPUT CHAR TO TTY
     STC   TTY              /PRINT COMMAND
          .
          .
          .
TTY       EQU   15B         /TTY = 15B
FCODE     OCT   130000      /PRINT FUNCTION CODE
```

Note: In Example 6-2 the computer checks the status of the Teletype upon *entering* the print routine. If the previously issued print command has been completed, there is no *computer* time wasted in executing the next print. Most importantly, the computer exits from the routine shortly after initiating the print cycle on the Teletype, and at least 0.1 sec of useful programming can be executed before it returns to print the next character. Of course, greater than 0.1-sec intervals between character prints can be allowed if efficient use of *Teletype* time is not required.

An important point to be made regarding the flowchart of Fig. 6-4 and Examples 6-1 and 6-2 is that although the clearing (initialization) of I/O logic after each output (or input) operation is suggested, it is not always necessary. For example, the control bit might be cleared automatically whenever the flag bit is set by the peripheral device as shown in Fig. 6-3. This relieves the programmer from incorporating a CLC instruction into the program when the character printing is completed. (This is the actual situation for the computer system used here.) However, the flag bit must still be cleared before one tries to carry out another print operation. From Fig. 6-3, it can be seen that the CLF instruction will clear the flag bit in the interface module as well as in the Teletype logic. The buffer flip-flop registers themselves do not have to be cleared in this case because gating is used which replaces the contents of one register with the contents of the other. (Refer to Chap. 7.)

Input through the teleprinter keyboard   The programming involved to input digital information to the computer through the teletype keyboard is determined by the hardware characteristics of the Teletype and the interface module in analogous fashion to output programming. A flowchart for keyboard input operations is given in Fig. 6-5. A subroutine to take in a single keyboard entry is given in Example 6-3.

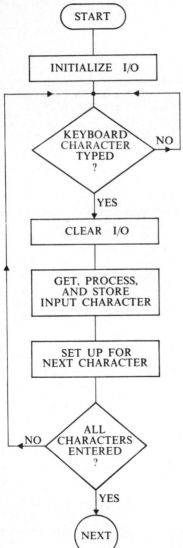

Fig. 6-5   Flowchart for typical Teletype printer keyboard input operation.

Example 6-3   Program for Teletype keyboard input

```
                .
                .
                .
        LDA   FCODE        /TELL TTY IT IS TO READ IN FROM
                           KEYBOARD
        OTA   TTY
        STC   TTY
        CLF   TTY          /ACTIVATE KEYBRD TO ACCEPT ENTRY
        SFS   TTY          /WAIT FOR FLAG INDICATING KEYBRD
                           ENTRY COMPLETE.
        JMP   *-1
        CLF   TTY          /CLEAR FLAG
        LIA   TTY          /BRING ENTRD ASCII CHARCTR INTO AC
                .
                .
                .
TTY     EQU   15B          /EQUATE "TTY" TO 15B, THE TTYP
                           /I/O CHANNL
FCODE   OCT   170000       /FUNCTION CODE FOR TTY KEYBRD
                           INPUT
```

The program of Example 6-3 reflects the assumption that both input and output Teletype operations can be handled with a single I/O channel. This will be true if the interface module has two buffer registers, one for input and one for output.

Punched paper-tape I/O with the teleprinter   One additional aspect of the Teletype teleprinter—and a tribute to its versatility—is the fact that paper tape can be punched or read in either an off-line or on-line mode. That is, when the Teletype printer is in the LOCAL mode, the operator can prepare a punched paper tape corresponding to the character string entered on the keyboard. (The specific coding is discussed below.) Also, in the LOCAL mode a punched paper-tape segment can be read in through the tape reader attachment with the coded characters being simultaneously printed by the Teletype. In addition, if the tape punch unit is on while a punched-tape segment is being read, the punch will duplicate the tape being read in.

In the on-line mode, the Teletype can be used to enter digital information stored on paper tape through the tape reader. The coding does not have to correspond to that representing the type-print characters available on the Teletype, but can represent any binary-coded format. In addition, punched paper-tape segments can be prepared under computer control in the on-line mode using the paper-tape punch of the Teletype. Again, the output can be any binary-coded format. The one restriction is that only 8 bits can be transferred at a time.

6-2   CODING OF INPUT/OUTPUT INFORMATION

In order to read or write information into or from the computer, there usually must be a translation of the information to a different form of coding. For example, a character struck on the Teletype keyboard generates an 8-bit coded value. If the character represented is a decimal number, it must be decoded and translated into binary form before the computer can do anything with it arithmetically. Conversely, if the computer is to output numerical information, that information must be translated from binary to some other coding which will be accepted by the output device, and the data must be transferred in such a manner as to be readily recognizable (such as a sequence of decimal digits with spaces separating discrete integer values).

For ordinary type-print characters (like A, B, C, 1, 2, 3, +, −, =, !, @, etc.) a standard 7-bit code has been developed. This coding is referred to as ASCII code. Examples of the coding are:

Character	Code	Character	Code	Character	Code
A	101_8	0	60_8	CR	15_8 (carriage return)
B	102_8	1	61_8	LF	12_8 (line feed)
C	103_8	2	62_8		
.		.			
.		.			
.		.			
Z	132_8	9	71_8		

Note that the code for digits 0 to 9 can be generated by adding 60_8 to the octal equivalent of each digit. (See Appendix A for the complete ASCII code listing.) The ASCII code can be used to represent type-print characters on paper tape as shown in Example 6-4.

When a character is typed on the Teletype keyboard, the Teletype hardware converts that character into ASCII code, and an 8-bit byte of information is generated. (It is also punched as an eight-level code on paper tape if the punch is on.) Thus, when the information is loaded into the computer, it is ASCII-coded. It must be decoded to be useful. Conversely, in order to output numerical data, they must be ASCII-coded in the computer before transmission to the Teletype.

For example, to type out the character 7, the result of a binary computation which is contained in location SUM, the following sequence is followed: (1) Load AC with SUM. (2) Add 60_8 to AC. (That is, convert to ASCII code for the character.) (3) Go to I/O routine to print out contents of AC. The I/O routine will output the ASCII-coded value 67_8 to the interface buffer register and tell the Teletype to print the character; the Teletype responds with a printed 7. [Also, if the tape punch is on, an ASCII-coded 7 (67_8) will be punched on tape.] The program could be written as shown in Example 6-5.

Example 6-4   Punched paper-tape format for various ASCII-coded characters

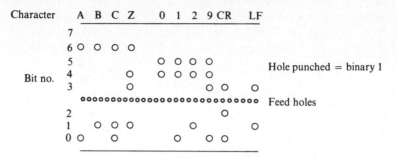

Note: The eighth bit (bit 7) is usually punched for ASCII coding. This allows one to easily identify an ASCII portion of punched tape. However, bit 7 is inessential for translation, can be ignored, and is sometimes not punched except as a parity check. That is, some punches will punch the eighth bit to cause an even (or odd) number of bits to always result for each word. Thus, a reader program could easily check for punch errors by checking for parity on each word.

Example 6-5   Routine to print digit with ASCII code

```
              .
              .
              .
       LDA  SUM      /GET NUMBER FROM 0 TO 9 (BINARY CODED)
       ADA  N60      /GENERATE ASCII CODE FOR NUMBR CHARACTR
       JSB  TYPE     /GO TO TTYP PRINT SUBROUT.
              .
              .
              .
TYPE   NOP           /ENTER W/ ASCII-CODED CHARCTR IN AC
       LDB  FCODE    /TELL TTY TO BE IN PRINT MODE
       OTB  TTY
       OTA  TTY      /OUTPUT ASCII CODE TO TTYP
       STC  TTY      /TELL TTYP TO PRINT
       CLF  TTY      /CLEAR FLAG
       SFS  TTY      /WAIT FOR FLAG WHEN PRINT COMPLETED
       JMP  *-1
       CLF  TTY      /CLEAR FLAG
       JMP  TYPE,I   /RETRN TO MAIN PROGRM
TTY    EQU  15B      /EQUATE "TTY" TO 15B, TTYP I/O CHANNL #
N60    OCT  60
FCODE  OCT  130000   /CODE FOR PRINT FUNCTION
              .
              .
              .
```

Note: The first instruction of the subroutine is an NOP, a *no operation*, whose only function is to use one memory cycle of time. Because it is a useless instruction, it is often used as the first in subroutines simply to reserve the linkage location.

If one wanted to type out the six-digit octal equivalent of a binary number stored in core memory, a more difficult task is involved. (1) The number must be broken down into parts corresponding to each octal digit. (2) Each of these must be coded in ASCII code. (3) The ASCII characters must be stored or generated sequentially for reference by the I/O routine. (4) The I/O routine must be called.

Two parts of this sequence need further explanation at this point. First, regarding the storage of ASCII characters, because these are 8-bit words, the most efficient storage of these is to pack them *two to a word*. Thus, if the number 17_8 were to be output, it could be packed as shown below:

	MSH		LSH	
Bit number 15		8	7	0
Contents	0 011 000	1	00 110 111	
	61_8		67_8	
	ASCII 1		ASCII 7	

Thus, the I/O routine written by the programmer might include not only *octal-to-ASCII* or *ASCII-to-octal conversion* but also an *ASCII packing* or *unpacking* routine in order to allow for efficient handling of ASCII-coded data.

A second consideration is how one can isolate the various octal digits contained in a binary word prior to octal-to-ASCII conversion. This can be done by taking advantage of the various shift-rotate instructions and the logical AND instruction. (See Appendix A.) The AND instruction performs a logical AND operation between the AC and the contents of location m; i.e., the result of the operation, found in the AC, is a binary number containing 1's in each bit location where *both* words on which the AND operation was performed had 1's prior to the operation. Thus, the value in location m can be used as a *mask* to *erase* all but those specific bits of interest in the AC.

Example 6-6   Logical AND memory reference instruction

Contents:	m	0	000	000	000	000	111	(7_8)
	AC	1	001	010	110	111	010	(112672_8)
Operation:		AND m						
Result in AC:	AC	0	000	000	000	000	010	(2_8)

Note: Only bits 0 to 2 of the AC come through the mask.

Example 6-7   Inclusive OR (IOR) memory reference instruction

Contents:	m	0	000	100	111	001	010
	AC	1	001	100	010	000	001
Operation:		IOR m					
Result in AC:	AC	1	001	100	111	001	011

Note: The logical AND and IOR instructions can only be performed with the A register in the H.P.2100 computer series.

Another useful logical memory reference instruction is the *inclusive* OR, IOR. It can be used in the packing routine. An IOR   *m* operation between the AC and location *m* results in a 1 in any bit of the AC where either the AC *or* the contents of location *m* had had a 1 previously.

6-3   USEFUL INPUT/OUTPUT SUBROUTINES

When one uses the Teletype for I/O, several fundamental operations are executed repeatedly. The simple operation of taking a single ASCII-coded character from the computer and causing that particular character to be printed on the Teletype, the reading in of a single character entered on the Teletype keyboard, the typing of a complete six-digit octal number corresponding to the 16-bit contents of a memory location, and the carriage-return-line-feed operation at the end of each printed line are all examples of simple operations executed repeatedly in I/O routines. It is convenient to consider each of these fundamental operations and prepare subroutines for each to provide efficient programming.

Print a character   The first subroutine to consider is one which simply takes an ASCII-coded value and executes a print operation on the Teletype. Such a subroutine was given in Example 6-5 (the Type subroutine). In order to use such a routine conveniently, it should be callable from anywhere in core memory.

Example 6-8   Type subroutine and zero-page linkage

```
            ORG  100B     /POINTERS TO SUBROUTS.
  PTYPE     DEF  TYPE
              .
              .
              .
            ORG  6000B
  TYPE      NOP            /CHARACTR PRINT. ENTR W/ASCII CHAR IN
                              ACCUMULATOR
            LDB  PCODE     /TELL TTY TO BE IN PRINT MODE
            OTB  TTY
            AND  MASK
            OTA  TTY       /OUTPUT ASCII CODE TO TTYP
            STC  TTY       /TELL TTYP TO PRINT
            CLF  TTY       /CLEAR FLAG
            SFS  TTY       /WAIT FOR FLAG WHEN PRINT COMPLETED
            JMP  *-1
            CLF  TTY       /CLEAR FLAG
            JMP  TYPE,I    /RETRN TO MAIN PROGRM
  TTY       EQU  15B       /EQUATE "TTY" TO 15B, TTYP I/O CHANNL #.
  N60       OCT  60
  PCODE     OCT  130000    /FUNCTION CODE FOR PRINT
  MASK      OCT  377       /MASK TO LET THROUGH ONLY LOWER 8-BITS
```

Thus, it should either be located on the zero page, or else the pointer to the subroutine should be stored on the zero page, and the subroutine may be called indirectly as shown in Example 6-8.

To call the subroutine of Example 6-8, the program gets the ASCII-coded character to be typed into the AC and then executes a JSB   PTYPE,I instruction. The appropriate character will then be printed on the Teletype printer. Note that the subroutine uses a mask to erase the upper 8-bits of the word containing the coded character to be printed. Thus, if any information was contained there, it will be lost when the routine is exited.

Read a character from the keyboard   A subroutine designed to read in one character from the keyboard can be written as suggested earlier in Example 6-3. This routine is shown in Example 6-9.

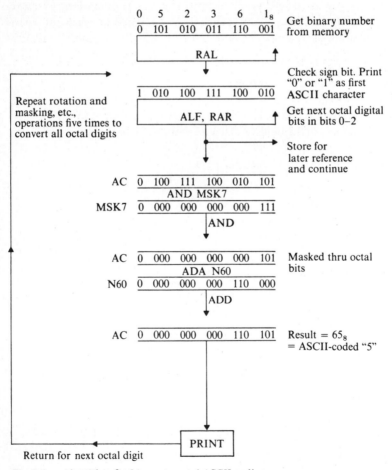

Fig. 6-6   Algorithm for binary-to-octal ASCII coding.

Example 6-9   Subroutine to read one keyboard entry

```
          ORG   101B
PINPT     DEF   INPUT
           .
           .
           .
          ORG   6020B
INPUT     NOP                 /SUBROUT. FOR KEYBRD ENTRY
          LDB   KCODE         /TELL TTY TO BE IN KEYBRD MODE
          OTB   TTY
          STC   TTY           /ACTIVATE KEYBRD TO ACCEPT ENTRY
          CLF   TTY
          SFS   TTY           /WAIT FOR FLAG INDICATING KEYBRD
                                ENTRY COMPLETE
          JMP   *-1
          CLF   TTY           /CLEAR FLAG
          LIA   TTY           /BRING ENTRD ASCII CHARCTR INTO AC
          JMP   INPUT,I       /RETRN MAIN PROGRAM W/ CHARACTR IN
                                ACCUMULATOR
KCODE     OCT   170000        /FUNCTION CODE FOR TTY KEYBRD ENTRY
```

Print six-character, octal equivalent   As illustrated in Fig. 6-6, a 16-bit binary value stored in a given memory location can be printed as a six-digit octal value. The subroutine which follows the algorithm of Fig. 6-6 is given in Example 6-10.

Example 6-10   Subroutine to take a 16-bit binary value and print the six-digit octal equivalent

```
          ORG   102B
PT6DG     DEF   T6DG          /POINTR TO T6DG SUBROUTINE
           .
           .
           .
          ORG   6040B
T6DG      NOP                 /ENTER W/ BINARY # IN AC
          RAL                 /GET BIT 15 INTO BIT 0
          STA   TEMP          /STORE PRTLY ROTTD WORD
          AND   MSK1          /MASK THRU BIT 0
          ADA   N60           /CNVRT TO ASCII-CODED CHAR
          JSB   PTYPE,I       /GO TO SUBROUT. TO PRINT ONE CHARACTR
LOOP      LDA   TEMP          /GET PARTLY ROTTD WORD
          ALF,RAR             /ROTATE NXT 3 BITS INTO BITS 0-2
          STA   TEMP          /STORE PRTLY ROTTD WORD
          AND   MSK7          /MASK THRU BITS 0-2
          ADA   N60           /CONVRT TO ASCII CODED CHARACTR
          JSB   PTYPE,I       /GO TO SUBROUT TO PRINT ONE CHRCTR
          ISZ   CNT5          /CONVRTD & PRNTD ALL 16 BITS?
          JMP   LOOP          /NO
          LDA   CT5ST         /YES--RESET FOR RE-ENTRY
```

Example 6-10   *(continued)*

```
            STA  CNT5
            JMP  T6DG,I      /RETRN MAIN PROGRM
CT5ST       OCT  −5
CNT5        OCT  −5
TEMP        OCT  0
N60         OCT  60
MSK1        OCT  1
MSK7        OCT  7
             .
             .
             .
```

Carriage-return-line-feed subroutine   One of the most implemented operations in any output routine for the Teletype is the carriage-return-line-feed (CR,LF) operation. This must be executed whenever a new line of print is to be started. Since the carriage-return and line-feed operations can be effected by outputting the particular 7-bit ASCII-coded values corresponding to each (15_8 and 12_8, respectively), the CR,LF function can be handled just like the printing of an alphanumeric character. A possible subroutine is illustrated in Example 6-11.

Example 6-11   Carriage-return-line-feed (CR,LF) subroutine

```
            ORG  103B
PCRLF       DEF  CRLF
             .
             .
             .
            ORG  6070B
CRLF        NOP
            LDA  NCR          /GET ASCII-CODED CARRIAGE RETRN
            JSB  PTYPE,I      /PRINT
            LDA  NLF          /GET ASCII-CODED LINE FEED
            JSB  PTYPE,I      /PRINT
            JMP  CRLF,I       /RETRN MAIN PROGRM
NCR         OCT  15
NLF         OCT  12
             .
             .
             .
```

General-purpose subroutine to print string of ASCII characters   A useful subroutine is one which can be used to type out the contents of an ASCII-coded buffer, where the location and the number of characters in that buffer may vary. Such a subroutine would be particularly useful for typing out messages. Example 6-12 illustrates this type of subroutine. (Note: The subroutine assumes ASCII-coded characters are stored in memory packed two to a word.)

Example 6-12   General-purpose ASCII print routine

```
            ORG  104B
PPRNT       DEF  PRINT
            .
            .
            .            Call sequence
            JSB  PPRNT,I /CALL PRINT SUBROUT
            DEF  ABFFR   /ASCII BFFR ADRSS
            DEC  N       /ASCII BFFR WORD LNGTH
            .
            .
            .
            ORG  6100B
PRINT       NOP
            LDA  PRINT,I /GET ASCII BFFR ADRSS
            STA  PBUF
            ISZ  PRINT
            LDA  PRINT,I /GET ASCII BFFR LNGTH
            CMA,INA
            STA  WCNTR   /STORE NGATED BFFR LNGTH IN WORD CNTR
            ISZ  PRINT   /GENERATE RETRN ADRSS
PRNT2       LDA  PBUF,I  /GET 2-CHRCTR ASCII WORD
PRNT1       ALF,ALF      /ROTATE MSH TO LSH
            STA  PBUF,I  /SAVE
            JSB  PTYPE,I /PRINT CHRCTR IN LOWER 8 BITS
            ISZ  CNT2    /PRNTD BOTH CHRCTRS THIS WORD?
            JMP  PRNT2   /NO
            LDA  CT2ST   /YES--RESET
            STA  CNT2
            ISZ  PBUF    /INCRMNT ASCII BFFR POINTR
            ISZ  WCNTR   /PRNTD ALL CHARACTRS?
            JMP  PRNT2   /NO--GET NXT ASCII PACKED WORD
            JMP  PRINT,I /YES--EXIT
PBUF        OCT  0
WCNTR       OCT  0
CNT2        OCT  -2
CT2ST       OCT  -2
```

One fact that should be pointed out here is that the subroutines based on two ASCII-coded characters per word do not fall apart when there is an uneven number of characters to be printed. In such a case, either the lower or upper 8 bits of a word in the ASCII buffer will contain all 0's. When the Teletype receives this character to type, it will not generate any character since no character has that code.

The use of the subroutine of Example 6-12 to print out an alphanumeric message is shown in Example 6-13.

The program of Example 6-13 has the Teletype print out the message THIS IS SIMPLE!?! with a CR,LF before and after. The coding for each

Example 6-13   Printing out messages

```
                    .
                    .
                    .
         JSB    PCRLF,I          /GET CR,LF FIRST
         JSB    PPRNT,I          /PRINT MESSAGE
         DEF    ABFFR            /ADRSS OF ASCII-CODED MESSAGE
         DEC    9                /DECIML LNGTH OF ASCII BFFR
         JSB    PCRLF,I          /END MESSAGE W/ CR,LF
                    .
                    .
                    .
ABFFR    OCT    052110           /ASCII-CODED "THIS IS SIMPLE!?!"
         OCT    044523
         OCT    020111
         OCT    051440
         OCT    051511
         OCT    046520
         OCT    046105
         OCT    020477
         OCT    020400
```

character is determined, and they are packed two to a word for access by the print routine. The coding for each character can be found in Sec. A-2 with coding for either the upper or lower 8-bits included. It is necessary that the programmer generate the ASCII-coded buffers for all messages upon assembly of the program. Illustrated in Example 6-14 is the actual generation of the ASCII buffer from the individual character codes.

6-4   NUMERICAL CONVERSION SUBROUTINES FOR I/O

Now that we have generated the fundamental I/O subroutines for communication between the Teletype printer and the computer, it is a relatively simple matter to write programs which provide for the convenience of Teletype I/O. However, the repertoire of subroutines is not yet complete. One fundamental problem that remains is to provide for conversion from one type of information coding to another. For example, in Example 6-10 it was pointed out that in order to print out the octal equivalent of the 16-bit binary value stored in a given memory location, it is necessary to convert that binary value into a string of ASCII-coded characters. Many other types of numerical conversion routines are necessary: binary-to-decimal (ASCII), decimal (ASCII)-to-binary, octal (ASCII)-to-binary, BCD-to-decimal (ASCII), and BCD-to-binary conversion. The first four conversions are required either to take in ASCII-coded information and transform it to a representation in the computer which can be useful arithmetically or to take numerical information represented in the computer and convert it to an appropriate ASCII-coded format for printout. The last

Example 6-14   Generation of packed-ASCII buffer

Letter	T	H		I	S		I		S	
ASCII code	124	110		111	123		040	111	123	040
16-bit word		052110			044523			020111		051440

S	I		M	P		L	E		!	?		!	
123	111		115	120		114	105		041	077		041	000
	051511			046520			046105			020477		020400	

Note: Spaces must be included in the ASCII-coded message, and the octal equivalent of the packed 16-bit word is not a straightforward combination of the octal equivalents of both ASCII words. (See Sec. A-2.) It should also be pointed out that the Hewlett-Packard Assembler has an instruction ASC which directs the Assembler to automatically code and pack designated alphanumeric messages included in programs. But not all Assemblers have this feature.

conversion routine is designed to take information represented in the computer in a BCD (binary-coded decimal) format and convert it to a binary format. No Teletype I/O is involved, however. BCD-coded information usually results from previous input from some data-acquisition device. It must be converted to binary before the computer can work with it conveniently.

Binary-to-octal (ASCII) conversion   A flowchart for a program designed to take a binary value from core memory and convert it to a six-digit octal (ASCII) printout is given in Fig. 6-7. The corresponding subroutine is given in Example 6-10, and the algorithm is illustrated in Fig. 6-6.

Octal (ASCII)-to-binary conversion   It is often convenient to enter information through the Teletype printer in octal notation since that notation is so readily correlated with binary values such as memory locations or machine instructions. With a 16-bit computer the logical format would be to enter data as six-digit octal integers. A flowchart for a program to handle numerical data input in this format is given in Fig. 6-8. Note that the assumption is made that all six digits are entered, including leading 0's. Also, it is assumed that the six ASCII-coded values are stored in a 3-word buffer, packed two to a word; this is not a necessary assumption, and a program could be written where the conversion is made as the six-digital octal value is typed.

Binary-to-decimal (ASCII) conversion   Most often the output of numerical information from the computer is desired in a decimal notation. In this case, it is necessary to convert the normal binary values to a decimal representation. The conversion of the decimal representation to ASCII-coded values can be accomplished simultaneously. The flow chart for such a conversion routine is given in Fig. 6-9.

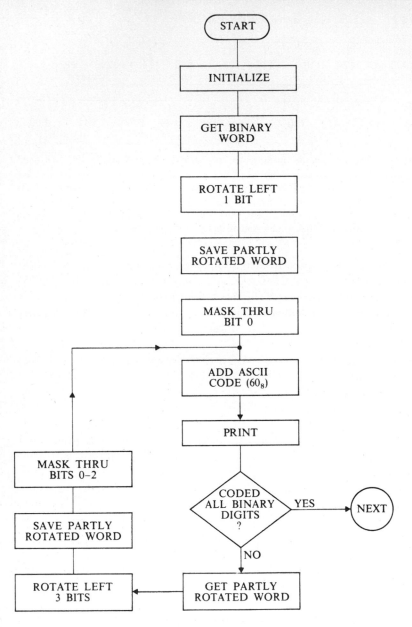

Fig. 6-7 Flowchart for binary-to-octal (ASCII) conversion.

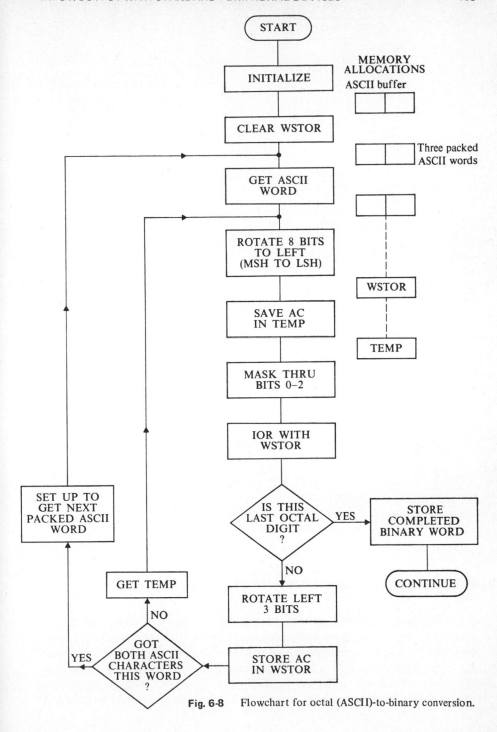

Fig. 6-8   Flowchart for octal (ASCII)-to-binary conversion.

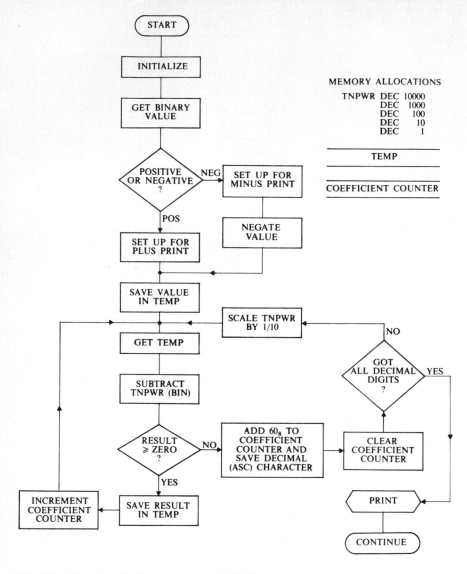

Fig. 6-9 Flowchart for binary-to-decimal (ASCII) conversion.

The basic algorithm defined by Fig. 6-9 is simply to divide the binary number by successively decreasing integral powers of 10, starting with the largest power of 10 which can be represented with the 16-bit word (10^4). [The division algorithm used is a simple repetitive subtraction operation which is faster than the divide subroutine. (Refer to Chap. 5.) However, with a hard-wired integer-divide function available, a simple divide instruction should be used.] A coefficient counter is used to represent the quotient for each division. This quotient can become as large as 9_{10} for division by any given TNPWR. However, 9_{10} is equivalent to 11_8 or 1001 in binary. When the quotient for each division is added to 60_8, the appropriate decimal (ASCII) character code is generated and stored for a later printout. (For example, $11_8 + 60_8 = 71_8$ = the ASCII code for 9_{10}.) The program will also provide for printing either a "+" or "−" before each numerical output.

Decimal (ASCII)-to-binary conversion   The input of numerical information through the Teletype printer is usually accomplished more conveniently in a decimal notation. However, the computer must convert that input information to binary values in order to utilize it in any computations. The flowchart of Fig. 6-10 outlines such a conversion routine.

The algorithm for the process outlined in Fig. 6-10 is analogous to that for binary-to-decimal (ASCII) conversion except that the process is reversed. The decimal (ASCII) characters are taken one at a time, starting with the most significant digit; the ASCII-coding bits are stripped (subtract 60_8), and the remaining bits (representing values from 0 to 9) are multiplied (in binary) by the appropriate integral power of 10. The products associated with each decimal digit are then summed to provide the binary representation of the original decimal number. Note, however, that the algorithm given assumes that the input decimal values are entered in a five-digit format which must include leading 0's plus a sign. Thus, 99 would be entered as +00099. This restriction could be altered readily to provide *free-field* input if the input program simply kept track of the number of sequential digits entered and selected the highest power of 10 to be used on that basis. It should be pointed out again here that if available, a hard-wired multiply instruction should be substituted for the repetitive addition algorithm suggested.

BCD-to-decimal (ASCII) and BCD-to-binary conversions   The conversion routines for BCD information are represented by the flowcharts of Figs. 6-11 and 6-12. The algorithms are fairly straightforward and require little comment here. Two points should be emphasized, however: First, it is assumed that the storage of BCD information in memory is packed four BCD values per word, with the most significant digit in the upper bits; for example,

$$9723_{10} = 1 \quad 001 \quad 011 \quad 100 \quad 100 \quad 011 = 1001 \quad 0111 \quad 0010 \quad 0011$$
$$\qquad\qquad\qquad\qquad\qquad\qquad\qquad\qquad 9 \qquad 7 \qquad 2 \qquad 3$$

Second, the ASCII character code for each decimal value coded in BCD can be generated simply by adding 60_8 to the BCD value. Again, note that a repetitive addition and multiplication is used in the algorithm of Fig. 6-12 and the same comments as above apply here.

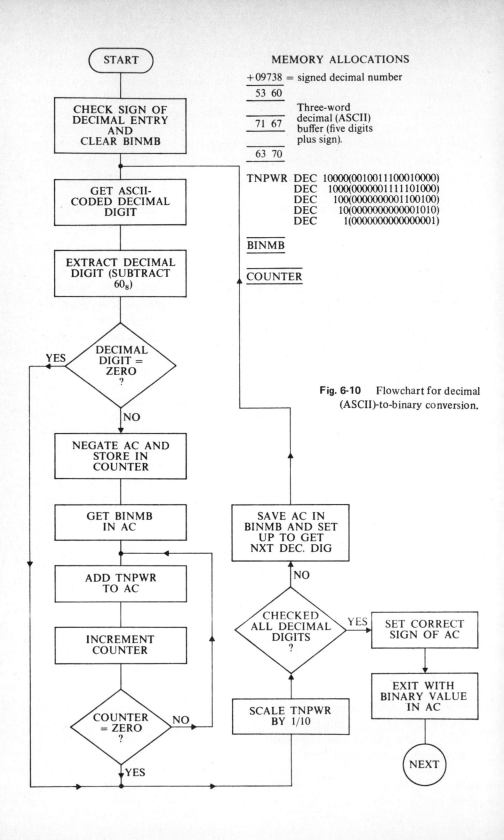

Fig. 6-10   Flowchart for decimal (ASCII)-to-binary conversion.

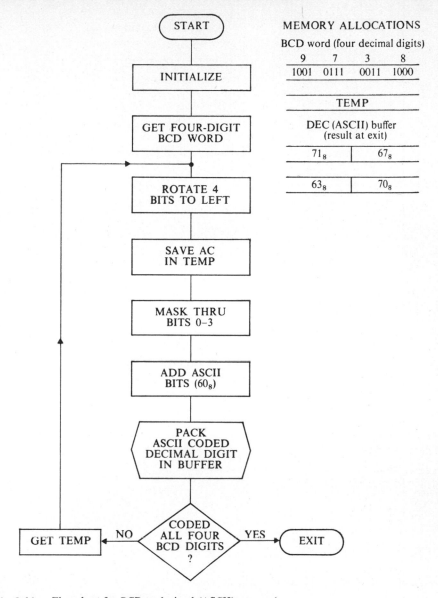

Fig. 6-11 Flowchart for BCD-to-decimal (ASCII) conversion.

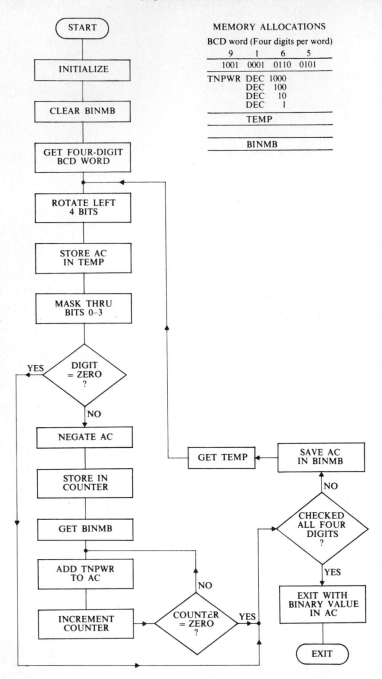

Fig. 6-12 Flowchart for BCD-to-binary conversion.

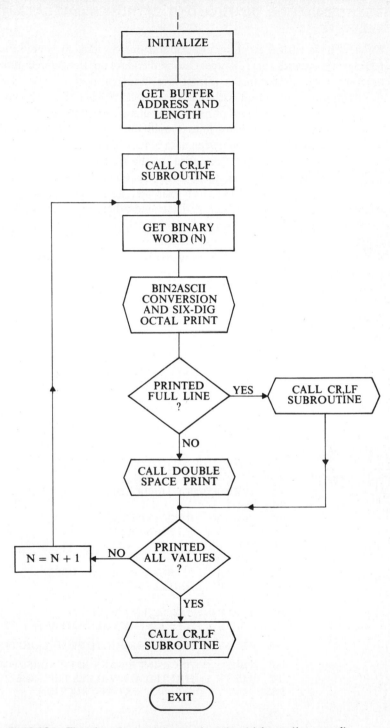

Fig. 6-13   Flowchart for octal-dump subroutine (eight per line spaced).

6-5 FORMATTING INPUT/OUTPUT OF NUMERICAL DATA WITH THE TELETYPE

We now have all the programming tools at our disposal to handle fundamental I/O operations with the Teletype. One of the tasks not considered in detail yet is how the programmer might *format* the input or output data. For example, should the data be output with one value per line? Four? Six? Should there be a CR,LF when all the data are printed? How many spaces should there be between numerical values? How many lines should there be between data rows? How many lines should there be between data blocks? And so forth.

The same sort of questions arise for the formatting of input information. In addition, the problem arises of how to communicate with the computer in other than numerical fashion. For example, how does one select alternative options through a keyboard entry? These and other questions will be considered here.

Output formatting   The easiest way to discuss output formatting of numerical data is to provide an example of a formatted output. See Example 6-15 and Figure 6-13.

Example 6-15   Octal Dump routine: program to list memory contents in octal, 8 words per line

```
                  .
                  .                 Call sequence
                  .
          JSB   ODUMP     /CALL OCTAL DUMP ROUTINE
          DEF   BUFFER    /ADDRESS OF MEMORY BLOCK TO BE LISTED
          DEC   N         /LNGTH OF MEMORY BLOCK
                  .
                  .
                  .
ODUMP     NOP
          LDA   ODUMP,I   /GET BFFR ADRSS
          STA   PBUF
          ISZ   ODUMP     /INCRMNT LINKAGE ADRSS
          LDA   ODUMP,I   /GET BUFFR LENGTH
          CMA,INA
          STA   DCNTR
          ISZ   ODUMP     /GENERATE RETRN ADRSS
          LDA   LCTST     /INITLZE LCNTR
          STA   LCNTR
          JSB   PCRLF,I   /INITIATE W/ CR,LF
GET       LDA   PBUF,I    /GET BINARY VALUE TO BE PRINTED

          JSB   PT6DG,I   /CALL SUBROUT TO PRINT 6-DIG OCTAL

          ISZ   PBUF      /INCRMNT BINARY BFFR ADRSS POINTR
          ISZ   LCNTR     /PRNTED ALL VALUES THIS LINE?
          RSS             /NO–SKIP NXT INSTRUCTION
```

Example 6-15 (*continued*)

```
              JMP   NWLIN    /YES--GO TO NEW LINE GENERATION
              LDA   SPACE    /GET ASCII-CODED "SPACE"
              JSB   PTYPE,I
              JSB   PTYPE,I  /PRINT TWO SPACES
BACK          ISZ   DCNTR    /PRNTD ALL DATA?
              JMP   GET      /NO--GET NXT VALUE AND PRINT
              JSB   PCRLF,I  /YES--GIVE CR,LF AND EXIT
              JMP   ODUMP,I  /EXIT SUBROUTINE
NWLIN         LDA   LCTST    /RESET WORD/LINE COUNTER
              STA   LCNTR
              JSB   PCRLF,I  /GIVE CR,LF TO START NEW LINE
              JMP   BACK     /CHECK TO SEE IF ALL DATA PRNTD
DCNTR         OCT   0
PBUF          OCT   0
SPACE         OCT   40
LCNTR         DEC   -8
LCTST         DEC   -8
                .
                .
                .
```

Many other variations of output formatting could be considered. For example, the program could be formatted to list only one numerical value per line; the formatting might involve four values per line with several spaces between; etc. In addition, the program might provide for nonnumerical statements, such as column headings or informational statements.

An example of a more complex output formatting problem arises if one considers the task of listing several columns of data on the Teletype, where data for each particular row of the columns correspond to different parameters measured under identical experimental conditions and are fetched from different blocks of core memory. That is, consider the problem of listing data, where all the data are not found in a single continuous block of core memory. Such a situation is considered in Example 6-16.

The flowchart corresponding to the output formatting defined by

Example 6-16 Output formatting of coherent data columns   *Problem*: Consider the case where gas chromatographic (GC) data are stored in memory for a series of observed peaks. These data include retention times in seconds, peak heights, and peak areas. Output these data in three distinct columns on the Teletype, with appropriate headings and spacing as shown below:

RET. TIME (SECS)	PEAK HEIGHT	PEAK AREA
10	101	567
35	639	3457
68	249	1048
97	358	1677

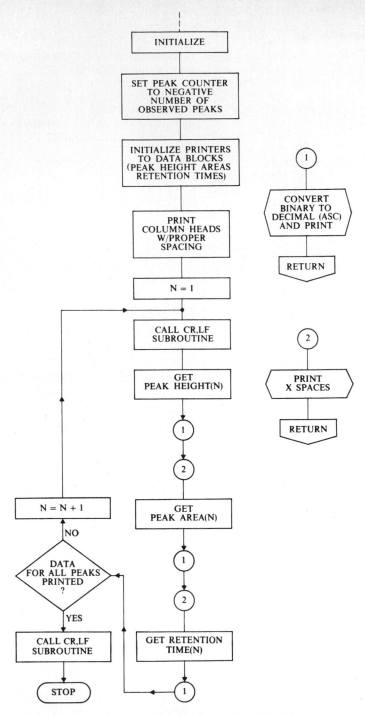

Fig. 6-14 Flowchart to format columnar output of GC data.

Example 6-16 is given in Fig. 6-14. It is left as an exercise for the reader to write the actual program involved. As guiding principles, simply remember that there are only 72 spaces available per line of Teletype output; the Teletype must always be told when to execute a CR,LF; multiline spacings can be accomplished by multiple LFs or CR,LFs, and it is possible to condense programming by the use of several subroutines to handle repetitive operations.

Input formatting There are two basic types of input information usually entered through the Teletype: [1] Numerical information to be implemented in some part of the program. That is, the operator can make numerical modifications of the program by providing for the input of certain critical parameters (like loop counters, threshold values, etc.) on the Teletype printer, with the program putting the input information in the appropriate core memory locations. In this fashion, a single program can be extremely versatile. [2] The operator may select alternative options existing within the program by the input of certain key words or characters at the appropriate time during the execution of the program.

The formatting of these types of operations involves primarily the coordination of input and output such that the computer interprets the input in the desired manner. For example, consider the case where a particular program which has three critical loops in it is to be executed. The program is to be designed such that the operator may choose to modify the counters for each loop by Teletype input. In addition, the program is to provide for the operator to select the option of retaining the program intact without modification. A typical dialogue between computer and operator on the Teletype might appear as:

Computer: MODIFY PROGRAM? (PRINT "YS" OR "NO")
Operator: YS
Computer: PRINT CNTR1, CNTR2 CNTR3:
Operator: 10
3
100

Several aspects of the simple dialogue above need further discussion. For example, how did the computer recognize the YS input, interpret it properly, and then respond with the next statement requesting the input data? Why were the input data entered on three successive lines rather than all on one line? How does the computer handle the different types of input data? What if an erroneous input is made?

Let us answer these questions one at a time. First, the ability of the computer to recognize and interpret nonnumerical inputs is simply related to the fact that alphanumeric characters are binary-coded in an ASCII format. Thus, the computer must simply be programmed to identify a particular ASCII-coded character or string of characters upon input in order to interpret that input. For example, the dialogue cited above requires that the computer be able to distinguish between a YS and a NO input at some particular point in the program in order to determine which direction the program is to take from that point. When a YS is entered at that point, the computer branches to a Parameter Input subroutine; when a NO answer is given, the computer bypasses the Parameter Input subroutine and goes directly to the execution of the program with the existing values of the parameters. The branching aspect of the program is not difficult. The more critical aspect is the computer's ability to distinguish between YS and NO inputs. This can be accomplished readily by recognizing that our keyboard input subroutine can be written to take two sequentially input, ASCII-coded characters and pack them two to a word. Thus, upon the input of YS, the binary, packed-ASCII word which results is 0 101 100 001 010 011 or 054123_8. The packed-ASCII equivalent of a NO input is 0 100 111 001 001 111 or 047117_8. Therefore, the computer can distinguish between these two particular inputs simply by a numerical comparison routine. This procedure is demonstrated by the program of Example 6-17.

One of the questions answered by the program of Example 6-17 is what happens if an erroneous input is made. For example, if the Teletype operator mistakenly types YES instead of YS, the computer will see the entry of the YE and will immediately return to the beginning of the routine—giving a CR,LF and then printing the initial question again. This is just one alternative way of handling erroneous inputs. What happens when erroneous numerical data are input? In this case, the computer has no way of knowing that the numerical values entered are in error. One way to handle such a problem would be to format the keyboard input routine such that an entire line of ASCII-coded input is interpreted at a time, rather than character by character. Thus, the computer program might include steps which involve looking for the CR,LF accompanying the end of a line of entry and then taking the entire line and processing it as required. One advantage of this approach is that the interpretation part of the input program could involve looking for a key character, which would indicate that there is an error in the line of input and that it should be ignored. In this case, if the Teletype operator recognizes that he has made an erroneous entry, he might type a predetermined character, which when seen by the computer will cause that line of entry to be ignored. Useful nonprinting characters for this function might be RUBOUT, ESC, or DEL.

One final question to be answered is why require that the numerical values be input one per line. This is not a rigid requirement at all. It depends only on how the input routine is designed. The programmer could have formatted the input routine such that multiple data entries per line are handled. However, this can be a more difficult program. Assignments at the end of this chapter require the reader to write programs with a variety of I/O formats, and the details of the problems involved will become quite obvious when these assignments are attempted.

Example 6-17   Interpretation of alphanumeric input

```
              .
              .
              .
BRNCH    JSB   PCRLF    /START W/ CR,LF
         JSB   PPRNT,I  /PRINT FIRST STMNT (3-WRD CALL SEQ)
         DEF   ABFRI
         DEC   N        /NUMBR WRDS IN ASCII-CODED STMNT
         JSB   PCRLF    /GIVE TWO CRLF'S
         JSB   PCRLF
         JSB   PINPT,I  /WAIT FOR KYBRD INPUT
         ALF,ALF        /RTRN W/ FRST CHAR IN AC. ROTATE
                          TO MSH OF AC
         STA   WSTOR    /PACK INTO WSTOR
         JSB   PINPT,I  /GET SECND CHARACTR
         IOR   WSTOR    /PACK 2ND CHARACTR INTO WSTOR
         STA   WSTOR
         CPA   WRD1     /COMPARE "YS" TO PACKED INPUT CHAR-
                          ACTRS–SAME
         JMP   PRMTR    /YES–GO TO PARAMETER INPUT ROUTINE
         CPA   WRD2     /NO--SEE IF INPUT = "NO"
         JMP   CONT     /YES--CONTINUE PROGRM W/ CURRENT
                          PARMTR VALUES
         JMP   BRNCH    /NO–WRONG ANSWRS GIVEN. ASK AGAIN
              .
              .
              .
WRD1     OCT   054123   /PACKED ASCII-CODED "YS"
WRD2     OCT   047117   /PACKED ASCII-CODED "NO"
WSTOR    OCT   0
ABFRI                   /ASCII-CODED STATEMENT--"MODIFY
                          PROGRAM?–ETC."
              .
              .
              .
```

6-6   HIGH-SPEED PUNCHED PAPER-TAPE INPUT/OUTPUT

High-speed tape input   The data transfer rates of the standard Teletype are slow enough that the input of long programs or data lists stored on punched paper tape becomes a very time-consuming process. Thus, most laboratory mini-computer systems are equipped with high-speed paper-tape readers. These devices are based on photoelectric sensing of the punched-hole patterns and parallel information transfer. The maximum read rates can be of the order of 300 to 500 eight-bit characters per second. This enhanced reading speed can provide a considerable saving in time. For example, a long program might require 60 sec to be read in completely through a high-speed photoreader; whereas that same program would require 30 min to be read in on the Teletype tape reader. Figure 6-15 illustrates the high-speed photoreader's mechanical operation.

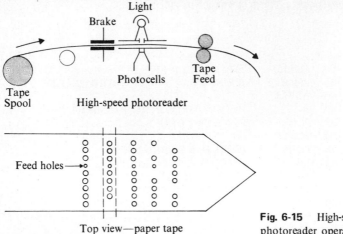

Fig. 6-15 High-speed punched-tape photoreader operation.

The details of the control and interfacing of a high-speed photoreader will vary with the manufacturer and the computer. However, the general principles of operation will be very similar. For the discussion here, we have made some basic assumptions regarding the electronic and mechanical characteristics of the high-speed photoreader which are generally true. That is, we can assume that the electromechanical features are such that a complete read cycle includes the turning on of the tape drive power, the reading of the status of the photocells at the appropriate time during the tape feed process, the setting of a flip-flop flag

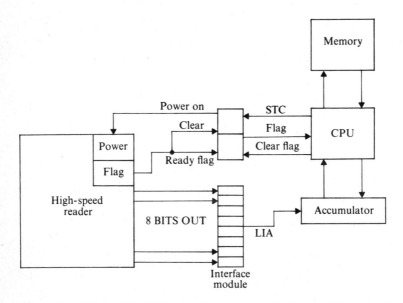

Fig. 6-16 Details of high-speed punched-tape photoreader interfacing.

which causes the 8-bit value to be strobed into the interface buffer register when the reading is complete, and the subsequent turning off of the tape feed power. Thus, the tape reader is oriented toward reading one 8-bit character at a time. The schematic details of the interfacing involved are given in Fig. 6-16.

Because the computer can operate on a microsecond time scale whereas the high-speed photoreader operates on a millisecond time scale, the computer can keep way ahead of the photoreader. Thus, even though the tape feed power is apparently going on and off repeatedly, the paper-tape input can proceed as smoothly as though the power were on continuously. On the other hand, the fact that the power automatically turns off after the input of each character allows for the possibility that the computer may be processing each character or string of characters as it is input. The continued input of characters can be determined by the result of the processing of the preceding input characters. Thus, the reading of each individual character is under computer control and may proceed as slowly as required by the simultaneous processing or as rapidly as the mechanical characteristics of the photoreader allow.

A program—based on the interface characteristics defined by Fig. 6-16— which will read in a string of 8-bit characters from the photoreader and pack them two to a word is given in Example 6-18.

Example 6-18 Program to read 8-bit bytes from a high-speed punched-tape photoreader and pack them in 16-bit words

```
              .
              .
              .
READ    STC   RDR             /POWR ON H.S. READR
        CLF   RDR             /CLR FLAG I/O CHANNL 13 (H.S.
                                READR CHANNL)
        SFS   RDR             /WAIT FOR FLAG
        JMP   *-1
        CLF   RDR
        LIA   RDR             /INPUT 8-BIT BYTE
        ALF,ALF               /MOVE TO MSH OF AC
        STA   TEMP            /STORE TEMPORARILY
        STC   RDR             /GET NXT BYTE
        SFS   RDR
        JMP   *-1
        CLF   RDR
        LIA   RDR
        IOR   TEMP
        STA   PBFFR,I         /STORE PACKED WORD
        ISZ   PBFFR
        ISZ   DCNTR           /ALL CHARACTRS READ?
        JMP   READ            /NO
              .               /YES, CONT.
              .
              .
RDR     EQU   13B             /EQUATE "RDR" TO 13B, I/O CHANNL
```

Two assumptions implicit in the program of Example 6-18 are that the program knows how many characters are to be read in and that the tape is positioned in the reader such that the very first 8-bit byte of information will be read in when the program begins to be executed. (That is, the program does not account for any "leader" on the paper tape.) If this is not the case, then the program will have to be rewritten accordingly, but these are formatting problems similar to those discussed in the previous section.

High-speed tape output   High-speed paper-tape punches which can operate at speeds on the order of 100 punched 8-bit characters per second are commercially available. If we assume that the interface characteristics are analogous to the photoreader interface of Fig. 6-16 except that an STC instruction causes the tape to be fed and punched rather than fed and read, then a program to output a string of 8-bit characters can be given by Example 6-19.

Example 6-19   Program to punch a string of 8-bit bytes with a high-speed paper-tape punch

```
              .
              .
              .
PNCHR    LDA  PBUF,I      /GET OUTPUT WORD (16-BIT)
OUT1     ALF,ALF          /ROTATE MSH TO LSH
         OTA  PNCH        /OUTPUT LOWER 8 BITS TO PUNCH CHNNL
         STC  PNCH        /POWR ON TO PUNCH
         CLF  PNCH
         SFS  PNCH        /WAIT FOR PUNCH COMPLETION
         JMP  *-1
         CLF  PNCH
         ISZ  CNT2        /PUNCHED BOTH BYTES THIS WORD?
         JMP  OUT1        /NO
         LDA  CT2ST       /YES–SET UP FOR NXT WRD
         STA  CNT2
         ISZ  PBUF
         ISZ  DCNTR       /PUNCHED ALL DATA?
         JMP  PNCHR       /NO
              .           /YES–CONT.
              .
              .
PNCH     EQU  14B         /EQUATE "PNCH" TO 14B, I/O CHNNL
```

6-7 INPUT/OUTPUT USING INTERRUPT CONTROL

Standard I/O peripheral devices generally have data transfer rates much slower than the computer's speed of operation. Thus, when program-controlled data transfers are employed, the computer can spend most of its time waiting for electromechanical I/O operations to be completed. For example, the Teletype output routines of Examples 6-1 and 6-2 are program-controlled data transfer routines where the computer can sit in a wait loop continuously checking the status of the Teletype until one print operation is completed and the next print operation can be initiated.

Of course, if the computer has no other useful work it can do while I/O data transfers are taking place, the inefficient routine of Example 6-1 is perfectly adequate. On the other hand, if there is useful work to be done, the I/O routine of Example 6-2 can be used, and the computer is able to execute other jobs in other parts of the program between print operations. However, although the programming of Example 6-2 is more efficient, the possibility of wasting a large percentage of the computer's time still exists unless it can be made aware of the precise moment when each print operation is completed. In this manner, its useful nonoutput programming can continue uninterrupted for the longest possible fraction of time available. Such a capability exists. It involves using the computer's *interrupt system* as described below.

Interrupt-controlled data transfers   An alternative means by which the computer can control I/O operations is through the interrupt system. Although varying in the details of operation from one computer to another, the interrupt system of a computer allows the computer to be made instantaneously aware of the needs of a peripheral device. That is, when the interrupt system is activated, an external-device flag can cause the computer to leave whatever part of the program it is currently executing and attempt to service the peripheral device causing the interrupt. Thus, the interrupt system allows a peripheral device to notify the computer that it is ready to be serviced regardless of what the computer might be engaged in currently. In most computers, the interrupt request from a peripheral device is recognized instantaneously, and the next machine operation involves servicing of the interrupt. The manner of service allowed varies with the computer. Nevertheless, such operations can provide for the most efficient use of machine time. Moreover, there are some types of computer operations—such as time sharing—which would be impossible without an interrupt system.

There are basically two types of interrupt systems: the *party-line* system and the *multichannel* system. In the party-line interrupt system, all the peripheral stations are on a single interrupt line. The computer must determine which device has caused the interrupt by questioning each device until the one requesting service is found. The questioning routine is determined by the user's software. Thus, the user must determine the priority of devices by the sequence in which they are queried upon interrupt. The first device in the list found requiring service is serviced, regardless of the needs of lower-priority devices. In addition, the user may write a program which allows the servicing routine for a lower-priority device to be interrupted by a higher-priority device. This must all be done by software and can get quite involved, not to mention the computer time used in determining the source of an interrupt. The advantages lie primarily in minimizing the amount of hardware inherent in the basic processor and I/O system, and this reduces the basic cost of the computer.

The computer used for the discussions in this text employs a multichannel priority interrupt system. This means that each external station is assigned a particular I/O channel through which an interrupt request can be made to the CPU. The CPU knows instantaneously where the request comes from, and it initiates service. Priority is established by the channel number. The lower the channel number is, the higher the priority is. Priority can be altered simply by altering physically the channel to which the external device is connected. An interrupt request is serviced by causing the immediate execution of the instruction found in the *memory location whose address is equivalent to the I/O channel* on which the interrupt was seen. The program counter is not affected by an interrupt-fetched instruction. Thus, if the interrupt-fetched instruction is JSB   m or JSB   m,I, the address of the next instruction in the *main program* will be stored at the first address of the subroutine. In this way, the computer can return to the main program. However, the programmer must accomplish the return by software design. Also, the programmer must provide for the storage of the contents of the various working registers upon interrupt so that they may be restored when control is returned to the main program.

An example of the programming involved in operating the high-speed photoreader under interrupt control is given in Example 6-20.

Several details should be pointed out regarding the illustration of Example 6-20. First of all, the interrupt system had to be turned on by a specific instruction STF   0 executed at the beginning of the main program. As soon as the interrupt system is activated, any external flag for a device connected to the computer will cause an interrupt. For this example, the next instruction is STC   RDR, which will start the read cycle on the high-speed tape reader. From that point, the computer proceeds merrily on, executing whatever program happens to be resident. The computer is oblivious to the progress of the high-speed reader until the read cycle is completed. At that point, the device flag is set, and this event is immediately sensed by the interrupt system. The computer responds by completing its current instruction execution and then executing the instruction in the location whose address corresponds to the channel number of the device causing the interrupt (channel 13_8 in this example). [Should more than one device be on-line and should simultaneous interrupts occur, the device having the lowest channel number would be serviced first. However, the programmer would have to design his program to take advantage of the hardware characteristics of the computer to maintain the desired priority arrangements. (This is discussed in detail below.)]

The interrupt instruction stored in location 13_8 is a JSB instruction which transfers control to a subroutine designed to service the high-speed reader at the completion of the read cycle. The address of the next location of the main

Example 6-20   Program for operating a high-speed tape reader under interrupt control

Location	Instruction	
Location	*Instruction*	
(13B)	JSB PRDSV,I	
	•	
	•	
	•	
PRDSV	DEF RDSVC	
	•	
	•	
	•	
START	NOP	/MAIN PROGRAM
	CLF RDR	/CLR FLAG ON H.S. READR
	STF 0	/TURN ON INTERRUPT SYSTEM
	STC RDR	/START READ CYCLE ON H.S. READR
	•	
	•	
	•	
RDSVC	NOP	
	STA ASAVE	/SAVE CONTENTS OF ACCUMULATOR(S)
	CLA	
	ELA	/SAVE LINK CONTENTS
	STA ESAVE	
	ISZ CNT2	/IS THIS FRST 8-BIT BYTE THIS WORD?
	JMP MSH	/YES--GO TO ROUT TO HANDLE PACKING-- IN MSH OF WORD.
	LDA CT2ST	/NO--RESET CNTR
	STA CNT2	
	LIA RDR	/BRING IN 8-BIT BYTE
	IOR TEMP	/PACK INTO LSH OF PACKED WORD
	STA PBFR,I	/SAVE PACKED WORD
	ISZ PBFR	
	ISZ DCNTR	**/ALL DATA IN?**
	JMP READ	/NO--GET MORE INPUT
	CLC RDR	/YES--CLR I/O DEVICE. REMOVE FROM INTERRUPT LINE
READ1	LDA ESAVE	/GET SAVED LINK VALUE FOR RETRN TO MAIN PROGRM
	ERA	/ROTATE AC BIT 0 INTO LINK
	LDA ASAVE	/GET SAVED AC VALUE FROM INTERRUPTED PROGRM
	CLF RDR	
	JMP RDSVC,I	/RETRN TO MAIN PROGRM WITH WRKING REGISTRS RESTORD
MSH	LIA RDR	/GET 8-BIT BYTE FROM H.S. READR
	ALF,ALF	/ROTATE TO MSH OF AC
	STA TEMP	/SAVE IN TEMPORARY STORAGE LOCATION
READ	STC RDR	/RESTART H.S. READR TO GET NEXT DATUM
	JMP READ1	/GO TO ROUT TO RESTORE REGS & RETRN TO MAIN PROGRAM
RDR	EQU 13B	

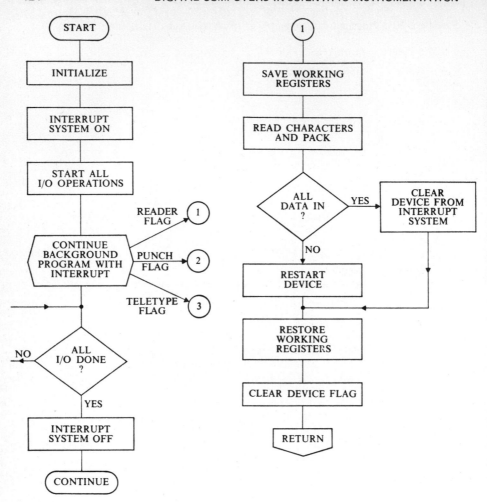

Fig. 6-17   Flowchart for program to time-share three standard peripherals.

program is stored in the first location of the service subroutine (RDSVC). The 8-bit byte is taken in from the reader and packed appropriately. Then, if all data are not in, the reader is told to start the next read cycle, and the subroutine is exited, with the computer returning to the main program to take up where it left off. In order to ensure that the execution of the main program can be interrupted and resumed later without any loss or distortion of information, the contents of the working registers are saved upon entry of the interrupt service routine. The working registers are then restored to the values existing at the time of interrupt when the service routine is exited. Because only the A or E

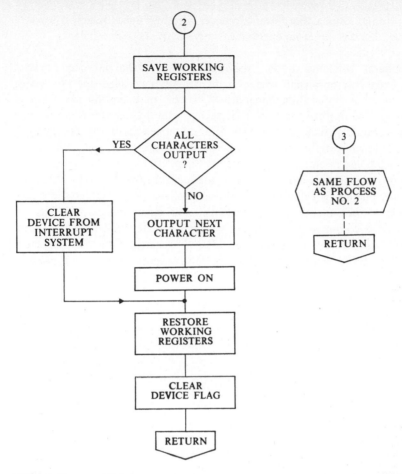

Figure 6-17   (*continued*)

registers might be altered by the RDSVC subroutine, only these are saved and restored.

It should be pointed out that the interrupt system can be deactivated at any time within the program by executing a CLF   0 instruction.

To appreciate the increased efficiency gained in the use of computational time by operating the high-speed reader under interrupt control, consider that only about 15 to 20 machine instructions need be executed to service the reader after each read cycle. Assuming an average execution time of about 2 μsec per instruction, only 30 or 40 μsec of computer time are required per character read-in. Thus, although the high-speed reader may operate at a rate of 300 characters per second, the computer can be utilizing about 99 percent of its time for other computational purposes! Moreover, the high-speed reader is serviced upon demand and is therefore operating at very nearly its peak efficiency.

Multiperipheral servicing under interrupt control   The interrupt system becomes extremely important whenever the computer is called upon to service several peripheral devices simultaneously. This sort of operation is a form of time sharing. The objective is to provide service with a computer response time for each peripheral such that to all outward appearances the computer is dedicated to each peripheral. Some aspects of this approach will be discussed here. More detailed discussion is presented in Chap. 10.

To provide a simple example of time sharing, consider the case where three standard peripherals—a Teletype, a high-speed tape reader, and a high-speed tape punch—are to be operating independently and simultaneously during the execution of a computational program which is executed independently from the operations of the three peripherals. The flowchart for such a program is given in Fig. 6-17. Some details for the programming involved are given in Example 6-21, where at least the Teletype printer service routine is illustrated.

Example 6-21   Some aspects of a time-sharing program servicing a Teletype output, a high-speed tape punch, and a high-speed tape reader.

```
              ORG  13B
              JSB  PRDSV,I          /STORAGE OF INTERRUPT SERVICE
                                     INSTRUCTIONS FOR DEVICES
                                     #1 THRU #3

              JSB  PPNCH,I
              JSB  PTYSV,I
               .
               .
               .
PRDSV         DEF  RDSVC
PPNCH         DEF  PNSVC
PTYSV         DEF  TYSVC
               .
               .
               .
              ORG  2000B
STRT          NOP                    /MAIN PROGRM
              CLF  RDR               /CLR ALLDEVICE FLAGS
              CLF  PNCH
              CLF  TTY
              STF  0                 /TURN INTRUPT SYSTEM ON
              JSB  TTLST             /GO TO ROUT TO START TTYP OUTPUT
              JSB  PLIST             /GO TO ROUT TO START PUNCH OUTPUT
              JSB  RDIN              /GO TO ROUT TO START H.S. READR
               .                     /CONTINUE WITH MAIN PROGRAM
               .
```

Example 6-21   (*continued*)

```
TTLST   NOP
        LDA  TCODE
        OTA  TTY
        LDA  PBUF1,I        /GET 8-BIT BYTE TO OUTPUT
        OTA  TTY            /OUTPUT TO TTYP
        STC  TTY            /POWER ON--PRINT
        CLF  TTY
        JMP  TTLST,I        /RETRN MAIN PROGRM
TYSVC   NOP                 /TTYP INTRRUPT SERVICE ROUTINE
        STA  ASAV1          /SAVE CONTNTS OF AC FOR RETRN TO
                              MAIN PROGRM
        CLA
        ELA                 /GET LINK INTO L.S.B. OF AC
        STA  ESAV1          /SAVE LINK FOR RETRN TO MAIN
                              PROGRM
        ISZ  PBUF1          /SET UP TO OUTPUT NXT DATUM
        ISZ  DCNT1          /ALL DATA OUTPUT?
        RSS
        JMP  EXIT1          /YES--EXIT W/O MORE PRINTS
        LDA  PBUF1,I        /NO--SET UP TO OUTPUT NEXT BYTE
        OTA  TTY
        STC  TTY            /TELL TTY TO PRINT
XIT     LDA  ESAV1          /RESTORE LINK REGISTER
        ERA
        LDA  ASAV1          /RESTORE AC
        CLF  TTY            /CLR DEVICE FLAG JUST BEFORE EXIT-
                              ING SERVICE ROUT
        JMP  TYSVC,I        /RETRN TO MAIN PROGRAM W/WRKING
                              REGS. RESTORED, GLG CLRD.
EXIT1   CLC  TTY            /REMOVE DEVICE FROM INTERRUPT
                              SYSTEM W/"CLC"
        CLF  TTY
        JMP  XIT            /EXIT SERVICE ROUT W/DEVICE
                              FINSHED & OFF
             .
             .
             .
PNSVC   NOP
             .
             .
             .
RDSVC   NOP
             .
             .
             .
RDR     EQU  13B
PNCH    EQU  14B
TTY     EQU  15B
TCODE   OCT  130000
```

Several aspects of the program of Example 6-21 should be considered in some detail. First of all, it is obvious that the highest-priority device is the high-speed reader since it was assigned to the lowest-number I/O channel of the three. This makes sense since it operates at the fastest rate of speed of the three devices and therefore needs servicing much more frequently. Similarly, the Teletype is assigned the lowest priority because it requires servicing least frequently. (A detailed analysis of the timing considerations is given in the next section.) A second detail to be pointed out here is that once the device flag is *cleared*, the device loses its place on the priority list. Thus, if the flag were cleared immediately upon entering the service routine, that service routine could be interrupted by a lower-priority device! Thus, in order to maintain the priority of the device being serviced, the flag is not cleared until the interrupt service routine is about to be exited. This ensures that the device service routine can be interrupted only by a higher-priority device. In this regard, it is also important to note that *not* clearing the device flag immediately does *not* cause the computer to be interrupted continuously. The computer services an interrupt flag from a given device only once and will not service it again unless that flag has been cleared and then set again.

One final detail to be noted is that the device can be removed from the priority interrupt line by a CLC instruction. This ensures that extraneous generation of the device flag does not cause an unintentional interrupt.

(*Note*: The entire discussion above is pertinent only to one specific computer system. A large number of variations in the hardware associated with a computer's interrupt system exists. Some of the more common variations are discussed in Chap. 13. However, the general principles described here are consistent with existing computers having hardware multichannel priority interrupt systems.)

Timing considerations for multiperipheral servicing   It is difficult to visualize timing involved in the time-sharing servicing of more than one peripheral device. Figure 6-18 attempts to show how device service time is allocated among three devices operating asynchronously at different speeds and corresponding priorities. Obviously the lower-priority devices can have their service routines segmented and/or delayed. The degree of seriousness of this situation depends on three primary factors: (1) whether the device operation is computer-controlled or externally controlled, (2) the frequency with which each device is to be serviced, and (3) the amount of service time—including overhead—required at each interrupt. The *overhead* is simply that amount of computer time required to determine which device is causing the interrupt, save and restore registers, maintain priority, etc., in addition to the actual servicing of the device. Obviously, with a hardware multichannel priority interrupt system the overhead can be small compared to a party-line system.

If we consider the three standard peripheral devices discussed already in

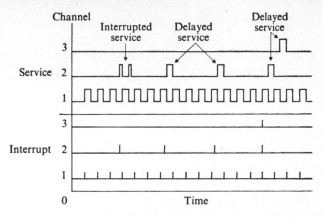

Fig. 6-18   Timing diagram for time sharing on priority interrupt basis: high-speed reader, channel 1 (highest priority), high-speed punch, channel 2 (next highest priority), Teletype printer, channel 3 (lowest priority).

this chapter, an estimate of the extent of service delays that might be anticipated for the lower-priority devices can be made. The Teletype printer service routine can be interrupted by both the reader and punch at some given time. Since the service routines for each of these devices require less than about 50 μsec, the maximum delay in completion of any given Teletype service routine will be about 100 μsec. For a device which can operate at a maximum speed of 10 characters per second, occasional delays on the order of 100 μsec are negligible. For the high-speed punch, which can be interrupted only by the high-speed reader, a servicing delay of about 50 μsec can occur on a given interrupt. Because the punch can operate at a maximum rate of about 100 characters per second—a punch cycle of about 10 msec—frequent delays on the order of 50 μsec are still negligible.

Consider what would happen if the priorities were reversed. The high-speed reader could then conceivably be interrupted for as long as 100 μsec during a given service cycle. For a device which can operate at a maximum speed of 300 characters per second, this begins to be significant. However, particularly since it would not occur on every cycle, it is nowhere near critical.

A case where a critical situation might arise would be where the high-speed reader were being operated under *external control* at its maximum rate. Then, the computer would have to be able to ensure the completion of the service routine during each complete read cycle. Thus, if the total service time for all higher-priority devices should *exceed* the time to complete one read cycle of the high-speed reader, some input information would be lost. However, with the reader situated properly as the highest-priority device, there is no problem.

It should be pointed out here that the previous discussion of timing considerations has been directed toward the worst possible case, i.e., for several devices operating asynchronously. If external arrangements can be made such that all time-sharing peripherals are operating synchronously and the computer program can be made aware of the synchronization characteristics, the need for the priority interrupt hardware and much of the overhead software disappears. The only consideration left is whether all service routines can be executed rapidly enough to avoid missing data from any given device.

Foreground/background computing   The main computational programs which can be executed while the interrupt system is activated and several peripheral devices are being serviced in a time-sharing environment are called *background* programs. The programs operating under interrupt control to service the peripherals are called *foreground* programs. The computer can appear to be executing foreground and background programs simultaneously.

To estimate the fraction of the computer's computational time available for background computing, it is necessary simply to evaluate the fraction of its time dedicated to foreground programming. This can be evaluated for each peripheral by multiplying the service request frequency by the amount of service time per request (including overhead). These fractions for each peripheral can be summed to give the total fraction of CPU time devoted to foreground programming. This is illustrated below:

Device	Service frequency	Service time	CPU time/second
High-speed reader	300 per sec	50 μsec	15 msec
High-speed punch	100 per sec	50 μsec	5 msec
Teletype printer	10 per sec	50 μsec	0.5 msec
Total foreground time			20.5 msec

Thus, for the three devices considered here and the service routines involved, only about 2 percent of the computer's time is tied up. Therefore, the computer can be executing background programs with 98 percent efficiency even though these three peripheral devices are operating continuously at their maximum rates.

EXERCISES

6-1. Prepare a flowchart, and write a program which will take a numerical message input on the keyboard and output the list of six-digit values corresponding to the packed ASCII-coded characters typed. Arrange it so that the list is printed after each CR,LF typed.

6-2. Write a program to print the following statement on the Teletype printer:
THESE ARE USEFUL CHARACTERS—*@$!?

6-3. Write programs corresponding to the flowcharts of Figs. 6-9 and 6-10.

6-4. Modify the flowcharts and programs of Figs. 6-9 and 6-10 to use integer MPY or DIV instructions. From information in Chap. 5 regarding time/space requirements of the subroutines, compare the execution time and the program space required for programs written in Exercises 3 and 4. (Assume that each numerical conversion is for a 5.)

6-5. Prepare a flowchart, and write a program to take an ASCII-coded character string input from the Teletype keyboard, pack two per word, and store.

6-6. Modify the Octal Dump program (Example 6-15) to print the memory address of the first location whose contents are printed on each line as shown below:

| (002000) | 177121 | 154432 $\cdots$ 133432 | 122332 | 177765 |
| (002010) | 176543 | 123234 $\cdots$ 145456 | 176567 | 154546 |

   .
   .
   .

6-7. Write an output routine which corresponds to Example 6-16 and Fig. 6-14.

6-8. Prepare a flowchart, and write a keyboard input subroutine which accepts five-digit decimal (ASCII) input values one per line and stores the binary equivalents in a buffer.

6-9. Do Exercise 6-8 with provision for the input of four five-digit decimal values per line, where at least one space must separate each value, but with no fixed spacing specified.

6-10. Prepare a flowchart, and write a routine to print the 10-digit decimal values corresponding to binary numbers stored in double precision.

6-11. Prepare a flowchart, and write the corresponding program for printing out values stored in floating-point notation.

6-12. Prepare a flowchart, and write the corresponding program for reading in a paper tape on the high-speed reader and exactly duplicating that tape on the high-speed punch simultaneously. (Use program control.)

6-13. Prepare a flowchart and program for a foreground/background application as follows: Foreground: Take in ASCII-coded data from the high-speed tape reader; simultaneously punch every third data value on the high-speed tape punch, *and* type every thirtieth data value on the Teletype in some appropriate form; use a 1,024-word buffer for the high-speed input data, but *reuse* this buffer so that there is no limit to the amount of data that may be input. (Be sure that no data will be lost to the punch or Teletype printer!) Background: The main program should involve repeatedly counting to 1 million with the computer recording the number of times this count is reached.

If the above program is executed, a *cyclic* paper-tape input should be used (ends taped together to allow repeated input of same data block). Execution should be allowed for some exact time interval (e.g., 100 sec), and the progress of the background program determined. Then, execute *only* the background program for the same time interval, and compare the progress with that attained previously.

7
Introduction
to Digital Logic

The design and operation of on-line computer systems generally involve three areas of specialization: software development, interfacing, and the experimental system to which the computer is connected. In previous chapters, we have discussed basic small-computer configurations and programming techniques. In this and the following chapter, we will discuss the hardware aspects important for interface design.

Interfacing refers to the electrical and logical connections necessary to put an experiment in direct communication with a computer. It may be as simple as plugging in a cable or as complex as the design of data-acquisition and control hardware. Much of interfacing involves the use of digital electronic devices and digital logic. Digital logic is introduced in this chapter to lay the foundations necessary for the more general discussion of interfacing presented in Chap. 8.

7-1 DIGITAL LOGIC STATES

Digital logic devices are the foundation used to build digital instrumentation including computers, interfaces, etc. They consist of a set of electronic circuits that perform simple logic operations. They can be connected to make more complex building blocks to perform the necessary logic, storage, arithmetic, interface, and timing operations that result in digital instrumentation. Logic operations are performed by the logic states generated with various digital devices. The logic states are usually represented by the binary number system.

In the simplest case, a digital logic function can be simulated with conventional switches and a battery. The output states can be indicated with a light bulb. Such a simple logic circuit is presented in Fig. 7-1. When both switches A and B are open, no current flows to the lamp. However, when either switch A or B or both are closed, current flows to light the lamp.

The circuit presented in Fig. 7-1 is performing the basic OR digital logic function. The output states, i.e., whether the lamp is on or off, can correspond to the numbers of the binary number system. The on condition can be defined as the binary number 1 and the off condition as the binary number 0. The input functions or switch positions can be defined in the same manner, with the open switch condition as a binary 0 and the closed condition as a binary 1.

A circuit for performing another basic logic function is presented in Fig. 7-2. In it, the two switches A and B are connected in series rather than in parallel as in Fig.7-1. In this case, both switches A and B must be closed for the lamp to light. This is a basic digital logic circuit to perform the AND function.

The operation of the circuits presented in Figs. 7-1 and 7-2 can be described in terms of the binary number system in tabular form as presented in Table 7-1. The output states are listed for all combinations of the input-switch conditions. If the circuits presented in Figs. 7-1 and 7-2 are expanded with more input switches, similar but more complex tables can be constructed. In addition,

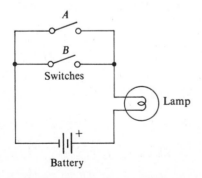

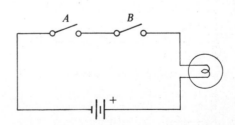

Fig. 7-1 Circuit to perform the basic
OR logic function.

Fig. 7-2 Circuit to perform the basic AND
logic function.

Table 7-1   Binary-number representation of logic-circuit operation

Input conditions		
Switch A	*Switch B*	*Output condition*
OR function[a]		
1	1	1
1	0	1
0	1	1
0	0	0
AND function[b]		
1	1	1
1	0	0
0	1	0
0	0	0

[a]For Fig. 7-1.
[b]For Fig. 7-2.

the AND and OR circuits presented above can be connected together to perform more complex logic operations. A complex logic circuit performing both AND and OR logic is presented in Fig. 7-3. A table defining the operation of that circuit can also be constructed and would be a worthwhile exercise at this time.

The logic circuits presented above are useful approximations for defining the simple nature of digital logic. They are not, however, used in modern electronic digital computers. Modern electronic digital logic uses transistors to perform the switching operations and is available in integrated-circuit form. *Integrated circuits* are miniaturized solid-state devices that may contain several

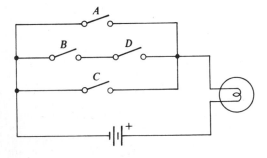

Fig. 7-3   Complex logic circuit.

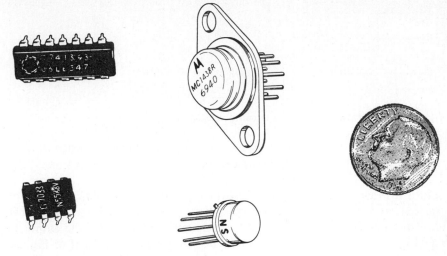

Fig. 7-4   Common integrated-circuit packages.

complete electronic circuits. Typical integrated-circuit packages are presented in Fig. 7-4. A single integrated-circuit logic package can perform extremely complex logic functions. The logic states are usually represented by voltage levels.

A common set of voltage levels for integrated-circuit, electronic, digital logic consists of 0 and +5 V for both inputs and outputs. For example, the basic OR logic function presented in Fig. 7-1 can be represented in terms of the box with two inputs and an output presented in Fig. 7-5. Instead of opening and closing switches as before, voltage levels can be applied to the inputs, and the voltage measured at the output. If +5 V corresponds to a binary 1 and 0 V corresponds to a binary 0, the operation of the box in Fig. 7-5 can be defined in the same way as the operation of the circuit in Fig. 7-1. Complex logic operations can be represented by simply connecting boxes together as in Fig. 7-6.

The circuit shown in box format in Fig. 7-6 performs the same logic function as the one represented in Fig. 7-3. Oftentimes it is necessary to define the operation of a logic system in a sentence format. This can easily be done and is illustrated below for the circuits presented in Figs. 7-3 and 7-6.

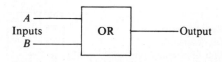

Fig. 7-5   Box representation of basic OR function.

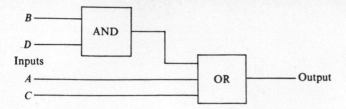

Fig. 7-6   Box representation of the circuit presented in Fig. 7-3.

In Figs. 7-3 and 7-6, the logic output is equal to a binary 1 when either A OR (B AND D) OR C is equal to a binary 1.

Notice that logic diagrams can be constructed from sentence descriptions. For the sentence description above, we know first that one of three variables, either *A, C,* or the combination of *B* and *D*, must equal a binary 1 for the output to equal a binary 1. Thus, the three-input OR box presented as a part of Fig. 7-6 could be drawn. From the sentence descriptions, two of the input variables *A* and *C* are connected directly to the box. However, the other two input variables *B* and *D* must be ANDed together before being connected to the output OR box. This is easily accomplished by connecting *B* and *D* as inputs to a two-input AND box, the output of which is connected as one of the input variables to the three-input OR box. Notice that we have worked backward from the outputs to the inputs in reconstructing the logic diagram. This is often easier than working from the inputs to the outputs.

7-2 SIMPLE LOGIC ELEMENTS, INTRODUCTION TO GATES

Actual modern, electronic digital logic comes in several microelectronic integrated-circuit forms, some of which are tabulated in Table 7-2. In addition, there is a class of digital logic called *metal-oxide-silicon field-effect transistor logic* (MOS) that is commonly used for extremely complex integrated-circuit logic functions and whose electronic circuits are quite different from those types listed in Table 7-2. A more complete discussion of integrated-circuit logic types

Table 7-2   Common integrated-circuit logic types

Logic type	Common abbreviation	Logic performed by	Logic voltage levels
Resistor-transistor	RTL	Driver transistors	0, +3.6 V
Diode-transistor	DTL	Input diodes	0, +5 V
Transistor-transistor	T²L or TTL	Input transistor	0, +5 V
Emitter-coupled	ECL	Input transistors	−1.55, −0.75 V
High-threshold	HTL	Input diodes	0, 8-30 V

including electronic-circuit descriptions and a comparison of operating characteristics is presented in Appendix B. Suffice it to say here that several different types of logic exist with different electronic operating characteristics and requirements. They differ in their electronic circuits, not the logic functions which they perform. The principles discussed here apply equally well to all.

There are five basic, simple logic functions from which all others are usually constructed. They are presented along with their common symbols in Fig. 7-7. A complete listing of standard logic symbols is presented in publication MIL–STD–806B available from the U.S. Government Printing Office. The actual hardware electronic device used to perform a particular basic logic function is called a *gate*. The symbols presented in Fig. 7-7 are often referred to as the *basic positive logic gate symbols*.

Of the logic functions presented in Fig. 7-7, the AND and OR gates have been considered above using switches and box diagrams. Their operation is the

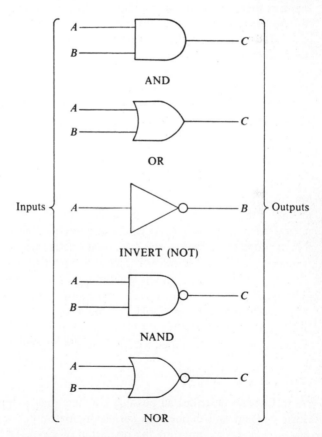

Fig. 7-7   Basic logic gates.

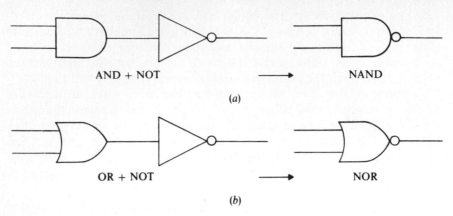

AND + NOT   ⟶   NAND

(*a*)

OR + NOT   ⟶   NOR

(*b*)

Fig. 7-8   Definitions of NAND and NOR gates.

same as that already discussed. The symbols presented in Fig. 7-7 will, however, be used from now on.

The INVERT or NOT gate has not been presented before. It does exactly what its title suggests; that is, negates the input. When a binary 1 is placed on the input, the output is a binary 0. In a like manner, when a binary 0 is placed on the input, the output is a binary 1. The inverter has been introduced here because it is a fundamental part of two more basic gates, the NAND and NOR gates. Because of the characteristics of the electronic circuits in many integrated-circuit logic types, the most basic gates do not perform AND and OR functions, but rather perform NAND and NOR functions. As a result, an understanding of these gates is extremely important.

The term NAND is derived from AND and NOT and functionally refers to an AND gate followed by an inverter as is illustrated in Fig. 7-8*a*. In fact, the NAND function can be generated in such a manner. The logical operating characteristics of the NAND gate can be defined in terms of a table using the binary number system in the same manner as was done for the AND and OR functions previously described. These characteristics along with those for the NOR gate derived from an OR gate and an inverter are presented in Table 7-3. The NOR gate equivalent is also illustrated in Fig. 7-8. The logical operation tabulations presented in Table 7-3 are commonly called *truth tables*. Truth tables can be used to define the logic operation of most digital logic. They are by far the most commonly used road map to the operation of logic systems.

7-3   TRUTH TABLES

A truth table describes all possible combinations of input variables for a given logic element and defines the corresponding output values. Truth tables are a useful way of summarizing the operation of simple and complex logic systems and provide a road map of operation. The truth tables and logic symbols for the simple logic functions considered thus far are tabulated in Fig. 7-9.

Table 7-3   Truth tables for NAND and NOR

Inputs		Output
A	*B*	*C*

NAND function

1	1	0
0	1	1
1	0	1
0	0	1

NOR function

1	1	0
0	1	0
1	0	0
0	0	1

The truth tables presented in Fig. 7-9 are straightforward for the AND, OR, and INVERT gates. However, at first they may not seem so for the NAND and NOR gates. Notice, though, that the NAND and NOR gate truth tables are easy to construct if we remember that they can be considered as AND and OR gates followed by inverters. As a result, one can arrive at NAND and NOR gate output values by simply inverting the corresponding AND and OR gate output values.

The real significance of truth tables lies in the fact that they simplify the design of complex logic systems. One can easily construct logic systems by working from requirements tabulated in truth-table form. The first problem is to construct a truth table from the known input parameters. Suppose we want a control signal Y to be a binary 1 whenever inputs A AND $B = 1$, OR inputs C AND D AND $E = 1$, OR input $F = 1$. In addition, input G must always equal 1 for Y to equal 1. Under all other conditions, we want Y to equal 0. We can summarize these statements in the following manner:

$Y = 1$ only when

$$
\begin{bmatrix}
\text{level 1} & A \text{ AND } B = 1 \\
 & \text{OR} \\
\text{level 2} & C \text{ AND } D \text{ AND } E = 1 \\
 & \text{OR} \\
\text{level 3} & F = 1
\end{bmatrix} \quad \text{AND } G = 1
$$

Notice that for convenience we have written all AND links horizontally and all OR links vertically and have numbered each vertical level within the brackets. The *long-form* truth table corresponding to the stated logic is given in Table 7-4.

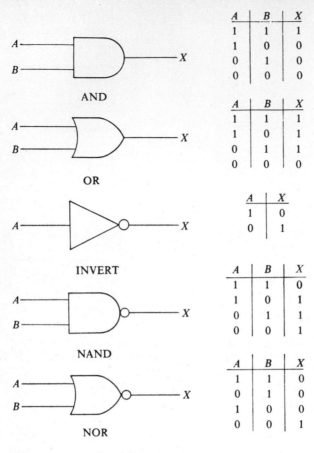

A	B	X
1	1	1
1	0	0
0	1	0
0	0	0

AND

A	B	X
1	1	1
1	0	1
0	1	1
0	0	0

OR

A	X
1	0
0	1

INVERT

A	B	X
1	1	0
1	0	1
0	1	1
0	0	1

NAND

A	B	X
1	1	0
0	1	0
1	0	0
0	0	1

NOR

Fig. 7-9   Logic gates and truth tables.

Table 7-4   Long-form truth table for the function Y = 1 only when A AND B = 1, OR C AND D AND E = 1, OR F = 1 AND G = 1.

	Inputs						Output
Case	A	B	C	D	E	F	G	Y
1	0	0	0	0	0	0	0	0
2	0	0	0	0	0	1	0	0
3	0	0	0	0	1	0	0	0
4	0	0	0	0	1	1	0	0
5	0	0	0	1	0	0	0	0
6	0	0	0	1	0	1	0	0

Table 7-4   (*continued*)

| | | | | Inputs | | | Output |
Case	A	B	C	D	E	F	G	Y
7	0	0	0	1	1	0	0	0
8	0	0	0	1	1	1	0	0
9	0	0	1	0	0	0	0	0
10	0	0	1	0	0	1	0	0
11	0	0	1	0	1	0	0	0
12	0	0	1	0	1	1	0	0
13	0	0	1	1	0	0	0	0
14	0	0	1	1	0	1	0	0
15	0	0	1	1	1	0	0	0
16	0	0	1	1	1	1	0	0
17	0	1	0	0	0	0	0	0
18	0	1	0	0	0	1	0	0
19	0	1	0	0	1	0	0	0
20	0	1	0	0	1	1	0	0
21	0	1	0	1	0	0	0	0
22	0	1	0	1	0	1	0	0
23	0	1	0	1	1	0	0	0
24	0	1	0	1	1	1	0	0
25	0	1	1	0	0	0	0	0
26	0	1	1	0	0	1	0	0
27	0	1	1	0	1	0	0	0
28	0	1	1	0	1	1	0	0
29	0	1	1	1	0	0	0	0
30	0	1	1	1	0	1	0	0
31	0	1	1	1	1	0	0	0
32	0	1	1	1	1	1	0	0
33	1	0	0	0	0	0	0	0
34	1	0	0	0	0	1	0	0
35	1	0	0	0	1	0	0	0
36	1	0	0	0	1	1	0	0
37	1	0	0	1	0	0	0	0
38	1	0	0	1	0	1	0	0
39	1	0	0	1	1	0	0	0
40	1	0	0	1	1	1	0	0
41	1	0	1	0	0	0	0	0
42	1	0	1	0	0	1	0	0
43	1	0	1	0	1	0	0	0
44	1	0	1	0	1	1	0	0
45	1	0	1	1	0	0	0	0
46	1	0	1	1	0	1	0	0
47	1	0	1	1	1	1	0	0
48	1	0	1	1	1	1	0	0
49	1	1	0	0	0	0	0	0
50	1	1	0	0	0	1	0	0

Table 7-4   (*continued*)

			Inputs				Output
Case	A	B	C	D	E	F	G	Y
51	1	1	0	0	1	0	0	0
52	1	1	0	0	1	1	0	0
53	1	1	0	1	0	0	0	0
54	1	1	0	1	0	1	0	0
55	1	1	0	1	1	0	0	0
56	1	1	0	1	1	1	0	0
57	1	1	1	0	0	0	0	0
58	1	1	1	0	0	1	0	0
59	1	1	1	0	1	0	0	0
60	1	1	1	0	1	1	0	0
61	1	1	1	1	0	0	0	0
62	1	1	1	1	0	1	0	0
63	1	1	1	1	1	0	0	0
64	1	1	1	1	1	1	0	0
65	0	0	0	0	0	0	1	0
66	0	0	0	0	0	1	1	1
67	0	0	0	0	1	0	1	0
68	0	0	0	0	1	1	1	1
69	0	0	0	1	0	0	1	0
70	0	0	0	1	0	1	1	1
71	0	0	0	1	1	0	1	0
72	0	0	0	1	1	1	1	1
73	0	0	1	0	0	0	1	0
74	0	0	1	0	0	1	1	1
75	0	0	1	0	1	0	1	0
76	0	0	1	0	1	1	1	1
77	0	0	1	1	0	0	1	0
78	0	0	1	1	0	1	1	1
79	0	0	1	1	1	0	1	1
80	0	0	1	1	1	1	1	1
81	0	1	0	0	0	0	1	0
82	0	1	0	0	0	1	1	1
83	0	1	0	0	1	0	1	0
84	0	1	0	0	1	1	1	1
85	0	1	0	1	0	0	1	0
86	0	1	0	1	0	1	1	1
87	0	1	0	1	1	0	1	0
88	0	1	0	1	1	1	1	1
89	0	1	1	0	0	0	1	0
90	0	1	1	0	0	1	1	1
91	0	1	1	0	1	0	1	0
92	0	1	1	0	1	1	1	1
93	0	1	1	1	0	0	1	0

Table 7-4 *(continued)*

			Inputs				Output
Case	A	B	C	D	E	F	G	Y
94	0	1	1	1	0	1	1	1
95	0	1	1	1	1	0	1	1
96	0	1	1	1	1	1	1	1
97	1	0	0	0	0	0	1	0
98	1	0	0	0	0	1	1	1
99	1	0	0	0	1	0	1	0
100	1	0	0	0	1	1	1	1
101	1	0	0	1	0	0	1	0
102	1	0	0	1	0	1	1	1
103	1	0	0	1	1	0	1	0
104	1	0	0	1	1	1	1	1
105	1	0	1	0	0	0	1	0
106	1	0	1	0	0	1	1	1
107	1	0	1	0	1	0	1	0
108	1	0	1	0	1	1	1	1
109	1	0	1	1	0	0	1	0
110	1	0	1	1	0	1	1	1
111	1	0	1	1	1	0	1	1
112	1	0	1	1	1	1	1	1
113	1	1	0	0	0	0	1	1
114	1	1	0	0	0	1	1	1
115	1	1	0	0	1	0	1	1
116	1	1	0	0	1	1	1	1
117	1	1	0	1	0	0	1	1
118	1	1	0	1	0	1	1	1
119	1	1	0	1	1	0	1	1
120	1	1	0	1	1	1	1	1
121	1	1	1	0	0	0	1	1
122	1	1	1	0	0	1	1	1
123	1	1	1	0	1	0	1	1
124	1	1	1	0	1	1	1	1
125	1	1	1	1	0	0	1	1
126	1	1	1	1	0	1	1	1
127	1	1	1	1	1	0	1	1
128	1	1	1	1	1	1	1	1

Referring back to the statement of requirements, notice that for Y to equal 1, G must equal 1, and for all the combinations of $G = 0$, Y will equal 0. Thus, an abbreviated truth table can be constructed. One method involves the use of a symbol X indicating that the state of an input variable makes no difference in the output. The case where $G = 0$ can be defined in one line in the following manner:

	Inputs							Output
Case	A	B	C	D	E	F	G	Y
1	X	X	X	X	X	X	0	0

This single line defines that $Y = 0$ whenever $G = 0$ regardless of the states of A, B, C, D, E, and F.

   Looking back at the list of requirements, notice that when either level 1, level 2, or level 3 is 1, we can have Y equal to 1 if $G = 1$. Notice that if level 1 is a binary 1, the states of levels 2 and 3 are unimportant. Therefore, for A AND $B = 1$ along with $G = 1$, another case can be added to our truth table.

	Inputs							Output
Case	A	B	C	D	E	F	G	Y
1	X	X	X	X	X	X	0	0
2	1	1	X	X	X	X	1	1

   Consider now the effect of level 2. If $G = 1$ and level 2 is 1, Y will equal 1 regardless of the states of levels 1 and 3. Level 2 will be 1 when C, D, and E are 1. In a similar manner, if level 3 is 1, or $F = 1$, the states of levels 1 and 2 are unimportant. These facts are summarized below.

	Inputs							Output
Case	A	B	C	D	E	F	G	Y
1	X	X	X	X	X	X	0	0
2	1	1	X	X	X	X	1	1
3	X	X	1	1	1	X	1	1
4	X	X	X	X	X	1	1	1

   To this point, all parameters have been taken care of except for one set. These are the cases where $G = 1$ and all other levels equal 0. For level 1, this means that either A or B or both are 0. For level 2, this means that any or all of C, D, and $E = 0$. For level 3, this means that $F = 0$. These conditions can be added, and the completed short-form truth table is presented in Table 7-5. Notice that it contains all the essential information given in the long-form truth table presented in Table 7-4. Notice also that it is much less cumbersome and easier to understand than the long-form truth table. It has only 10 cases to consider rather than the 128 cases in the long-form table.

At this stage, one might ask why make a truth table when the information was already given by our tabulation of input parameters. The answer is that often the tabulation of parameters is ambiguous to all except the author. The truth table is, however, unambiguous to all. It also makes logic design and check-out procedures simpler. Last, and maybe most important, its construction oftentimes points out conceptual logic errors.

Table 7-5   Short-form truth table

			Inputs				Output
Case	A	B	C	D	E	F	G	Y
1	X	X	X	X	X	X	0	0
2	1	1	X	X	X	X	1	1
3	X	X	1	1	1	X	1	1
4	X	X	X	X	X	1	1	1
5	0	X	0	X	X	0	1	0
6	X	0	0	X	X	0	1	0
7	0	X	X	0	X	0	1	0
8	X	0	X	0	X	0	1	0
9	0	X	X	X	0	0	1	0
10	X	0	X	X	0	0	1	0

7-4   SIMPLE LOGIC DESIGN

Having generated a truth table, we can now design a logic system around it. From Table 7-5 when $Y = 1$, G must equal 1; and when $G = 0$, $Y = 0$. Working back from the output, we can deduce that G and something else will have to be 1 to make $Y = 1$. The last gate will then have to be an AND gate as illustrated in Fig. 7-10a. Also, when A and B equal 1 and $G = 1$, then $Y = 1$. Thus, A and B can be put into an AND gate as is illustrated in Fig. 7-10b. This completes level 1. Notice that in case 3, C, D, and E equal 1. They are thus fed into the AND gate presented in Fig. 7-10c. Notice that when either A and $B = 1$, C, D, and $E = 1$, or $F = 1$, along with $G = 1$, then $Y = 1$. Thus, the outputs of the gates presented in Fig. 7-10b and c can be placed into an OR gate along with input F.

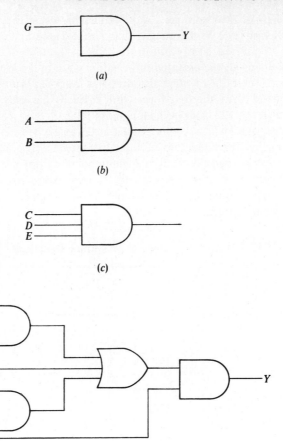

Fig. 7-10   Logic system generated from Table 7-5.

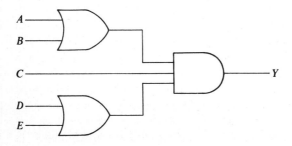

Fig. 7-11   Logic diagram for Table 7-6.

The output of this gate is connected to an input of the AND gate presented in Fig. 7-10a to complete the system. The completed logic diagram is presented in Fig. 7-10d.

Let us now turn the problem around and attempt to generate a meaningful truth table from a logic diagram. When trying to interpret existing diagrams, this problem often occurs. Consider the logic diagram presented in Fig. 7-11. The construction of the truth table can begin by noting that for Y to equal 1, C must equal 1. Also, note that when $C = 0$, $Y = 0$. Therefore, we have as a first approximation:

		Inputs				Output
Case	A	B	C	D	E	Y
1	X	X	0	X	X	0

For Y to equal 1, the outputs of both OR gates must equal 1, in addition to C being equal to 1. That means that either A or B or both must equal 1 for the top OR gate and either D or E or both must equal 1 for the bottom OR gate. These conditions add four more cases to the table.

		Inputs				Output
Case	A	B	C	D	E	Y
1	X	X	0	X	X	0
2	1	X	1	1	X	1
3	X	1	1	1	X	1
4	1	X	1	X	1	1
5	X	1	1	X	1	1

Now consider the cases where $C = 1$ but either one or both of the OR gate outputs are 0. For this condition, both A and B or D and E must be 0. They add the last two cases, and the short-form truth table (Table 7-6) defining the operation of the logic diagram presented in Fig. 7-11 is now complete.

For the simple logic functions and truth tables so far considered, an intuitive approach to their operation has been presented. This approach has been employed in an attempt to illustrate the techniques involved in constructing and using simple logic diagrams and truth tables without becoming involved in a defining mathematics that is usually not needed in interfacing problems. There

Table 7-6   Short-form truth table for logic diagram in Fig. 7-11

	Inputs					Output
Case	A	B	C	D	E	Y
1	X	X	0	X	X	0
2	1	X	1	1	X	1
3	X	1	1	1	X	1
4	1	X	1	X	1	1
5	X	1	1	X	1	1
6	0	0	1	X	X	0
7	X	X	1	0	0	0

is, however, a well-established branch of mathematics called *Boolean algebra* that can be helpful for more complex logic design. There are also mapping techniques developed to make complex logic design more efficient and less costly through the elimination of redundant logic gates. These techniques will not be extensively covered in the text which follows because they are not usually needed for computer-experiment interfacing. For those readers who wish to know more about Boolean algebra and mapping techniques, a discussion is presented in Appendix C. Some basic and necessary Boolean algebra considerations follow in the next section.

7-5   BOOLEAN ALGEBRA

Thus far, we have considered verbal descriptions of logic functions and truth tables to clearly define their operation. We will consider here some of the very basic concepts of Boolean algebra which provide a mathematical description of logic functions. Boolean algebra is the algebra of logic and consists of a symbolic method of studying logical operations. We have previously discussed AND, OR, and NOT or INVERT functions. The Boolean algebra symbols for these functions are presented in Table 7-7. The AND function can be represented by a dot or by placing the variables adjacent to each other. The OR function can be represented by a "+." The NOT or INVERT function can be represented by a prime beside or bar above the variable.

A complex logic function can, of course, be written in terms of Boolean algebra. For example, $X = AB + CD + F(\overline{G} + H)$ is read, "$X = 1$ whenever A AND $B = 1$, OR C AND $D = 1$, OR $F = 1$ when $G = 0$ OR $H = 1$." Notice that $G = 0$ is a requirement rather than $G = 1$. This is because of the complement bar above it in the Boolean expression. It implies that G must be equal to 0. $\overline{G}$ is usually expressed verbally as "not G."

Many theorems and postulates have been developed for Boolean algebra and

Table 7-7 Symbols for Boolean algebra operations

Function	Example	Symbol	Example
AND	$X = A$ AND B	$\cdot$	$X = A \cdot B, X = AB$
OR	$X = A$ OR B	$+$	$X = A + B$
NOT	NOT A	$^-, '$	$\overline{A}, A'$

are necessary for the design and evaluation of complex logic systems. They are presented and discussed in Appendix C. Only one theorem will be presented here. It is De Morgan's theorem, which is used to complement complex logic expressions. It is necessary for an understanding of NAND and NOR gates. For example, a NAND gate is equivalent to an AND gate followed by an inverter. In other words, it is an *output-inverted* or *complemented* AND gate. Thus, a Boolean expression can be written in the following manner for a NAND gate. It says that $X = 1$ when the complement of $(A$ AND $B) = 1$:

$$X = (AB)' \tag{7-1}$$

This is a correct expression, but it makes the construction of a complex truth table cumbersome because it requires first constructing the AND gate truth table and then complementing its output column. With the use of De Morgan's theorem, the expression can be placed in a more convenient form. De Morgan's theorem states that

$$(AB)' = \overline{A} + \overline{B} \tag{7-2}$$

and

$$(A + B)' = \overline{A} \cdot \overline{B} \tag{7-3}$$

The more general implications of De Morgan's theorem can be considered after defining two new terms, a *dual* and a *literal*. To obtain the dual of a Boolean expression, one must interchange all occurrences of a "+" and a "·" and of a 1 and a 0. Some examples are presented in Table 7-8. A literal is defined as any single *variable* within the dual expression. For example, in the expression

$$\overline{A}B + C \tag{7-4}$$

whose dual is

$$\overline{A} + BC \tag{7-5}$$

Table 7-8 Duality examples

Expression	Dual
$1 \cdot A + B$	$0 + AB$
$\overline{A}B + C$	$\overline{A} + BC$
$A (B + C)$	$(A + B)(A + C)$

Table 7-9   Complementing Boolean expressions

Expression	Dual	Complement
$1 \cdot A + B$	$0 + AB$	$\overline{AB}$
$\overline{A}B + C$	$\overline{A} + BC$	$A + \overline{BC}$
$A(B + C)$	$(A + B)(A + C)$	$(\overline{A} + \overline{B})(\overline{A} + \overline{C})$

the letters $\overline{A}$, B, and C are all literals. *In order to complement a Boolean expression, one must complement all literals in the dual expression.* This is illustrated in Table 7-9. The application of these rules for the NAND and NOR functions leads to the statements of De Morgan's theorem presented above.

To verify that the NAND gate is defined by $X = \overline{A} + \overline{B}$, a truth table can be constructed. This is done by substituting 1's and 0's for A and B in the equation and tabulating the results. It is presented in Table 7-10. Notice that the results are the same for the two expressions indicating that $X = (AB)' = A + B$.

In the case of the NOR function, it can be expressed by

$$X = (A + B)' = \overline{AB} \tag{7-6}$$

It is left to the reader to verify that this is correct.

Boolean algebra is often used to interpret and generate logic diagrams. Consider the diagram presented in Fig. 7-12. It has two inputs A and B. B is inverted by the upper inverter to $\overline{B}$, and A is inverted by the lower inverter to $\overline{A}$. The inputs to the upper NOR gate are then A and $\overline{B}$. From the Boolean

Table 7-10   Truth table for NAND using De Morgan's theorem

	Inputs		Outputs	
Case	A	B	AB	$(AB)'$
1	0	0	0	1
2	0	1	0	1
3	1	0	0	1
4	1	1	1	0
	$\overline{A}$	$\overline{B}$		$\overline{A} + \overline{B}$
1	1	1		1
2	1	0		1
3	0	1		1
4	0	0		0

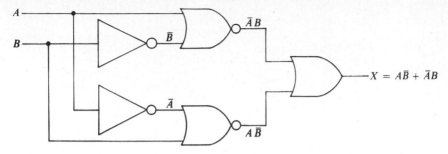

Fig. 7-12 Exclusive OR function.

expression for a NOR gate [Eq. (7-3)], the output of the upper gate is $\overline{A}B$. In a like manner, the output of the lower NOR gate is $A\overline{B}$. These outputs become inputs to the OR gate, resulting in the output expression

$$X = A\overline{B} + \overline{A}C \tag{7-7}$$

This particular function is called the *exclusive* OR function (XOR) and is the basis for binary arithmetic operations. It says that X is true if A or B but not both are true. The reader should at this point construct a truth table and verify the operation of this function. It has a defining symbol "$\oplus$" and is written as

$$X = A \oplus B \tag{7-8}$$

where

$$X = A\overline{B} + \overline{A}B = A \oplus B \tag{7-9}$$

There are other ways to generate the XOR function. If we start at the inputs, both A and B and $\overline{A}$ and $\overline{B}$ are required. These can be provided with the two inverters presented in Fig. 7-13a. Notice that the expressions $A\overline{B}$ and $\overline{A}B$ can be generated with AND gates, as illustrated in Fig. 7-13b. Notice also that the addition of an OR gate at the output completes the logic diagram shown in Fig. 7-13c. There are still more ways to generate the same function. The reader should work through the logic diagram presented in Fig. 7-14 and see if it will give the same results, viz., $X = A\overline{B} + \overline{A}B$. The XOR function is available in integrated-circuit form and is designated by the symbol presented in Fig. 7-15.

There is one more function which needs to be defined, the *coincidence* function. As the name implies, this function gives a 1 output whenever both inputs are the same, either all 0's or all 1's. Its symbol is illustrated in Fig. 7-16. The reader should construct its logic diagram using AND, OR, NAND, NOR, and INVERT gates in a manner similar to that just presented for the XOR function. A truth table should then be constructed to verify its operation.

One final aspect of Boolean algebra should be presented here. Thus far, we have said that electronic logic devices have two states, usually voltage levels, and

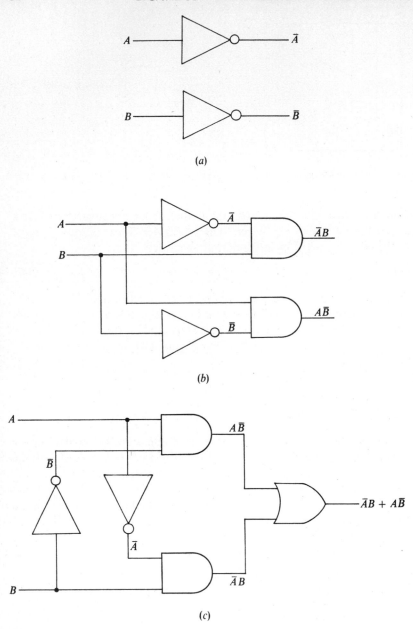

Fig. 7-13 Exclusive OR function.

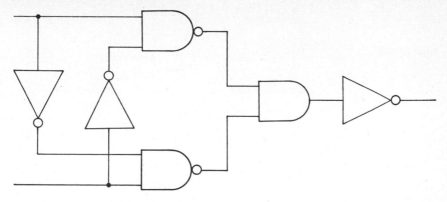

Fig. 7-14   Logic function.

$$X = A \oplus B = A\bar{B} + \bar{A}B$$

Fig. 7-15   Exclusive OR (XOR) function.

$$X = A \odot B = AB + \bar{A}\bar{B}$$

Fig. 7-16   Coincidence function.

that they may be represented by the binary numbers 0 and 1. To be more specific, a lower voltage was defined as a logical 0 and a higher voltage as a logical 1. However, the terms "true" and "false" have not yet entered the discussions. It is possible to define the validity of the binary numbers in two ways. A binary 1 can be defined as true and a binary 0 as false; this is called *positive logic*. Or a binary 1 can be defined as false and a binary 0 as true; this is called *negative logic*. Any logic diagram can be implemented in any of three ways using positive, negative, or mixed logic. It is important to note that positive-logic NAND, NOR, AND, and OR gates become negative-logic NOR, NAND, OR, and AND gates, respectively. The symbols used for both types of logic are summarized in Fig. 7-17.

7-6   ARITHMETIC APPLICATIONS OF LOGIC ELEMENTS

The logic elements discussed above can be used to perform arithmetic operations in digital computers. We will discuss here simple methods to perform both binary addition and subtraction. First, let us review those operations.

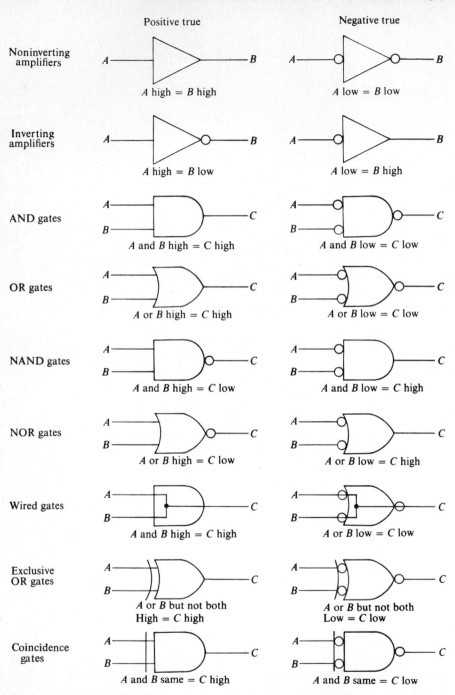

Fig. 7-17　　Logic gate symbols.

Table 7-11   Binary addition

Added numbers	Sum	Carry
0 + 0	0	0
1 + 0	1	0
0 + 1	1	0
1 + 1	0	1

Binary addition of two numbers is simple and straightforward. To add two numbers such as 1011 and 1010, one simply sums them remembering that $0 + 0 = 0$, $1 + 0 = 1$, and $1 + 1 = 10$. The addition is then

$$\begin{array}{r} 1011 \\ \underline{1010} \\ 10101 \end{array}$$

The addition can be broken down into the generation of a sum and a carry. For example, the addition of $1 + 1$ gives a sum of 0 and a carry of 1 for an answer of 10. In designing a logic system to perform addition, both the sum and the carry need to be considered. The conditions for the sum and carry are tabulated in Table 7-11. Notice that the sum is always 0 when the two numbers being added are the same. Or in other words, notice that the sum is always 1 when the two numbers are different. The XOR function previously discussed generates the addition function. This can be verified from a truth table of its operation. Notice also that the carry is 1 only when both inputs are 1. The AND function will perform this operation.

In its simplest form, then, the logic system presented in Fig. 7-18 can add

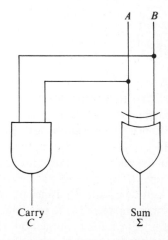

Fig. 7-18   Simple binary adder.

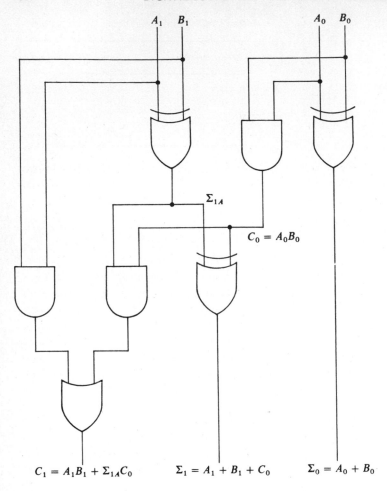

$$C_1 = A_1B_1 + \Sigma_{1A}C_0 \qquad \Sigma_1 = A_1 + B_1 + C_0 \qquad \Sigma_0 = A_0 + B_0$$

Fig. 7-19   Two-bit binary adder.

two binary numbers together. However, to add two larger numbers together, the carries from lower stages must be added to higher stages. The logic system to perform the addition of two 2-bit numbers is presented in Fig. 7-19. Notice that the most significant bits A_1 and B_1 are added together and their sum is added to the carry generated from the least significant bits A_0 and B_0 to produce the sum of the most significant bits. This is called a *ripple carry adder* because the carries ripple from stage to stage as the sums are generated. Notice also that the most significant carry C_1 is generated taking into account the state of C_0. Additional stages can be added to handle larger numbers.

For binary subtraction, the same technique can be employed except that one number must be complemented and provisions must be made to take care of the various kinds of negative binary numbers such as 1's-complement, 2's-complement, etc. In actual practice, one usually purchases complete 2-, 4-, or 8-bit arithmetic units in one integrated-circuit package to perform these operations.

7-7   FLIP-FLOPS

So far, we have discussed common logic elements used to perform arithmetic and logic operations. Now we will discuss some of the logic elements used for counting and storage operations. The basic device used for counting and storage is the *flip-flop* or *bistable multivibrator.* The flip-flop, in its many forms, can be constructed from gates, but it is usually purchased in integrated-circuit form.

The most basic flip-flop is called a *reset-set* or RS flip-flop. It can be constructed from two cross-coupled NAND gates as illustrated in Fig. 7-20. It has two inputs labeled S for set and C for clear and two outputs labeled Q and $\overline{Q}$. The Q output goes to a binary 1 and remains there when the S input momentarily goes from 1 to 0. In like manner, the $\overline{Q}$ output goes to a binary 1 and remains there when the C input momentarily goes from 1 to 0. Whenever Q is a 1, $\overline{Q}$ is a 0; that is, they are always complementary. As a result, the C input will also clear the Q output to a logical 0. The RS flip-flop will remain in whatever state it has been set or cleared to until its states are changed by applying negative-going pulses to one or the other of the inputs. It is thus a bistable device with two stable states that can be used for binary data storage. A truth table for the operation of the RS flip-flop is presented in Table 7-12. This is a somewhat different kind of a table than we have considered before in that it has time as a variable. t_n is the time before the specified input conditions have been imposed, and t_{n+1} is the time after they have been imposed. Since Q and $\overline{Q}$ always complement each other, only Q is given in the table. NC means that the output is unchanged from its previous state. U means that the output is undefined and may go to either state.

The RS flip-flop is commonly used for control and storage operations in

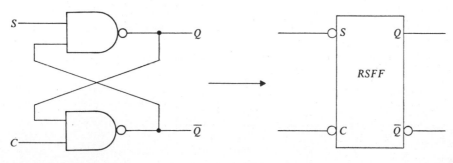

Fig. 7-20   RS flip-flop.

Table 7-12   RS flip-flop truth table

	Inputs		t_n	t_{n+1}
Case	S	C	Q	Q
1	1	0	X	0
2	0	1	X	1
3	1	1	X	NC
4	0	0	X	U

cases where two input signals occur. One signal is used to set its output and allow storage of a binary 1; while the other input is used to clear its output and store a binary 0. Notice in Table 7-12 that when $S = C = 0$, the output is undefined. When using the RS flip-flop as a storage element with two data inputs which always complement each other, this condition offers no hindrance to its use. However, for other operations, such as counting, where only one data input signal is provided, the RS flip-flop cannot be used directly.

Consider the case where the RS flip-flop is to be used to count a series of negative-going pulses. If the pulse train is connected to the S input only and the Q output is initially a binary 0, the first negative-going pulse, from 1 to 0, will set the Q output to 1. However, any further pulses will not change the state of the Q output. An analogous situation occurs if we use only the C input. If both the S and C inputs are connected together to the pulse train, the first negative-going pulse will apply a binary 0 to both inputs. The result will be undefined, as can be seen from the truth table. Thus, some other flip-flop will be needed for counting operations.

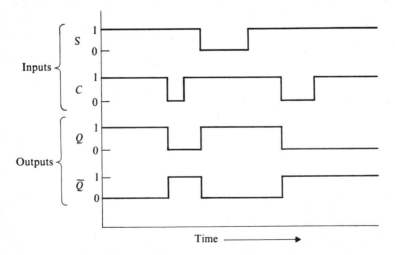

Fig. 7-21   RS flip-flop timing chart.

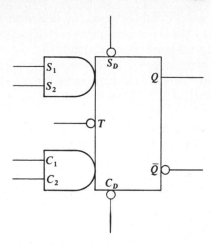

Fig. 7-22   Clocked flip-flop.

A timing chart for the RS flip-flop is presented in Fig. 7-21. Notice that the Q output is initially set at 1 and remains so until the C input goes to a binary 0. Q then changes to 0 and remains there until it is again set to 1 by the S input. It is then cleared back to 0. Notice that the $\overline{Q}$ output always complements the Q output state. Timing charts are useful for defining the dynamic operation of logic devices. Often they are used in place of, as well as with, truth tables.

A more versatile flip-flop than the RS type is called a *clocked* flip-flop. It is presented in Fig. 7-22 and can be used for counting operations. The clocked flip-flop has direct set S_D and direct clear C_D inputs that operate in the same manner as the S and C inputs of the RS flip-flop. This is commonly called *asynchronous* operation since no timing requirements are made. For this mode of operation, the truth table presented in Table 7-12 can be used. The clocked

Table 7-13   Clocked flip-flop truth table

Case	t_n				t_{n+1}
	S_1	S_2	C_1	C_2	Q
1	0	X	0	X	NC
2	X	0	X	0	NC
3	X	0	0	X	NC
4	0	X	X	0	NC
5	0	X	1	1	0
6	X	0	1	1	0
7	1	1	0	X	1
8	1	1	X	0	1
9	1	1	1	1	U

flip-flop has, however, another mode of operation called the *synchronous* mode. In the synchronous mode, information is entered into the flip-flop through the AND-gated S and C inputs (S_1, S_2 and C_1, C_2). The flip-flop does not change state, however, until a negative-going transition (from 1 to 0) occurs at the *clock* or T input. The truth table for the synchronous mode of operation is presented in Table 7-13. In the truth table, t_n refers to the input conditions prior to a timing pulse, and t_{n+1} refers to the time after a timing pulse has been applied to the T input. Notice, though, that there is still an undefined state. It occurs when all gated synchronous inputs are 1. As a result, the clocked flip-flop still cannot be used for counting in this mode. If, however, S_1 is connected to $\overline{Q}$, C_2 is connected to Q, and S_2 and C_1 are connected to a binary 1, as is shown in Fig. 7-23, a different condition exists. Since either Q or $\overline{Q}$ must always equal 0 and since Q is connected to C_2 and $\overline{Q}$ is connected to S_1, a condition where all gated inputs equal 1 can never be generated. As a result, there will never be an undefined output state. If a pulse train is applied to the T input, the flip-flop will change state on each negative-going pulse. This is called *JK* operation, and a truth table is presented in Table 7-14. Since the S_1 and C_2 inputs are connected to the outputs, they are not listed in the table. The timing chart is presented in Fig. 7-24. Notice that when S_2 and C_1 are 1, the Q output changes from 0 to 1 on every other input pulse, dividing the frequency in half. This flip-flop will serve for counting functions.

Normally, one would not have to connect a clocked flip-flop into the *JK* mode since many integrated-circuit *JK* flip-flops are available. A typical one is presented in Fig. 7-25. Notice that the gated S and C inputs are renamed J and K.

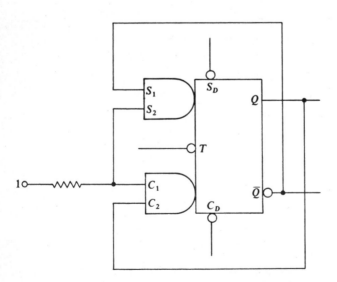

Fig. 7-23   Clocked flip-flop connected in *JK* modes.

Table 7-14   JK truth table

t_n		t_{n+1}
S_2	C_1	$\overline{Q}$
0	0	NC
1	0	1
0	1	0
1	1	$\overline{Q}_n$ (complements)

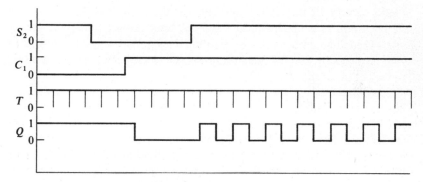

Fig. 7-24   Timing chart for *JK* flip-flop.

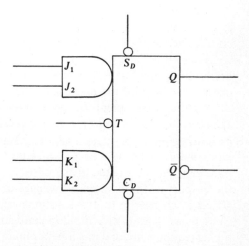

Fig. 7-25   *JK* flip-flop.

The connections from J and K to Q and $\bar{Q}$ are made internally and usually do not appear on the diagram.

There are, in addition to the RS, clocked, and JK flip-flops presented above, other popular configurations for many varied applications. They include many variations of input gating, including capacitor coupling for ac-only operation. Along with these, there is another general class of flip-flops called *master-slave* types. They have been developed to help overcome the critical nature of the timing requirements for the flip-flops already discussed.

When one actually uses clocked flip-flops such as the JK units described above, one particular problem often occurs. It is that the output changes resulting from a given clock pulse and set of input parameters applied to the flip-flop can change the input information during the life of the clock pulse. This results because of the connections between the outputs and inputs necessary for JK operation as is illustrated in Fig. 7-23. Very critical timing requirements are often needed to ensure that the flip-flop does not settle in the wrong state. This may involve very careful synchronization of clock pulses and input information and selection of the clock-pulse duty cycle.

A master-slave flip-flop is actually two flip-flops in one, with a master flip-flop that feeds data to a slave flip-flop. The resulting flip-flop configuration can be any one of those types already discussed, such as clocked or JK. A JK master-slave flip-flop and clock-input waveform are presented in Fig. 7-26. For the sake of illustration, the various internal gates and connections are presented. In actual practice, the symbols for master-slave flip-flops are not distinguished from those already presented. Almost all clocked types of flip-flops available in integrated-circuit form are of the master-slave type. All the counter and register circuits presented in the pages which follow were designed with master-slave flip-flops.

Referring to Fig. 7-26a, notice that both the master and the slave are gated RS flip-flops. The clock-pulse waveform presented in Fig. 7-26b has four points on it labeled A, B, C, and D. The operation of the master-slave JK flip-flop can be explained as follows: As the clock pulse goes positive from 0 to 1 past point A, the slave input gates 3 and 4 become disabled isolating the slave from the master. The state of the master prior to A is stored on the slave outputs. As the clock pulse passes point B, input gates 1 and 2 of the master are enabled allowing data to be transferred in through the S and R inputs. As the clock passes C, gates 1 and 2 are disabled, isolating S and R from the master. As the clock pulse further falls past D, inverter 5 enables gates 3 and 4, allowing data transfer from the master to the slave. Notice that with master-slave operation, the outputs of the slave do not change until the clock pulse is completed. The effects of changes in the slave outputs cannot reach the master inputs during a clock pulse. Master-slave flip-flop configurations thus exhibit much less critical timing requirements and much better noise immunity.

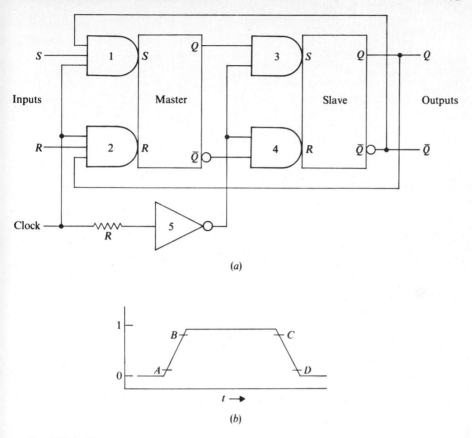

(a)

(b)

Fig. 7-26   Master-slave *JK* flip-flop and clock pulse.

7-8 DATA STORAGE LATCHES

One simple data storage device is called a data *latch*. It is often used for the temporary storage of binary information. A data latch can be built with either RS or synchronous flip-flops. The one presented in Fig. 7-27 is constructed from RS flip-flops. When the timing-pulse TP input is high, a binary 1, the outputs of all the input OR gates are high. No information from the data inputs D_0 and D_1 can be transferred to the latch. (The reader should refer to the OR gate truth table presented in Fig. 7-9 to verify that this is true.) However, when the TP input is a binary 0, the data inputs can transfer information to the flip-flops. For example, if a binary 1 is presented at D_0 the S input of *FFA* will be a binary 0, which will set Q to a binary 1. Notice that the C input will be a binary 1. As a

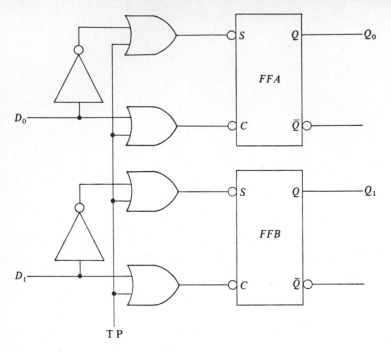

Fig. 7-27   Two-bit data latch using RS flip-flops.

result, when *TP* is low, the *Q* outputs will follow the *D* inputs. When TP is high, the *Q* outputs will not change regardless of what conditions occur at the *D* inputs. Data is thus stored in binary form until the TP input again goes momentarily to 0.

A data latch can also be built from synchronous flip-flops, as illustrated with the clocked flip-flops presented in Fig. 7-28. Again, this is a 2-bit latch. While two NOR gates and an inverter were added to each RS flip-flop to make a data latch, only one inverter need be added for each clocked flip-flop. The clocked flip-flop data latch operates in much the same way as the RS flip-flop data latch except for the operation of the TP input. In the clocked flip-flop data latch, the TP line must undergo a negative-going 1 to 0 transition each time data is transferred.

In other words, the clocked flip-flop data latch will only operate in a synchronous manner, that is, each time a negative-going clock pulse is present. This is in contrast to the RS flip-flop data latch where the outputs will follow the inputs whenever the TP input is 0. A data storage latch can be constructed from *JK* flip-flops. Latches can, of course, be constructed for any number of data bits. Integrated-circuit data latches are commonly available in 4- and 8-bit configurations.

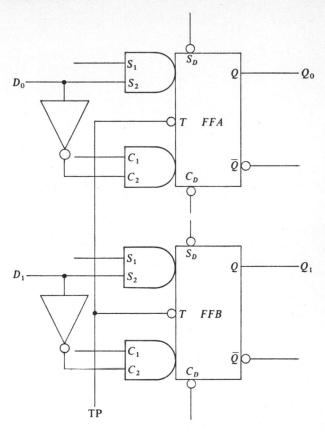

Fig. 7-28   Two-bit data latch using clocked flip-flops.

7-9   SHIFT REGISTERS

Shift registers are, like data latches, often used for data storage. They also have other features to make them more versatile. They can acquire and output data in both serial and parallel; data latches can acquire and output data only in parallel. Also, they can be used to move data within the register while it is being stored. They are sometimes used for digital signal delay lines for timing and synchronization problems. The simplest shift register is the serial I/O type (Fig. 7-29).

The 4-bit serial I/O shift register presented in Fig. 7-29 is constructed from clocked flip-flops. It can, however, also be constructed from JK flip-flops in exactly the same manner. It cannot be constructed from RS flip-flops. Notice that a shorthand notation with the AND gate symbols omitted is used for the clocked flip-flops. The operation of the register is quite simple. The C_D inputs are all tied together to provide a common clear line. This allows all the flip-flops

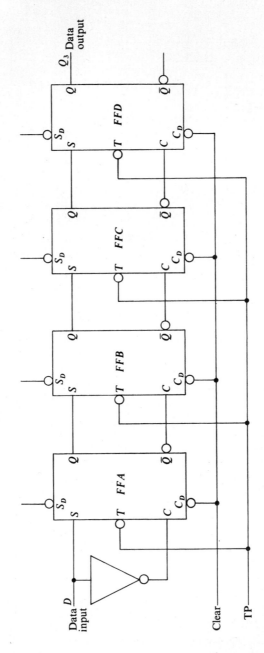

Fig. 7-29 Four-bit serial input/output shift register.

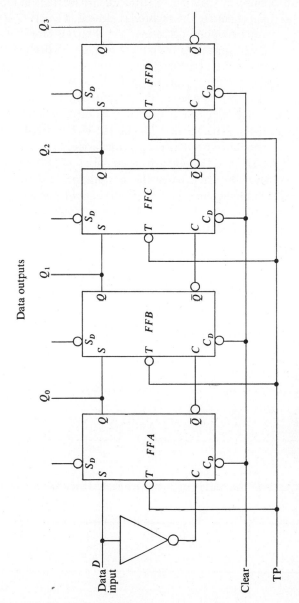

Fig. 7-30 Four-bit serial-input/parallel-output shift register.

to have their Q outputs set to 0 simultaneously. A common set line could also be used. The first stage inputs to *FFA* are connected in exactly the same manner as for the synchronous data latch previously presented. Successive stages have their outputs and inputs connected together to allow data to be transferred from one to the other in a serial manner. If the register is cleared, a binary 1 is presented at the data input, and a clock-pulse transition from 1 to 0 is provided on the TP input line, the binary 1 will appear on the output of *FFA*. Notice that a binary 1 is now applied to the S input of *FFB*. The next clock pulse will transfer the binary 1 to the output of *FFB* presenting it at the input of *FFC*. Thus, data present at the data input will appear at the output of *FFD* four clock pulses later. An n clock-pulse data delay can be generated by using n flip-flops.

In some applications, digital data might be received in serial but is needed in parallel for the computer or data-acquisition system. A serial-to-parallel converter can be built by providing enough clock pulses and flip-flops to fill the serial I/O shift register with data. Then the outputs of each flip-flop in the register can be read simultaneously. This is a *serial-input parallel-output* shift register and is illustrated in Fig. 7-30. Notice that except for the parallel data outputs it is exactly the same as the register presented in Fig. 7-29.

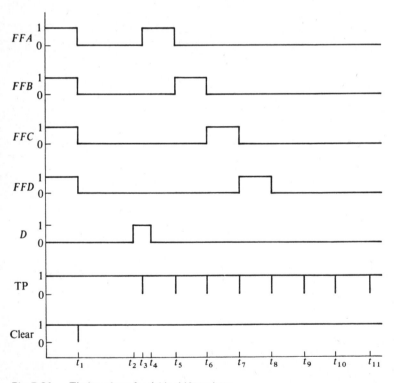

Fig. 7-31   Timing chart for 4-bit shift register.

The operation of the shift registers presented in Figs. 7-29 and 7-30 is easily described using the timing chart presented in Fig. 7-31. The outputs of all flip-flops are initially at a binary 1. A clear pulse at t_1 resets them to 0. At t_2, the data input goes to a binary 1, which is transferred to the output of *FFA* by a clock pulse TP at t_3. At t_4, the data input is reset to binary 0. The binary 1 data level is transferred through *FFB* and *FFC* by timing pulses at t_5 and t_6 and appears at the output of *FFD* at t_7 four clock pulses after it entered the register. Timing pulse t_8 shifts the binary 1 out of *FFD*, leaving the entire register contents equal to 0. Notice that any combinations of 1's and 0's can be entered into and shifted through the register providing that the proper timing sequences are met. Notice also that the data input level must be present before a timing pulse is initiated.

There exist also parallel-input registers with both serial and parallel outputs. A parallel-input shift register is presented in Fig. 7-32. For parallel outputs, the data bits are read from the outputs of each flip-flop. Notice that data can be entered either through S_D inputs in an asynchronous mode or through the gated inputs in a synchronous mode. When using the asynchronous mode, the S_D inputs must be returned to a binary 1 before data can be shifted through the register. When using the synchronous mode, the data inputs must be returned to a binary 0 before shifting can occur. Parallel-entry shift registers often have gates connected at the data inputs with a common control line to provide these characteristics. For data to be entered, a strobe signal or level change is required. When the strobe is released, the shifting can occur. It would be a good exercise at this point to design strobed inputs for both asynchronous and synchronous operation. The asynchronous design parameters are listed in Table 7-15. In order to design the logic to perform this operation, remember that one gate will be needed for each S_D input. The reader can look back at Figs. 7-27 and 7-28 to get a hint on how this may be accomplished. It is left to the reader to set up the design characteristics and logic system for strobed synchronous entry.

Table 7-15   Design characteristics for strobed data input

	Asynchronous mode
Data characteristics	A binary 0 at any S_D input results in $Q = 1$ for that flip-flop
Strobe characteristics	A common strobe controls all flip-flop inputs. When strobe = 0 (or 1), data are entered into the flip-flop. When strobe = 1 (or 0), data cannot pass, and S_D inputs are all held at 1.

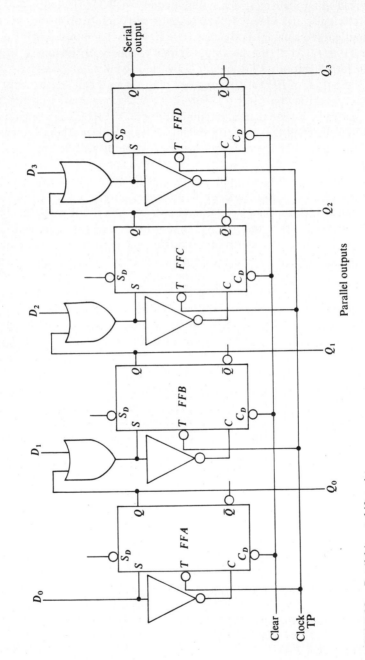

Fig. 7-32 Parallel-input shift register.

Since data have been entered in parallel into the register presented in Fig. 7-32, it can be read out of the corresponding flip-flops in parallel or in serial out of *FFD* by shifting it out 1 bit at a time. This provides either a parallel-to-parallel or parallel-to-serial converter. Notice also that data can be entered in parallel and shifted one or more positions; then the results can be read out again in parallel. This is an easy way to truncate a number, throwing away the least significant bits. For example, suppose the number 1011 is entered into the register via inputs D_0 to D_3. This can then be shifted one position and read out as 0101 from data outputs Q_0 to Q_4. In addition, the register can be used to multiply and divide numbers by powers of 2. If the number $1000_2 = 8_{10}$ is entered into the register and shifted once, the parallel-output number is $0100_2 = 4_{10}$. Shifting twice gives $0010 = 2_{10} = 8_{10} \div 4_{10}$; shifting three times gives $0001 = 1_{10} = 8_{10} \div 8_{10}$. Notice that if the register is turned around so that D_0 and Q_0 correspond to the least rather than the most significant bits, binary multiplication by powers of 2 can be performed.

In actual practice, many types of shift registers are available in integrated-circuit form and need not be constructed by the user. Wiring schemes are available from the manufacturers that allow all forms of serial and parallel operation, including shifting to the right, the left, or either direction.

7-10   ASYNCHRONOUS COUNTERS

Having considered flip-flops and shift registers, we are ready to discuss counters. Often one needs to provide counters in an interface to count events such as the number of data points taken and the number of times data exceed a predetermined threshold, to divide down the clock frequencies, etc. As for the registers just discussed, the flip-flop is the basic device used for counters. The most common flip-flop used in modern integrated-circuit counters is the *JK* flip-flop presented in Fig. 7-25. It can be used to construct two basic types of counters, asynchronous and synchronous, that will count up or down in a variety of counting schemes. The simplest counter is the asynchronous binary up-counter presented in Fig. 7-33.

The asynchronous binary up-counter presented in Fig. 7-33 counts from 0 to 15 and resets. Its timing chart is presented in Fig. 7-34. Initially all flip-flops in the counter are cleared to 0. When the first negative-going transition or clock pulse is presented at the count input, the output of *FFA* (Q_0) changes state from 0 to 1. When the second clock pulse occurs, it again changes the state of *FFA*, returning it from 1 to 0. This transition causes FF_2 (Q_1) to change state from a binary 0 to a binary 1. When *FF2* returns to a binary 0, it causes *FF3* (Q_2) to change state. In the same manner, *FF3* causes *FF4* to change state. Any number of flip-flops can be connected to count to larger numbers. The asynchronous counter is often called a ripple counter because of the way counts ripple through or are passed along from flip-flop to flip-flop. From the timing

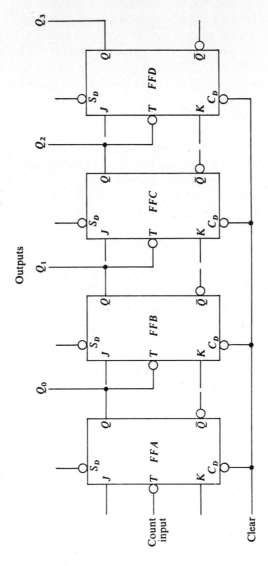

Fig. 7-33 Asynchronous binary up-counter.

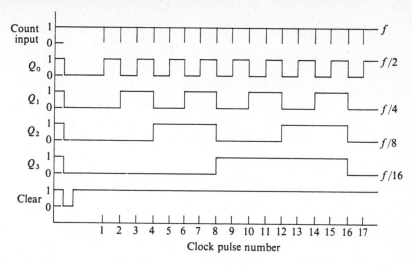

Fig. 7-34   Asynchronous binary up-counter timing chart.

chart, we can see that binary frequency division occurs with $Q_0 = f/2$, $Q_1 = f/4$, $Q_2 = f/8$, and $Q_3 = f/16$. This is then a convenient way to divide down clock frequencies in binary orders of magnitude. Notice also that we can read the binary number output of the counter from the timing chart after any given number of clock pulses has occurred. Consider the case after 10 clock pulses. Reading down from the top of the chart at clock pulse number 10, $Q_0 = 0$, $Q_1 = 1$, $Q_2 = 0$, and $Q_3 = 1$. The binary number is thus 1010 since Q_0 is the least significant bit. Notice that $1010_2 = 10_{10}$.

In addition to an asynchronous binary up-counter, an asynchronous binary down-counter can also be designed. For a 4-bit counter, it will count down from 15 to 0. This is accomplished by disconnecting each T input from the corresponding Q output and connecting it to the corresponding $\overline{Q}$ output. The reader should verify that this is true by constructing a binary down-counter and its timing chart.

Since we are accustomed to thinking in terms of the decade number system, decade counters are often used in interface systems. They not only count in powers of 10 but can be used to divide down clock frequencies in decade rather than binary steps. A simple decade counter is presented in Fig. 7-35. Recall that increasing the counting range or decreasing the frequency in binary steps involves only the addition of a single flip-flop. However, to increase the counting range or decrease the frequency range in decade steps involves adding complete decade counter assemblies, each of which is constructed from four flip-flops.

The decade counter presented in Fig. 7-35 follows the count sequence

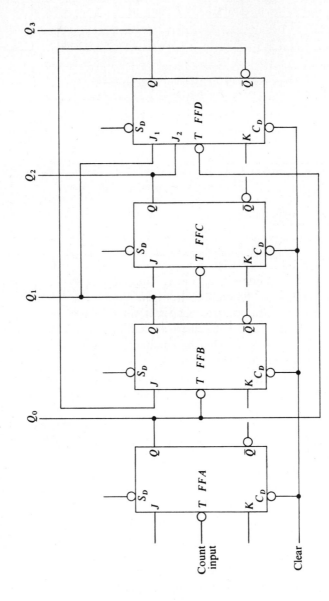

Fig. 7-35 Asynchronous decade up-counter.

Table 7-16   BCD number system

Decade	BCD
0	0000
1	0001
2	0010
3	0011
4	0100
5	0101
6	0110
7	0111
8	1000
9	1001

presented in Table 7-16. The BCD number system is a binary representation of the decimal number system.

The timing chart for the decade counter presented in Fig. 7-35 is presented in Fig. 7-36. Notice that it is the same as the binary up-counter timing chart presented in Fig. 7-34 up through clock pulse number 8. From there on, however, it differs. The chief requirement for the operation of the decade counter is that it must reset to 0 after a count of 9_{10} or 1001_2. This requires some additional connections over those needed for the binary counter. Notice in Fig. 7-35 that the $\overline{Q}$ output of FFD is connected to the J input of FFB. This allows FFB to change state only when the $\overline{Q}$ output of FFD = 1. This occurs for

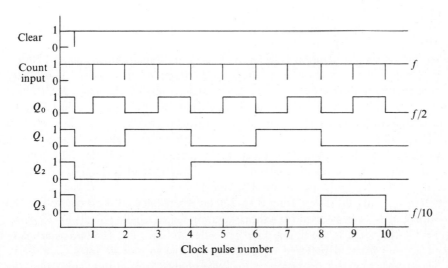

Fig. 7-36   Asynchronous decade up-counter timing chart.

counts 0 to 7. However, for counts 8, 9, and above, *FFB* is disabled. Notice also that *FFD* has two *J* inputs J_1 and J_2. They allow *FFD* to change state from 0 to 1 (*Q* output) only when $Q_1 = 1$, $Q_2 = 1$, and Q_0 goes from 1 to 0. These conditions are only present at a count of 7 and allow the count of 8 to occur. The count transition from 9 to 0 occurring on clock pulse number 10 in Fig. 7-36 occurs in the following way: For a count of 9, *FFB* is disabled because of the connection from $\overline{Q}$ of *FFD* to its *J* input. It cannot change state from the $Q_1 = 0$ state that it is in. Inputs J_1 and J_2 of *FFD* are also 0. Any 1-to-0 timing signal will thus set Q_3 back to 0. This can be verified by reviewing the *JK* flip-flop truth table presented in Table 7-14. Clock pulse 10 toggles Q_0 of *FFA* from 1 to 0. Q_0 is connected to the *T* input of *FFD* causing its *Q* output also to go from 1 to 0. The counter thus resets to 0 on clock pulse 10. It would be a good idea at this time to review the operation of the decade counter for each input clock pulse. The timing chart presented in Fig. 7-36 will help the reader to accomplish this easily.

The decade counter presented in Fig. 7-35 can be used as a down-counter rather than as an up-counter by reading the $\overline{Q}$ rather than the Q outputs. In addition, a down-counter can be constructed in which the Q outputs are used. This is accomplished by interchanging all J and K and Q and $\overline{Q}$ connections. The reader should at this point design such a counter and verify its operation by constructing a timing chart.

For many problems, the binary and BCD counters presented above are used. However, sometimes a counter that counts or divides frequencies by other values is wanted. A counter to divide by any number from 2 to 15 can be built by using the binary counter presented in Fig. 7-33 and a few gates. A counter designed to divide by 13 is presented in Fig. 7-37. Its timing chart is presented in Fig. 7-38. The principles used to describe this counter can be applied to the design of counters to divide by any number between 2 and 15 or, by using more flip-flops to divide by any number.

This type of counter is called a *divide-by-n* asynchronous counter. Notice that the divide-by-*n* counter, where *n* = 13, presented in Fig. 7-37 is a binary counter where the outputs have been fed back through a four-input NAND gate and two inverters to the C_D inputs of the flip-flops. For counts 0 to 12_{10}, the counter operates exactly like a binary counter. However, on count 13_{10}, which is 1101_2, all inputs to gate G_1 are at a binary 1. From the Boolean expression for a four-input NAND gate, the truth table presented in Table 7-17 can be constructed. Notice from Table 7-17 that the output of gate G_1 is a binary 0 only when all inputs are binary 1's. For all other cases, the output is 1. The counter counts in binary from 0 to 12 and resets back to 0 on count 13. The output of G_1 can be used for a short-duration negative-going divide-by-13 pulse. This is illustrated in the timing chart presented in Fig. 7-38. The two inverters I_1 and I_2 are inserted between G_1 and the C_D line to provide for a signal delay necessary for correct operation of the reset count. Without the signal delay, the

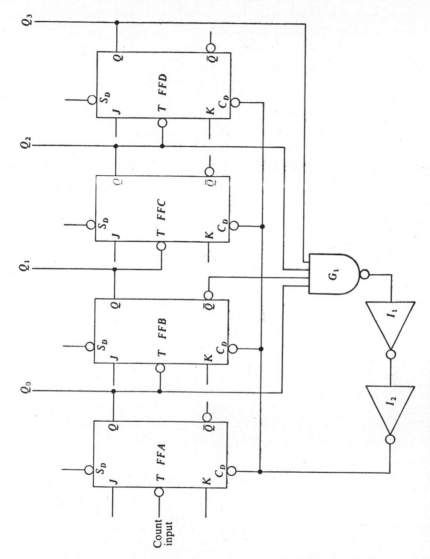

Fig. 7-37 Divide-by-13 asynchronous counter.

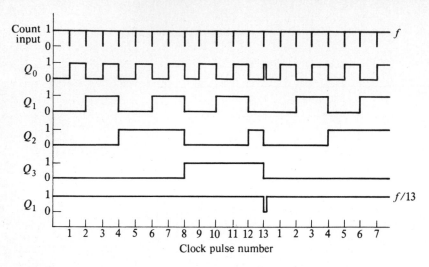

Fig. 7-38   Timing chart for asynchronous divide-by-13 counter.

flip-flops do not have enough time to settle, which results in resetting to numbers other than 0.

Consider now how the connections to gate G_1 are selected. For a reset to occur, all inputs must be at a binary 1. The dividing number n is first selected and written out in binary. For the divide-by-13_{10} counter, the binary equivalent is 1101_2. The correspondence of each bit in the number to each flip-flop is then made as is shown in Table 7-18. Whenever a binary 1 appears at the output of a given flip-flop for a given reset number n, the Q output is connected to G_1. Likewise, whenever a corresponding binary 0 appears for a given reset number n, the $\overline{Q}$ output is connected to G_1. This results in all input to G_1 being true for the reset number and produces the required false or binary 0 output on G_1 needed for counter reset. It would be a good exercise for the reader to design a divide-by-11 counter and verify its operation with a timing chart.

Table 7-17   Truth table for a four-input NAND gate

$E = \overline{A} + \overline{B} + \overline{C} + \overline{D}$

A	B	C	D	E
1	1	1	1	0
0	X	X	X	1
X	0	X	X	1
X	X	0	X	1
X	X	X	0	1

Table 7-18 G_1 connections for divide-by-13 asynchronous counters

Flip-flop	Bit	Output connection
FFA	1	Q_0
FFB	0	Q_1
FFC	1	Q_2
FFD	1	Q_3

7-11 SYNCHRONOUS COUNTERS

Unlike asynchronous counters, where the output change of one flip-flop is applied to the clock input and thus changes the state of a succeeding flip-flop, *synchronous*-counter flip-flop outputs set up the J and K inputs of succeeding flip-flops so that a common clock signal can cause the proper count sequence to occur.

A synchronous binary up-counter is presented in Fig. 7-39. Compare this counter with the asynchronous binary up-counter presented in Fig. 7-33. Notice that the synchronous counter requires external gating while the asynchronous counter does not. This is because the count sequence is generated by the external gates which set up the J and K inputs of each flip-flop. The timing chart presented in Fig. 7-34 for the asynchronous binary counter can also be used for the synchronous counter presented in Fig. 7-39.

The operation of the synchronous binary counter presented in Fig. 7-39 is quite simple. If we start with all flip-flops set to 0, the first pulse applied to the count input will change the state only of *FFA*. This is because the J and K inputs of all other flip-flops are at a logical 0, which disables the count input to those flip-flops. On each succeeding clock pulse, *FFA* will change state. When the second clock pulse is applied to the clock input, *FFB* changes state, with the Q output going to a logical 1. This occurs because its J and K inputs, which are connected to the Q output of *FFA*, were at a logical 1 after the first clock pulse. Notice now that *FFA* has its Q output at a logical 0, *FFB* has its Q output at a logical 1, and *FFC*'s and *FFD*'s Q outputs are both at a logical 0. On the third clock pulse, *FFA* changes state. It is now, along with *FFB*, at a logical 1. Since the outputs of *FFA* and *FFB* are connected to the J and K inputs of *FFC* through AND gate G_1, *FFC* now has both inputs at a logical 1. This means that the next clock pulse will change the state of *FFC*. And, in fact, on the fourth clock pulse *FFA* changes from 1 to 0, *FFB* changes from 1 to 0, and *FFC* changes from 0 to 1, giving a binary count of 4. In a like manner, when *FFA*, *FFB*, and *FFC* are all at a logical 1, *FFD* will change state. This occurs at a count of 8. The reader should work through the complete count sequence from 0 to 15 in binary for the synchronous binary up-counter.

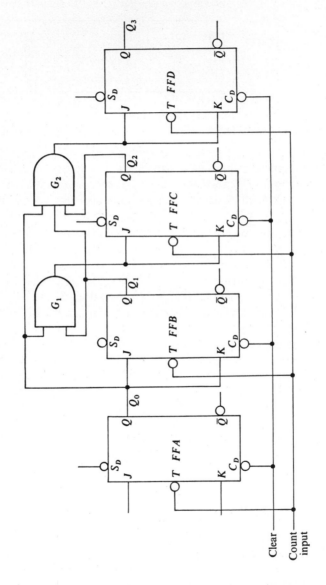

Fig. 7-39 Synchronous binary up-counter.

The synchronous binary counter presented in Fig. 7-39, like the asynchronous binary counter, can be used as either an up- or down-counter. A countdown sequence can be obtained by reading the $\overline{Q}$ outputs of each flip-flop or by connecting the $\overline{Q}$ outputs instead of the Q outputs on each flip-flop to the succeeding stages and reading the count numbers from the Q outputs.

A synchronous decade up-counter is presented in Figure 7-40. It should be compared to the asynchronous decade counter presented in Figure 7-35. The timing chart presented in Figure 7-36 applies equally well to both counters. Notice again that the synchronous counter involves more complex external gating. The evaluation of the operation of the synchronous decade up-counter is the same as for the binary synchronous counter if one remembers that a flip-flop will only change state when it receives a clock pulse and when either or both of the J and K inputs are a logical 1. The reader should work through the operation of the synchronous decade counter, paying especially careful attention to what occurs between the counts of 9 and 0.

The question arises of why one would use the more complex synchronous counters rather than the simple asynchronous counters. One important reason is that in asynchronous counters the various count sequences must ripple from one flip-flop to the next. This means that an incoming count on the first flip-flop that will eventually change the state of a flip-flop farther down the line must pass through all intermediate flip-flops. This operation takes a fairly long period of time and is equal to the sum of the signal *propagation delay times* of each succeeding flip-flop. The maximum count rate possible with asynchronous counters is determined, then, not only by the signal propagation delay of an individual flip-flop, but by the total propagation delay of all the flip-flops in sequence. However, in synchronous counting systems, the count sequences do not have to ripple down all the flip-flops. The propagation delay for a given synchronous counter is usually no more than that for one flip-flop and one gate. The result is that the synchronous counters can operate at much higher speeds. Notice, however, that as the number of stages increases in synchronous counters so does the number of inputs to the gates used between counter stages. In fact, if one looks at a synchronous binary up-counter of 15 stages, one sees that an input gate with 14 inputs will be required at the last stage. One way to get around this problem is to use semisynchronous operation. In this mode of operation, flip-flops are run synchronously up to perhaps eight or nine stages, and then these units are connected together asynchronously. When used in decade counters, each decade may be made up of four individual stages, all of which are synchronous internally, but each external stage or decade is connected to another decade asynchronously. This cuts down on the cost and complexity, while still allowing high-speed operation. The operation is not, however, at as high a speed as is a totally synchronous system.

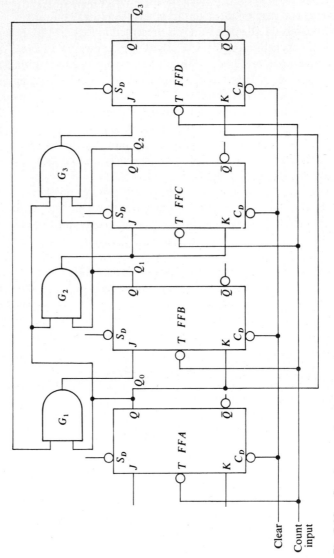

Fig. 7·40   Synchronous decade up-counter.

7-12 COMPLEX LOGIC ELEMENTS

The registers and counters presented so far have been discussed so that the approach taken to building more complex logical devices from simple gates and flip-flops can be understood. However, in actual use registers and counters are very seldom constructed from individual logic elements. The reason for this is that complex logic elements are available in integrated-circuit form. These include synchronous and asynchronous counters, shift and storage registers, memory systems, arithmetic elements, number converters, number-system decoders, digital multiplexers, digital comparators, and many others. These complex devices are usually referred to as either medium-scale integrated circuits (MSI) or large-scale integrated circuits (LSI). LSI technology, as the name implies, differs from MSI technology in that more complex logic functions requiring more gating and internal logic are performed in a single package. In addition, the MOS technology used in LSI devices is different from the conventional bipolar transistor technology used in MSI devices. (See Appendix B.)

The importance of complex logic elements cannot be overemphasized. They allow interface design to be done from a functional point of view using large functional blocks. Interface design complexity is reduced to the interconnection of these complex functional blocks along with the various clocks and timing elements necessary. It is important, however, for the designer to be able to implement the complex logical functions. The information necessary is usually provided on a data sheet and in application notes available from the manufacturer. Such a data sheet is presented in Fig. 7-41. The data sheet is for an MSI 4-bit shift register manufactured by Fairchild Semiconductor. Notice that under the general description it is stated that the register may be used in shift-left, shift-right, serial-serial, serial-parallel, parallel-serial, and parallel-parallel data transfers. It can also be used in many counter applications. Notice that the logic symbol is only a block diagram showing some inputs and outputs and control terminals. The data inputs are labeled P_0 to P_3, while the data outputs are labeled Q_0 to Q_3. Notice also that the complementary data output for the fourth bit is labeled $\overline{Q}_3$. The control inputs consist of a parallel-entry terminal labeled PE, a master-reset terminal labeled MR, a clock-pulse input labeled C_p, and the J and K inputs for the least significant stage. In order to provide parallel entry, the small circle at the PE input states that a binary 0 must be provided. In a like manner, the MR occurs when the MR terminal is at a binary 0. The actual shift transition occurs at a 0-to-1 transition at the clock pulse or C_p input. The second page of the data sheet is presented in Fig. 7-42. At the top of that figure, a functional description is provided on the operation of the shift register. This includes a truth table for serial entry and a list of the various operating requirements for that device. In addition, the electrical characteristics are presented.

In addition to providing data sheets describing the characteristics of MSI and LSI devices, most manufacturers provide applications information. For

184

GENERAL DESCRIPTION — The 9300 Four Bit Shift Register is a high speed multi-functional sequential logic block which is useful in a wide variety of register and counter applications. As a register it may be used in serial-serial, shift left, shift right, serial-parallel, parallel-serial, and parallel-parallel data transfers. The circuit uses TTμL for high speed and high fanout capability, and is compatible with all devices in the CCSL group of digital integrated circuits.

- 15 MHz shift frequency

- Synchronous parallel entry

- J, $\overline{K}$ inputs to first stage

- Asynchronous common reset

- Typical power dissipation of 300 mW

- The input/output characteristics provide easy interfacing with Fairchild DTμL, LPDTμL, and TTμL families (CCSL).

- All ceramic "HERMETIC" 16 pin Dual In-Line package.

- Input diode clamping

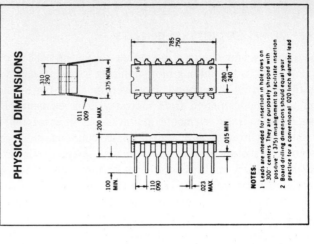

PHYSICAL DIMENSIONS

NOTES:

1 Leads are intended for insertion in hole rows on .300" centers. They are purposely shipped with "positive" (.375") misalignment to facilitate insertion

2 Board drilling dimensions should equal your practice for a conventional .020 inch diameter lead

Figure 1

ABSOLUTE MAXIMUM RATINGS (above which the useful life may be impaired)

Storage Temperature	$-65°C$ to $+150°C$
Temperature (Ambient) Under Bias	$-55°C$ to $+125°C$
V_{CC} Pin Potential to Ground Pin	$-0.5\,V$ to $+7\,V$
Voltage Applied to Outputs for high output state	$-0.5\,V$ to $+V_{CC}$ value
Input Voltage (D.C.)	$-0.5\,V$ to $+5.5\,V$

ORDER INFORMATION

Specify U6B9300XXX for 16 pin Dual In-Line package where XXX is 51X for the $-55°C$ to $+125°C$ temperature range, or 59X for the $0°C$ to $+75°C$ temperature range.

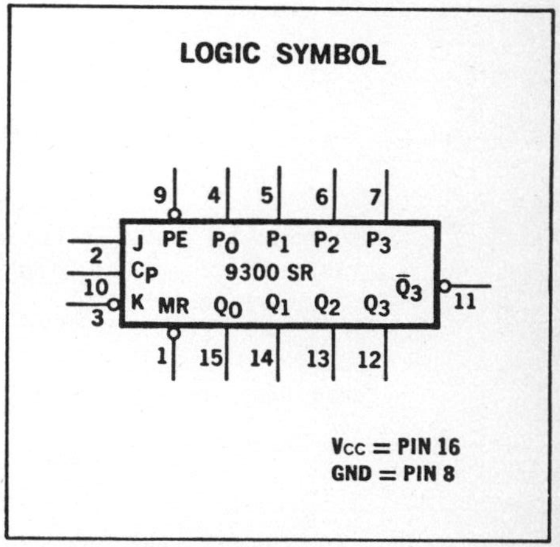

LOGIC SYMBOL

Figure 2

Fig. 7-41 MSI 4-bit shift register. (*Fairchild Semiconductor.*)

FAIRCHILD MEDIUM SCALE INTEGRATION · 9300

FUNCTIONAL DESCRIPTION

The logic symbol of Figure 2 provides an indication of the functional characteristics of the 9300 four bit shift register. Several special logical features of the 9300 design which provide a high degree of general usefulness are described below:

1. A J$\overline{K}$ input is provided to the first flip flop in the register. This type of input is the same as the more common JK input except that the low voltage level activates the $\overline{K}$ input. This provides the greater power of the JK type input for more general applications and at the same time the simple D type input that is most appropriate for a shift register can be easily obtained by simply tying the two inputs together.

2. There is no restriction on the activity of the J or $\overline{K}$ inputs for logical operation — except for the set up and release time requirements.

3. Parallel inputs for all four stages are provided. These will determine the next condition of the shift register synchronous with the clock input, whenever the Parallel Enable input is low. With the Parallel Enable input low the element appears as four common clocked D flip flops. When the Parallel Enable is high, or not connected, the shift register performs a one bit shift for each clock input. In both cases the next state of the flip flops occurs after the low to high transition of the clock input.

4. An internal clock buffer provides both reduced clock input loading, and the ability to gate the clock with only a single NAND gate.

5. The active high output is provided for all four stages and an active low output is provided for the last stage.

6. A master asynchronous clear input allows the setting to zero of all stages, independent of the condition of any other inputs.

TABLE I — TRUTH TABLE FOR SERIAL ENTRY

(PE = HIGH, MR = HIGH. (n + 1) indicates state after next clock)

J	$\overline{K}$	Q_0 at t_{n+1}
L	L	L
L	H	Q_0 at t_n (no change)
H	L	$\overline{Q}_0$ at t_n (toggles)
H	H	H

TABLE II — LOADING RULES (1 U.L. = 1 TTμL Gate Input Load)

INPUTS	LOADING
J, $\overline{K}$, $\overline{MR}$, P_0, P_1, P_2 & P_3	1 U.L.
$\overline{PE}$	2.3 U.L.
C_P	4 U.L.
OUTPUTS	FANOUT
Q_0, Q_1, Q_2, Q_3 & $\overline{Q}_3$	6 U.L.

TABLE III ELECTRICAL CHARACTERISTICS ($T_A = -55°C$ to $+125°C$, $V_{CC} = 5.0$ V $\pm 10\%$) (Part #U6B930051X)

SYMBOL	CHARACTERISTICS	LIMITS							UNITS	CONDITIONS & COMMENTS
		$-55°C$		$+25°C$			$+125°C$			
		MIN.	MAX.	MIN.	TYP.	MAX.	MIN.	MAX.		
V_{OH}	Output High Voltage	2.2		2.4	2.7		2.4		Volts	$V_{CC} = 4.5$ V, $I_{OH} = -0.36$ mA
V_{OL}	Output Low Voltage		0.4		0.2	0.4		0.4	Volts	$V_{CC} = 5.5$ V, $I_{OL} = 9.6$ mA $V_{CC} = 4.5$ V, $I_{OL} = 7.44$ mA
V_{IH}	Input High Voltage	2.0		1.7			1.4		Volts	Guaranteed input high threshold for all inputs
V_{IL}	Input Low Voltage		0.8			0.9		0.8	Volts	Guaranteed input low threshold for all inputs
I_F	Input Load Current* $J, K, MR, P_0, P_1, P_2 \& P_3$		-1.6 -1.24		-1.10 -0.97	-1.6 -1.24		-1.6 -1.24	mA mA	$V_{CC} = 5.5$ V $V_{CC} = 4.5$ V $\quad V_F = 0.4$ V
I_R	Input Leakage Current* $J, K, MR, P_0, P_1, P_2 \& P_3$				15			60	μA	$V_{CC} = 5.5$ V, $V_R = 4.5$ V

TABLE IV ELECTRICAL CHARACTERISTICS ($T_A = 0°C$ to $+75°C$, $V_{CC} = 5.0$ V $\pm 5\%$) (Part #U6B930059X)

SYMBOL	CHARACTERISTICS	LIMITS							UNITS	CONDITIONS
		$0°C$		$+25°C$			$+75°C$			
		MIN.	MAX.	MIN.	TYP.	MAX.	MIN.	MAX.		
V_{OH}	Output High Voltage	2.4		2.4	3.0		2.4		Volts	$V_{CC} = 4.75$ V, $I_{OH} = -0.36$ mA
V_{OL}	Output Low Voltage		0.45		0.2	0.45		0.45	Volts	$V_{CC} = 5.25$ V, $I_{OL} = 9.6$ mA $V_{CC} = 4.75$ V, $I_{OL} = 8.5$ mA
V_{IH}	Input High Voltage	1.9		1.8			1.6		Volts	Guaranteed input high threshold for all inputs
V_{IL}	Input Low Voltage		0.85			0.85		0.85	Volts	Guaranteed input low threshold for all inputs
I_F	Input Load Current* $J, \overline{K}, MR, P_0, P_1, P_2 \& P_3$		-1.6 -1.41		-1.0 -0.9	-1.6 -1.41		-1.6 -1.41	mA mA	$V_{CC} = 5.25$ V $V_{CC} = 4.75$ V $\quad V_F = 0.45$ V
I_R	Input Leakage Current* $J, \overline{K}, MR, P_0, P_1, P_2 \& P_3$				15			60	μA	$V_{CC} = 5.25$ V, $V_R = 4.5$ V

*For CP and PE input currents, use load factors in Table II

Fig. 7-42 Fairchild medium-scale integration. (*Fairchild Semiconductor.*)

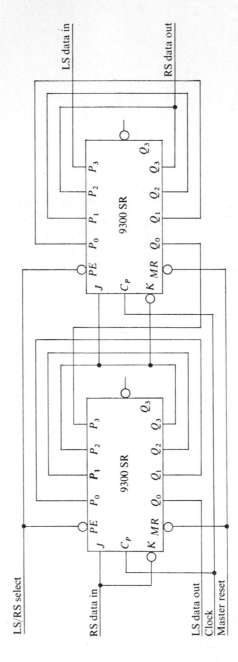

Fig. 7-43 Eight-bit left right shift register. This register shifts left or right on each shift clock, depending upon the condition of the shift-left shift-right select input. If this input is high, right shift occurs; and if low, left shift occurs.

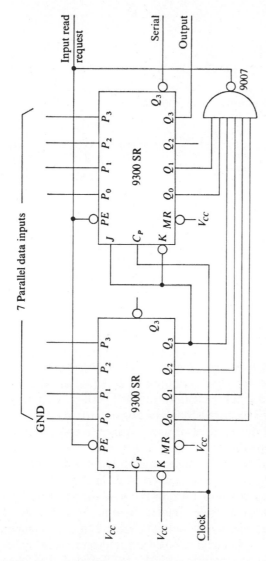

Fig. 7.44 Seven-bit parallel-to-serial converter. This parallel to serial converter uses a marker bit to count the data bits shifted out so that a parallel-load enable is generated to load the next parallel word for conversion at the correct time.

189

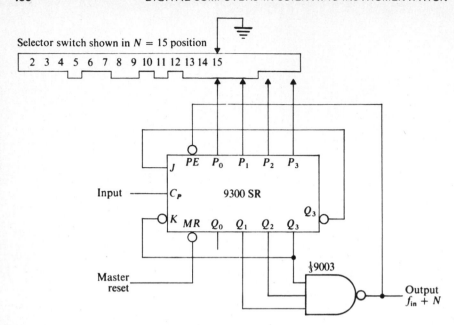

Fig. 7-45   Divide-by-n counter for n = 2 to 15. This counter produces an output pulse for every n input pulses, where the number n is determined by the setting of the slide selector switch as shown or by logic inputs to the parallel data lines from an external source.

example, several applications from the Fairchild Semiconductor data sheet presented in Figs. 7-41 and 7-42 are given in Figs. 7-43 to 7-46. In Fig. 7-43, two Fairchild Semiconductor type 9300s are shown as an 8-bit left- or right-shift register. The register shifts data entered in serial either left or right on each shift clock pulse, depending upon the condition of an input line called *shift-left* or *shift-right select.* If this input is high, a right shift occurs; and if low, a left shift occurs. In Fig. 7-44, a 7-bit parallel-to-serial converter is shown. It accepts data in parallel and shifts it out in serial. It uses a marker bit to count the data bits shifted out so that a parallel-load enable is generated to load the next parallel word for conversion at the correct time. In addition to the two type-9300 shift registers, a six-input NAND gate is required. A divide-by-n counter, where n is from 2 to 15, is presented in Fig. 7-45. In this application, the type-9300 shift register is used as a programmable counter. In other words, a switch-selected input can be used to provide any division ratio from 2 to 15. A computer output or any digital number output could also be used to determine the divide ratio. In Fig. 7-46, a two-decade programmable divider is illustrated. This circuit divides by any number n from 1 to 100. The selected number n is 1 greater than shown on the slide switches. As an example, the switches are showing 56. Therefore, the circuit will divide by 57 with this setting. These four examples serve to

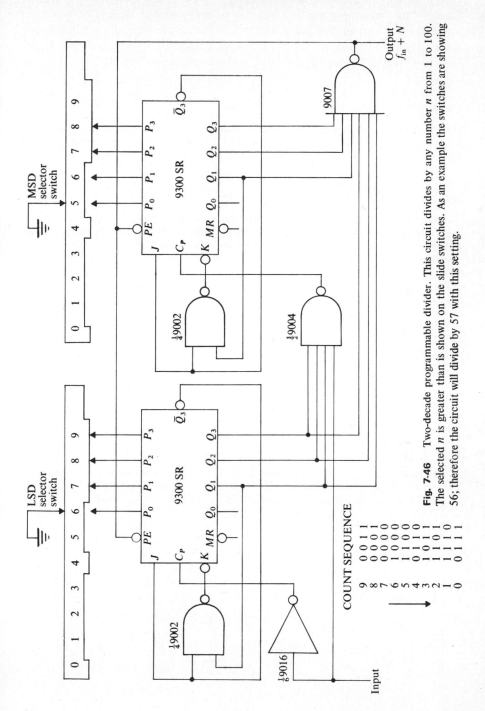

COUNT SEQUENCE

9	0	0	1	1	1
8	0	0	0	0	0
7	0	0	0	0	0
6	1	1	0	0	0
5	0	1	0	1	0
4	0	1	0	1	0
3	1	0	1	1	1
2	1	1	1	0	1
1	1	1	1	1	0
0	0	1	1	1	1

Fig. 7-46 Two-decade programmable divider. This circuit divides by any number n from 1 to 100. The selected n is greater than is shown on the slide switches. As an example the switches are showing 56; therefore the circuit will divide by 57 with this setting.

191

illustrate the versatility and ease of use of many MSI devices. The types of applications shown in the manufacturer's literature are generally applicable to many situations. They are often extensive enough that very complex logic operations can be designed and utilized with a minimum of experience.

EXERCISES

Various exercises were suggested in the text of this chapter, and the reader is urged to work these out. In addition, several exercises related to Boolean expressions are provided at the end of Appendix C.

BIBLIOGRAPHY

The following list of references is provided for the reader who desires more information on the topics presented in Chap. 7. Basic logic design, including Boolean algebra and mapping techniques, is covered well in Refs. 1 to 5. Latches, counters, and other more complex logic circuits are presented in Ref. 6; while Refs. 7 and 8 provide a good electronics background. Integrated-circuit devices, their operating characteristics, uses, etc., are given a superb overview in Ref. 9. In addition, one of the most useful sources of information on integrated-circuit logic devices is the application note sets published by the various manufacturers. Local electronics vendors can put you in touch with the manufacturers to obtain their sets.

1. Nashelsky, Louis: "Digital Computer Theory," Wiley, New York, 1966.
2. Chu, Yaohan: "Digital Computer Design Fundamentals," McGraw-Hill, New York, 1962.
3. Braun, Edward L.: "Digital Computer Design," Academic, New York, 1963.
4. Hoernes, G. E., and M. F. Heilweil: "Introduction to Boolean Algebra and Logic Design," McGraw-Hill, New York, 1964.
5. Maley, G. A., and John Earle: "The Logic Design of Transistor Digital Computers," Prentice-Hall, Englewood Cliffs, N.J., 1963.
6. Malmstadt, H. V., and C. G. Enke: "Digital Electronics for Scientists," Benjamin, New York, 1969.
7. Malmstadt, H. V., C. G. Enke, and E. C. Toren, Jr.: "Electronics for Scientists," Benjamin, New York, 1963.
8. Hunten, Donald M.: "Introduction to Electronics," Holt, New York, 1964.
9. Morris, R. L., and John R. Miller: "Designing with TTL Integrated Circuits," McGraw-Hill, New York, 1971.

8
Introduction to Interfacing

In the previous chapters, we have been concerned with the fundamentals of digital-computer operation and digital logic. We have considered machine- and assembly-language programming, simple chemical data processing algorithms and programming, the essential features of input/output (I/O) programming with standard peripheral devices, and basic digital logic design. In this chapter, we will consider the hardware fundamentals necessary to utilize the digital computer for on-line laboratory applications. In subsequent chapters, we will consider the software for on-line laboratory applications.

First of all, let us restate what is meant by on-line computer operation. Figures 8-1 to 8-3 provide a graphic description of the requirements of an on-line computer system. The computer is considered on-line to an experimental system when a direct electronic communication link exists between the experiment and the computer and, perhaps, between the computer and the experiment. In Chap. 6, we considered in some detail the hardware characteristics of the communication link—or interface—between the computer and standard peripheral devices. In this chapter, we will consider the characteristics of general hardware for interfaces between experimental systems and the digital computer. We will

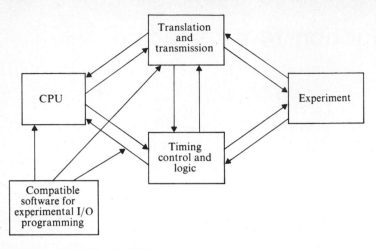

Fig. 8-1   Functional interfacing elements.

consider software approaches for communicating with experimental systems in the following chapter.

The various functions required for communication between an experiment and the digital computer are illustrated in Fig. 8-1. These functions are carried out by:

Translation and transmission elements.   These are illustrated in Fig. 8-2 and include analog and digital hardware which allow the appropriate conversion or handling of electronic information for communication between experiment and computer. Typical elements include analog-to-digital converters, digital-to-analog converters, voltage amplifiers, current-to-voltage converters, signal conditioners, sample-and-hold amplifiers, multiplexers, etc.

Timing, control, and logic elements.   These are illustrated in Fig. 8-3 and include such hardware as a digital clock, logic gates, flip-flops, counters, one-shots, Schmitt triggers, analog switches, level converters, etc.

Appropriate software to drive the interface hardware.   The I/O programs are necessary and important parts of the interface. Software often provides the necessary controls needed for the operation of the electronic elements in the interface. This will be discussed in Chap. 9.

8-1   TRANSLATION ELEMENTS

Digital-to-analog conversion (DAC)   A *digital-to-analog converter* (DAC) is used to change digital numerical information into a continuously variable analog

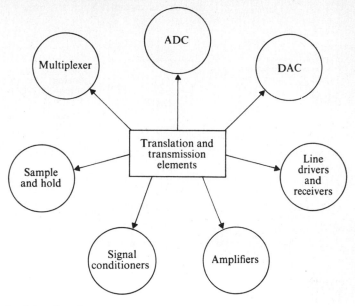

Fig. 8-2   Translation and transmission elements.

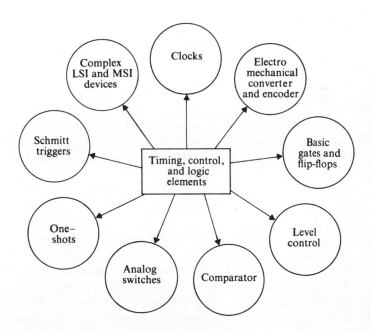

Fig. 8-3   Timing control and logic elements.

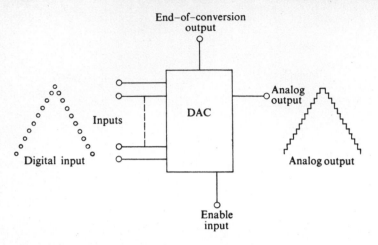

Fig. 8-4   Basic digital-to-analog converter (DAC).

output. DACs are often used as control devices in chemical experiments. For example, in a typical fast-sweep polarographic experiment, a DAC can be used to provide the electrochemical-cell control voltages and ramp functions. Because the computer can generate numbers in any sequence, nonlinear ramps can be generated which provide a wide variety of versatile control signals.

A basic DAC application is illustrated in Fig. 8-4. The DAC takes a digital input and converts it to an analog output which consists of a series of voltage steps. The minimum step magnitude is a function of the dynamic range and resolution of the converter. For example, if the converter has a 10-V output maximum and has a resolution of 1 part in 1,024, or 10 bits, the minimum voltage step on the output will be about 10 mV. Notice in Fig. 8-4 that there are digital inputs and an analog output, an *enable* input at which a conversion can be started, and an *end-of-conversion* output which indicates when a conversion is complete. The actual conversion from digital numbers to an analog output is accomplished within the converter by a series of resistors and switches called a *ladder network*, as is presented in Fig. 8-5.

The DAC ladder network presented in Fig. 8-5 consists of parallel resistors in binary orders of magnitude, a voltage source, and a current meter. The output current through the meter is a function of which of the switches S_1 to S_4 are closed. For example, if only switch S_4 is closed, the output current is equal to V/R, where V is the magnitude of the reference voltage. If, on the other hand, switch S_1 is closed and all other switches are open, the current is equal to $V/8R$. Thus, with S_4 closed the current is inversely proportional to $2^3 = 8$; while with only S_1 closed the current is inversely proportional to $2^0 = 1$. By closing each switch, the current can be varied in binary orders of magnitude. It would be worthwhile, at this point, for the reader to work through all the current values

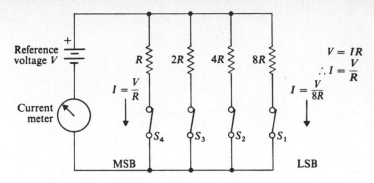

Fig. 8-5   Four-bit basic DAC ladder network.

for combinations of closed switches to satisfy himself that the device will produce currents proportional to the binary pattern represented by the open and closed switches. Notice that if the magnitudes of the resistors are some function other than R, $2R$, $4R$, and $8R$ or, in other words, correspond to some other number system, a different current-to-number system relationship will exist. In fact, ladder networks are manufactured to give currents corresponding to various number systems.

In an actual DAC digital control signals operate the switches and provide the digital input. The switches are electronic in nature and are closed by the digital electronic signals. An electronic digital ladder network is illustrated in Fig. 8-6. Notice that the only difference between the circuit presented in Fig.

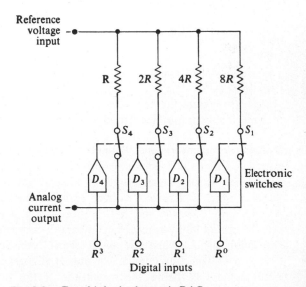

Fig. 8-6   Four-bit basic electronic DAC.

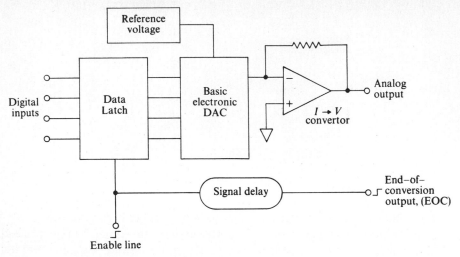

Fig. 8-7   Four-bit DAC with control elements.

8-6 and the one in Fig. 8-5 is the inclusion of digital input drivers D_1 to D_4 to control the electronic switches. We will refer to the ladder network presented in Fig. 8-6, including the electronic switches, as a basic electronic DAC in the remainder of this chapter.

Also required to complete the operation of converting a digital signal to an appropriate output voltage are a reference voltage and a current-to-voltage converter at the output. A practical 4-bit electronic DAC is presented in Fig. 8-7. Notice that the basic elements are all present. In addition, an enable input and an end-of-conversion output are included. In Fig. 8-7, a data latch precedes the basic electronic DAC ladder network. The digital information presented to the ladder switches can be changed with the enable line controlling the data latch. The end-of-conversion output consists of the enable-line input after it passes through a signal delay. The length of delay is long enough to allow the entire system to perform the necessary operations; it provides a status signal indicating that the analog output has settled.

Voltage comparators and Schmitt triggers   The next translation element considered is the voltage *comparator*. It is an analog device with two inputs and an output. It is generally used to compare one voltage with another and to indicate which is larger. A voltage comparator and its resulting waveforms are presented in Figs. 8-8 and 8-9. Notice in Fig. 8-9 that there are a reference voltage and an input signal. The comparator compares the reference voltage with the input signal. When the input signal rises above the magnitude of the reference voltage, the comparator output changes state from 1 to 0. When the input signal falls below the magnitude of the reference voltage, the output changes back from 0 to 1.

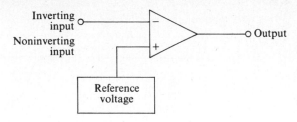

Fig. 8-8   Typical voltage comparator.

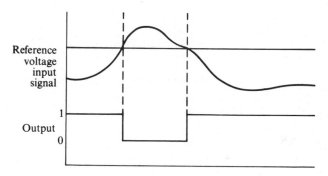

Fig. 8-9   Timing chart for comparator.

A device somewhat similar to a comparator is the *Schmitt trigger* presented in Fig. 8-10. In it, there are input, output, and upper and lower threshold terminals. Its typical waveform is presented in Fig. 8-11. The two reference levels V_{t+} and V_{t-} are the upper and lower thresholds. When the input signal goes above the upper threshold, the output changes from 1 to 0. However, when it goes below the upper threshold, the output signal does not change as long as it is above the lower threshold. When it goes below the lower threshold, the output changes. In a like manner, the increasing signal will not change the output when it only goes above the lower threshold. The difference between the upper threshold and lower threshold is commonly called the *hysteresis* or *backlash level* of the Schmitt trigger. It is used to provide noise immunity for the device.

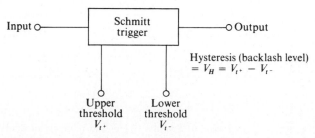

Fig. 8-10   Typical Schmitt trigger.

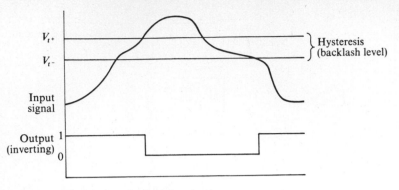

Fig. 8-11　Timing chart for Schmitt trigger.

In Fig. 8-12, both Schmitt trigger and comparator outputs are given where a noisy input signal exists. Notice in the case of the comparator that as the signal oscillates above and below the comparator reference voltage the output also oscillates. Notice in the case of the Schmitt trigger that the magnitude of the noise is less than the magnitude of the hysteresis; as a result, the Schmitt trigger output does not have noise spikes like the comparator. It is a device which can be used to extract control signals from noisy input signals and eliminate the noise by proper selection of the hysteresis level. It is often used as a level conversion device to take digital signals of one level and convert them to that of another level.

The lack of hysteresis in the comparator makes it more susceptible to noisy environments. It also makes it a more accurate switch.

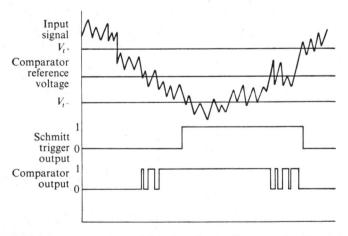

Fig. 8-12　Comparison of Schmitt trigger and comparator operating characteristics.

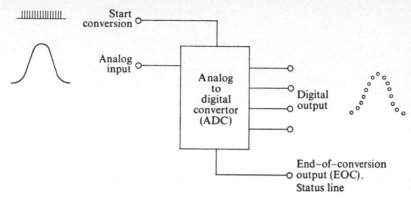

Fig. 8-13   Four-bit analog-to-digital converter (ADC).

Analog-to-digital conversion (ADC) A typical *analog-to-digital converter* (ADC) is presented in Fig. 8-13. It consists of an analog input, digital outputs, a start conversion input, and an end-of-conversion output. The ADC changes an analog or continuous voltage from an experimental system into a series of discrete digital values so that a computer can be presented with digital data in a format which it can handle. The most common output format is a binary digital representation of the analog input. There are many types of ADCs, some very fast and some very slow, some which require high-level voltage inputs in the range of from one to several volts, and some which require low-level voltage, current, or resistance inputs. We will, in this section, discuss several types of fast converters.

Counter converters   Probably the simplest ADC is the *counter converter* presented in Fig. 8-14. It is usually a fast high-level converter having conversion times of less than 1 msec. This makes it capable of providing greater than 1,000 data points per second. Basically, it consists of a comparator, a clock or pulse generator, a counter such as discussed in Chap. 7, and the DAC presented in Fig. 8-7. When an analog input signal is presented to one input of the comparator and a start-of-conversion signal is presented to the counter, the counter resets to 0 and starts counting up, it presents a digital input to the DAC. The DAC in turn provides a corresponding analog output voltage to the other input of the comparator. When the counter output number reaches a magnitude that provides a voltage to the comparator (through the DAC) equal in magnitude to the analog input signal, the comparator changes state, turning off the clock and stopping the counter. The digital output representation of the analog input is read in parallel from the counter outputs. A status or end-of-conversion signal can be obtained from the output of the comparator as it changes state upon completion of a conversion. The counter

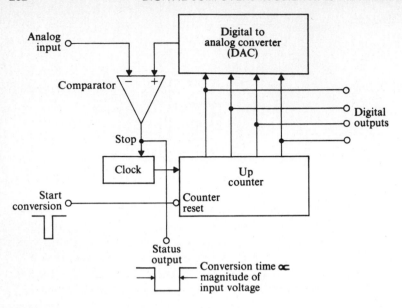

Fig. 8-14   Four-bit counter ADC.

converter is simple and inexpensive. The conversion time is proportional to the magnitude of the input voltage. That is, since the counter always starts counting from 0, the larger the analog input voltage, the longer it will take for the counter to count up to the value where the DAC output applied to the comparator is equal to the analog input voltage.

Continuous converters   The *continuous converter* presented in Fig. 8-15 is very similar to the counter converter presented in Fig. 8-14 with the exception

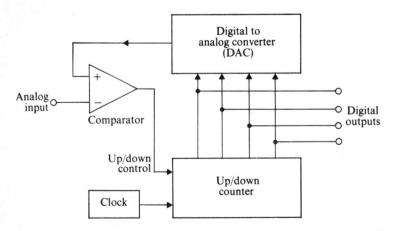

Fig. 8-15   Four-bit continuous ADC.

that the counter used can count both up and down; the comparator, instead of turning off the clock, controls the counting direction of the counter; and the counter counts all the time. When the analog input signal voltage has been exceeded by the output of the DAC, the comparator changes the direction of the counter from up to down. When the DAC output falls below the analog input, the comparator changes the direction of the counter to count up. The up-down counter tracks the analog input voltage if it is not changing faster than the counter can follow. When a new signal is applied to the analog input, the continuous converter locks onto it and follows it. Digital outputs can be read at intervals from the output of the counter. After it locks on a signal, the continuous converter is extremely fast.

Successive-approximation converters The 4-bit *successive-approximation* ADC presented in Fig. 8-16 differs from both the counter and the continuous converters in that in place of a counter it has a pattern generator. It consists of a comparator, a DAC, a buffer-register data latch in which the digital output is stored, the pattern generator, and some control logic. Its operation is most easily understood with reference to Fig. 8-17. All the possible number combinations generated by the pattern generator are presented for a 4-bit successive-approximation ADC. Notice that for the 4-bit conversion, it always takes four

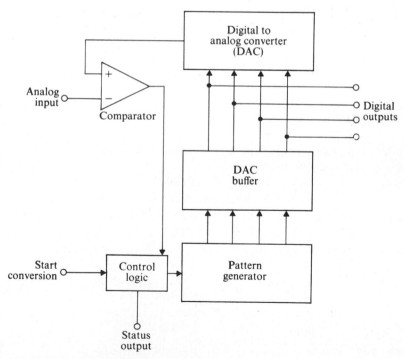

Fig. 8-16   Four-bit successive-approximation ADC.

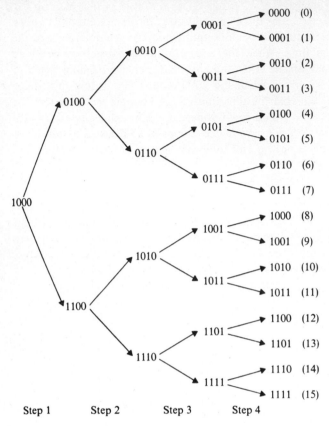

Step 1 Step 2 Step 3 Step 4

Fig. 8-17 Possible pattern generation for 4-bit successive-approximation ADC.

steps to completion. Unlike the counter converter, where the conversion time is a function of the magnitude of the number, a successive-approximation converter always has a fixed conversion time.

A typical conversion is illustrated in Fig. 8-18 for an input voltage that is equal to a binary output value between 10_{10} and 11_{10}. At the beginning of the conversion, the converter starts with 1000_2 or 8_{10}. Since the input voltage is greater than that, it divides the range from 1000_2 to 1111_2 in half, resulting at

Step 1 Step 2 Step 3 Step 4

Fig. 8-18 Four-bit successive-approximation ADC pattern for an analog input between 10_{10} and 11_{10} at the output.

the end of step 1 with the number 1100_2 or 12_{10}. In step 2, it divides the range between 1000_2 and 1100_2 in half since 12_{10} is greater than the magnitude of the input signal. This results at the end of step 2 with 1010_2 or 10_{10} being fed from the pattern generator through the buffer into the DAC and to the comparator. Since this number is smaller in magnitude than the input signal, the converter then divides the range between 10_{10} and 12_{10} in half, resulting at the end of step 3 with a digital output of 1011_2 or 11_{10}. Since this is now greater in magnitude than the input signal, the converter divides back to 1010_2 or 10_{10}, which is closest to the magnitude of the input signal.

Fast converters compared In comparing fast high-level converters, the counter converter is slower and much less expensive than the others. The conversion time, however, depends upon the input voltage magnitude. The continuous converter is slow for step signals. It is fast, though, for signals that change with a speed that the converter can lock on and track. The successive-approximation converter is the most versatile and the most expensive. Because the price of electronics has become low, it is still relatively inexpensive and is the most widely used fast converter.

In selecting high-level ADCs, several criteria are important. They include the input voltage range, format of the output, resolution of the output, logic voltage levels of the output, conversion speed, control signals, and power-supply requirements. Input voltages generally range from 1 to 10 V full scale. They may be positive, negative, or bipolar. Common ADC input voltage ranges are 0 to 1 V, 0 to -1 V, 0 to 5 V, 0 to -5 V, 0 to 10 V, 0 to -10 V, -1 to $+1$ V, -5 to $+5$ V, and -10 to $+10$ V.

The most commonly used ADC output format is the binary number system because of its direct compatibility with digital computers. Other codes such as BCD, Gray code, etc., are sometimes used. In addition, bipolar input converters such as the ± 1 V or ± 10 V units can have different forms of binary coding to account for the dual polarity. Some have a straight binary code setting 0 V at the input equal to half scale on the binary output. Others use different ways to distinguish the positive and negative outputs. Usually, they correspond to common binary arithmetic methods for handling negative numbers such as 2's-complement, 1's-complement, or sign-plus-magnitude notations.

The converter output resolution and dynamic range are determined by the number of data bits available. Common binary converters have 8-, 10-, or 12-bit output configurations, giving resolutions of 1 part in 256, 1,024, or 4,096, respectively. Converters using BCD output configurations often have three or four decimal digits represented, giving resolutions in the range of 10^3 to 10^4. Higher-resolution converters are available for applications requiring more resolution.

By far the most common output logic voltage levels are 0 and $+5$ V. They are the common integrated-circuit levels for DTL and TTL logic. There are, however, different logic voltage levels used including 0 and +12 V, 0 and −3 V, and others. The logic voltage levels should, of course, be compatible with those used on the computer or data system. If they are not, some sort of voltage level conversion will be necessary.

Conversion time is the time required after a start signal for a conversion to be completed. It determines the maximum data rate of a given converter. Fast converter conversion times range from 100 μsec to about 0.1 μsec, yielding conversion rates from 10 kHz to about 10 MHz.

Control signals for fast ADC units usually consist of a start command and an end-of-conversion or status signal. Other control signals are possible. In units with output storage buffers, controls may be available to load or store in the buffer.

Power-supply requirements, while they may seem trivial, are in fact very important. The presence or absence of an internal reference supply is a good example of this. The magnitude and stability of the general power supply are also important. ADC units with built-in voltage regulators have much less severe power-supply requirements than those with no internal regulation.

Low-level analog-to-digital conversion   For low-level converters, the input signal can be millivolts rather than volts in magnitude. Low-level converters are characterized by much slower operating speeds than those previously discussed. While fast converters can have conversion times of less than 1 μsec and conversion rates of better than 1 MHz, low-level converters very often have

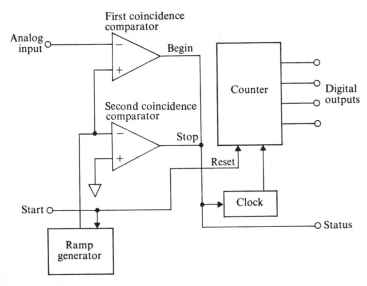

Fig. 8-19   Four-bit voltage-to-time converter.

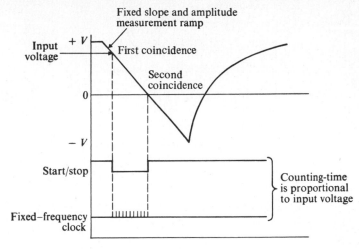

Fig. 8-20   Voltage-to-time converter timing chart.

conversion times on the order of milliseconds to seconds, with data rates far less than 1,000 conversions per second.

Voltage-to-time converter   Probably the simplest low-level converter is the *voltage-to-time converter*. It is illustrated in Fig. 8-19, and its associated waveforms are illustrated in Fig. 8-20. It consists of two comparators, a ramp generator, a counter, and a clock. When an input voltage is placed at the analog input and a conversion is initiated, a fixed-slope and fixed-amplitude ramp is generated. When the magnitude of this ramp falls below that of the input voltage, it starts a fixed-rate clock, which causes a counter to begin counting up. The counter counts up until the fixed-slope ramp crosses a threshold equal to the reference input of the second coincidence comparator. This is very often 0V. When this occurs, the clock is turned off, the counter stopped, and the digital output is read from the counter. This is called a voltage-to-time converter because the time over which the counter counts is a function of the magnitude of the input voltage.

Voltage-to-frequency converter   An example of a second type of low-level ADC is the 4-bit *voltage-to-frequency converter* presented in Fig. 8-21. In it, there are an integrator, a comparator with a reference voltage, a counter, a fixed-time control logic section, and a charge pump. When an analog input voltage is presented at the input, the integrator begins to integrate that voltage, forming a ramp at its output. When the output ramp exceeds the value of the reference voltage on the comparator, the comparator changes state. This causes the charge pump to discharge the integrator capacitor at the input. The capacitor is recharged and the cycle repeated. Thus, a pulse-train output from the comparator is generated. The rate at which the integrator capacitor charges up is

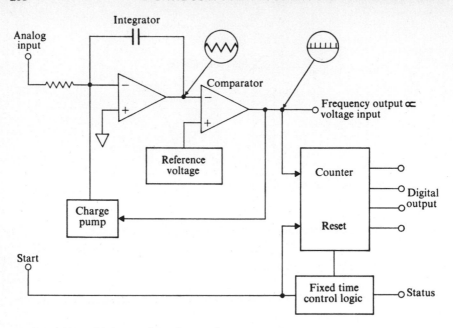

Fig. 8-21 Four-bit integrating voltage-to-frequency converter.

a function of the input voltage level. The larger the input voltage level, the faster the integrator will charge up and discharge. As a result, greater input signals cause a higher-frequency output than do smaller input voltage signals.

The output of the comparator can be used to drive a counter directly. A control-logic section provides a fixed time interval over which the variable rate counting occurs. The start signal resets the counter and initiates the counting time interval. At the end of that interval, the control logic puts out an end-of-conversion pulse on the status output.

Dual-slope integrating converter A third type of low-level converter is the *dual-slope integrating ADC* presented in Fig. 8-22. Its waveforms are presented in Fig. 8-23. In it, as in a voltage-to-frequency converter, there is an integrator. There are also a zero-crossing detector, a fixed-frequency clock, a counter, a constant current source, and some control logic. When an analog signal is presented at the input and the start signal initiates a conversion, the electronic switch at the input closes, and the integrator charges up for a fixed period of time. Since the charge rate of the integrator is a function of the magnitude of the input voltage, the level to which it charges will be in direct proportion to the magnitude of the input voltage. At the end of this time interval, which starts at t_0 and ends at t_1 in Fig. 8-23, the electronic switch at the input opens and the charge is held by the integrator. The integrator is then discharged at a fixed rate through a constant current source. During the

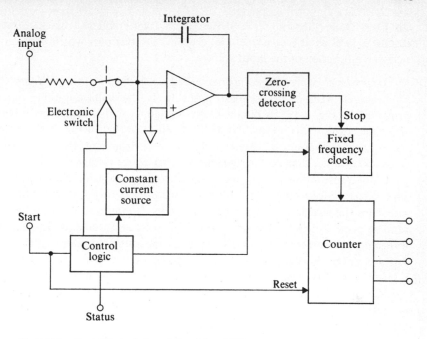

Fig. 8-22   Four-bit dual-slope integrating ADC.

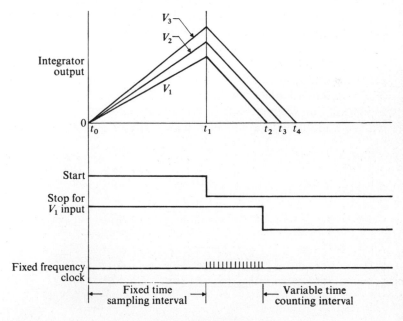

Fig. 8-23   Timing chart for dual-slope integrating ADC.

discharge time, the counter is driven by a fixed-frequency clock. Since the discharge is done at a constant rate, the time interval over which this counting occurs is a function of the input voltage.

Low-level converters compared   We have looked at three different types of low-level ADCs. The simplest probably is the voltage-to-time converter presented in Fig. 8-19. Its disadvantage, though, is that it in no way integrates the input signal. This means that as the ramp presented in Fig. 8-20 passes the first coincidence point, if there is a high-noise environment, it may trigger on a noise spike rather than on the signal itself. This causes voltage-to-time converters to be noise-susceptible and to give incorrect readings in high-noise environments. The voltage-to-frequency converter is less susceptible to this problem because it integrates the signals over short periods of time. It integrates out high-frequency random noise and gives truer readings in high-noise environments. The dual-slope integrating converter integrates over an entire measurement interval rather than a series of intervals. While the integrating types of low-level ADCs give data in high-noise environments, they are usually slower than nonintegrating types. Notice also that fast high-level converters, because of their very nature, will convert on noise spikes and give erroneous readings that do not occur in the integrating low-level converters.

Electrical-to-mechanical conversion   Having considered both ADCs and DACs, we can discuss other types of converters which translate either electrical signals into mechanical motion or mechanical motion into electrical signals. They are digital devices in that they are controlled by or yield digital signals. The first of these is the *digital stepping motor* presented in Fig. 8-24. It is essentially an incremental digital electric motor. If it is provided with a proper pulse train or pattern, it will move in either direction as a function of that input signal. It has a minimum resolution angle ϕ. In other words, there is a minimum angle that the motor will turn when presented with one pulse. For example, let us suppose that we had a motor with a minimum pulse angle of $60°$. It would then take six

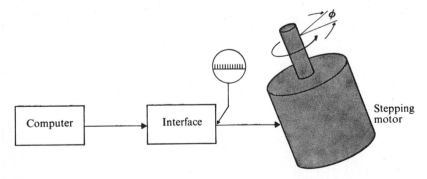

Fig. 8-24   Digital stepping motor.

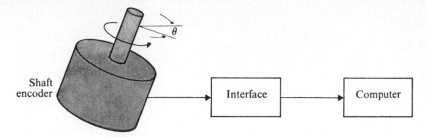

Fig. 8-25   Digital shaft position encoder.

pulses for the motor to turn one revolution. Stepping motors commonly have pulse angles less than 1° They are useful for driving mechanical components through a computer interface. Examples of this might be the movement of a grating in a spectrophotometer, the actuation of needle valves in gas chromatography, or the movement of the slits in a mass spectrometer.

Another useful device, which is the counterpart of the stepping motor, is the *shaft encoder* presented in Fig. 8-25. A shaft encoder yields a digital signal as a function of its shaft position. It, like a stepping motor, has a minimum resolution angle θ. Shaft encoders can divide a complete revolution into increments from a few up to several thousand. They have output bit configurations in binary and BCD as well as other number codes. Shaft encoders are used primarily to show the position of components in digital form. For example, the grating of a spectrophotometer might be connected to a shaft encoder to provide a digital representation of the wavelength scale.

8-2   ANALOG SIGNAL TRANSLATIONS

A common element used in most analog signal-handling situations is the *operational amplifier.* It is generally used with computer interface systems for analog signal handling and conditioning. The general characteristics of an operational amplifier are summarized in Fig. 8-26. Referring to this figure, one sees that it is a device with two input terminals and one output terminal. It has gain in that a small input signal will produce a large output signal. This is

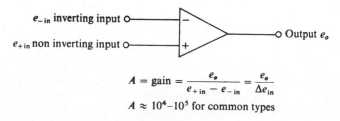

$$A = \text{gain} = \frac{e_o}{e_{+\text{in}} - e_{-\text{in}}} = \frac{e_o}{\Delta e_{\text{in}}}$$

$$A \approx 10^4 - 10^5 \text{ for common types}$$

Fig. 8-26   Basic operational amplifier.

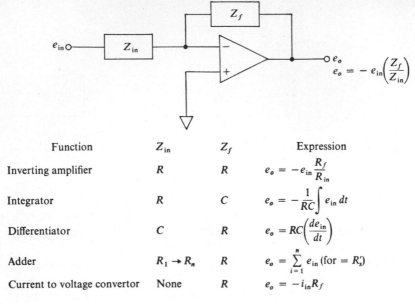

Function	Z_{in}	Z_f	Expression
Inverting amplifier	R	R	$e_o = -e_{in}\dfrac{R_f}{R_{in}}$
Integrator	R	C	$e_o = -\dfrac{1}{RC}\displaystyle\int e_{in}\,dt$
Differentiator	C	R	$e_o = RC\left(\dfrac{de_{in}}{dt}\right)$
Adder	$R_1 \rightarrow R_n$	R	$e_o = \displaystyle\sum_{i=1}^{n} e_{in}\ (\text{for} = R_s')$
Current to voltage convertor	None	R	$e_o = -i_{in}R_f$

Fig. 8-27 Basic inverting operational-amplifier configuration and common functions.

illustrated in equation form in Fig. 8-26. Common amplifier gain magnitudes range from around 10,000 to 100,000. Typical uses of the operational amplifier with appropriate feedback and input circuitry are summarized in Fig. 8-27. The configuration presented is called an *inverting amplifier* because the input signal polarity is reversed at the output. The output of the amplifier is a function of three variables: the input signal, the feedback element Z_f, and the input element Z_{in}. If the input and feedback elements are resistors, the amplifier provides inverting voltage amplification. If, on the other hand, the input and feedback elements are something other than resistors—for example, combinations of resistors and capacitors—the amplifier provides other functions.

Operational amplifiers can also be used in noninverting configurations, in which the output polarity is not inverted with respect to the input. Common noninverting configurations are presented in Fig. 8-28.

A useful application of the operational amplifier in digital interfacing is presented in Fig. 8-29. This is an *inverting ranging amplifier*. In it, a series of feedback resistors R_1 to R_4 can be placed in or out of the feedback loop by a series of electronic switches. If digital control signals are used to open and close the switches, the computer can control the gain of the amplifier. This is very useful when measuring signals which exceed the useful dynamic range of an ADC. A simple program can be written that changes the gain of the amplifier

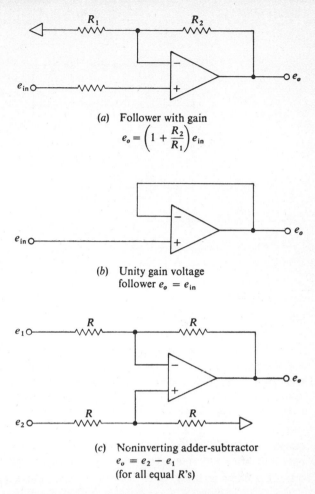

(a) Follower with gain

$$e_o = \left(1 + \frac{R_2}{R_1}\right) e_{in}$$

(b) Unity gain voltage
follower $e_o = e_{in}$

(c) Noninverting adder-subtractor
$e_o = e_2 - e_1$
(for all equal R's)

Fig. 8-28 Common noninverting operational-amplifier configurations.

when the signal reaching the computer is either too great or too small, thereby providing automatic amplifier ranging.

Another useful interfacing configuration is the *peak follower*, used to measure the peak magnitude of an incoming signal and to hold that peak magnitude until the data-handling system can read it. It consists of a voltage-follower operational amplifier, a capacitor, a reset switch, and an input diode. With the diode polarity as shown in Fig. 8-30, the response curves presented in Fig. 8-31 are obtained. Notice that the magnitude of the peak is stored on the capacitor until the reset switch is closed.

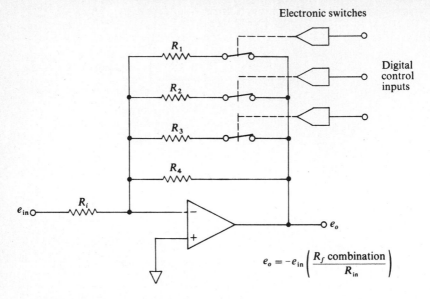

$$e_o = -e_{in}\left(\frac{R_f \text{ combination}}{R_{in}}\right)$$

Fig. 8-29 Inverting ranging amplifier.

Two other useful configurations consist of *track-and-hold* and *sample-and-hold* amplifier systems. While in practice, one would generally buy a ready-built track-and-hold or sample-and-hold system, it is useful here to review their operation. A track-and-hold amplifier is presented in Fig. 8-32. Its response curves and timing chart are presented in Fig. 8-33. An incoming signal is fed to the capacitor in Fig. 8-32 when the electronic switch is closed. In this mode, the output of the amplifier will continually follow the input signal. When the electronic switch is opened and isolates the input signal from the capacitor, the output remains at the voltage last seen by the input capacitor. This is illustrated

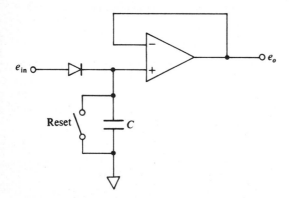

Fig. 8-30 Peak follower.

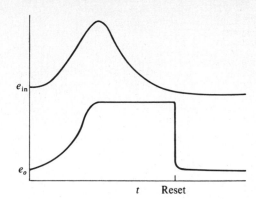

Fig. 8-31   Response curves for peak
follower.

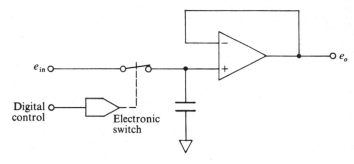

Fig. 8-32   Track-and-hold amplifier.

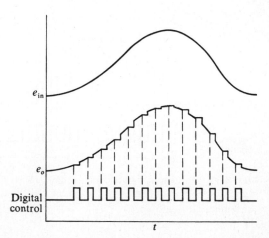

Fig. 8-33   Response curves and
timing chart for track-and-hold
amplifier.

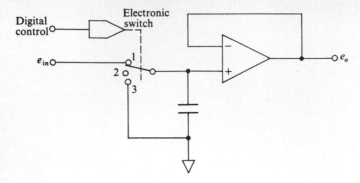

Fig. 8-34   Sample-and-hold amplifier.

in Fig. 8-33, where each time the digital control signal goes to a logical 1, the electronic switch opens, and the magnitude of the voltage at the time it opened is stored on the capacitor. Notice that the output waveform results in a series of levels stored on the capacitor and read from the amplifier output.

A sample-and-hold amplifier is presented in Fig. 8-34. It is much like a track-and-hold amplifier except that the electronic switch shorts out the capacitor in a reset mode. This switch actually has three positions. In one position, the input signal is connected to the capacitor. In the second position, the input signal is isolated from the capacitor. In the third position, the electronic switch shorts out the capacitor, resetting the output of the amplifier to 0. (The operation is illustrated in Fig. 8-35.) This device can be used not only

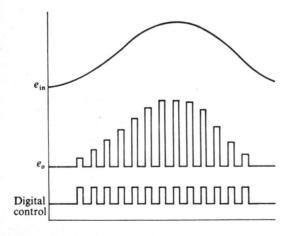

Fig. 8-35   Response curves and timing chart for sample-and-hold amplifier.

to track a signal and periodically store its magnitude for brief periods of time, but also when it is desired to reset to 0 the output of the sample-and-hold amplifier after each data point. This is useful when more than one input signal is fed to a single ADC. If after each data point we set the output of the sample-and-hold amplifier to 0, we can connect the output of all sample-and-hold amplifiers to the input of the inverting adder presented in Fig. 8-27 without interference between sample-and-hold amplifiers. The adder is connected to the analog input of the ADC.

Sample-and-hold and track-and-hold amplifiers are used to briefly store incoming signals for an ADC. Oftentimes, this is necessary when several signals need to be stored simultaneously and then sequentially switched into the input of the converter. It is also useful when converting extremely short transient signals as they can be stored and converted later.

Another useful application of operational amplifiers is the construction of *active filters*. Active filters are used to restrict the bandwidth of signals so as to reject unwanted noise. There are several basic types of active filters, including the high-pass and low-pass types illustrated in Fig. 8-36. A low-pass filter is

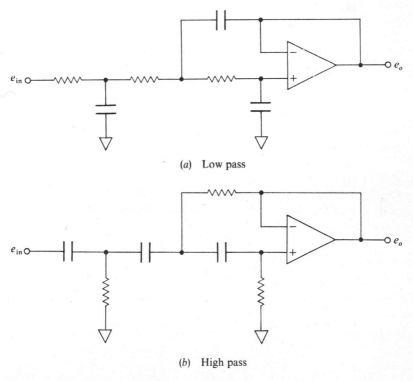

(*a*) Low pass

(*b*) High pass

Fig. 8-36   Active-filter amplifier.

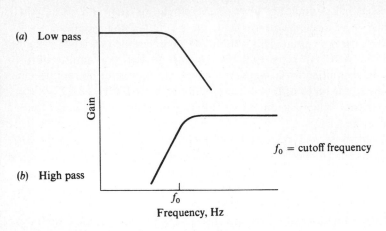

(a) Low pass

Gain

f_0 = cutoff frequency

(b) High pass

f_0

Frequency, Hz

Fig. 8-37   Response curves for active filters.

designed to reject high-frequency noise yet pass dc or low-frequency signals. In contrast, a high-pass filter is designed to reject dc and low-frequency signals and pass high-frequency signals. The response curves for high- and low-pass filters are illustrated in Fig. 8-37. Notice, also, that we can have *bandpass* and *band-reject* or *notch* filters. Their output characteristics are presented in Fig. 8-38. A bandpass filter is designed to reject frequencies both above and below a particular frequency. In so doing, it rejects both high- and low-frequency noise, while passing a relatively fixed frequency signal. A band-reject filter, on the other hand, is designed to pass all frequencies except for a particular one. For example, a band-reject filter set at 60 Hz may be designed to reject 60-Hz line frequency noise.

Analog multiplexers   When one has several signals that must be connected to the input of one ADC, a common device used is the *analog-signal multiplexer* presented in Fig. 8-39. With it, several input signals can be sequentially switched

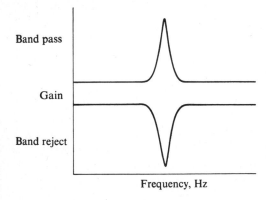

Band pass

Gain

Band reject

Frequency, Hz

Fig. 8-38   Bandpass and band-reject (notch) response curves.

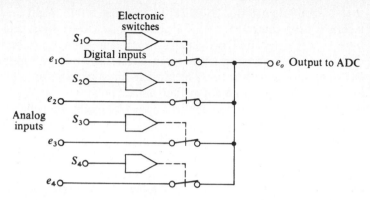

Fig. 8-39   Analog signal multiplexer.

to one output. Often, the input signals will come from the output of sample-and-hold or track-and-hold amplifiers. Analog-signal multiplexers are a series of electronic switches with digital control inputs to allow a computer or interface to control this operation.

Analog switches   Thus far, several uses for electronic analog switches have been discussed, including resistor ladder-network switches, autoranging amplifier feedback switches, sample-and-hold and track-and-hold amplifier switches, and analog-signal multiplexer switches. At this time, we will discuss in general the characteristics of various types of analog switches.

Analog switches can generally be divided into three categories: mechanical, electromechanical, and electronic. Common toggle, slide, and rotary switches are examples of mechanical types. Their contacts are opened or closed manually. Electromechanical switches are usually relays of one sort or another. In them, an electromagnetic coil is energized to open or close the contacts. The contacts may be either dry or coated with mercury. Mercury-wetted contacts usually exhibit somewhat better switching characteristics.

Electromechanical analog switches exhibit some very desirable characteristics. They have essentially zero resistance when the contacts are closed and infinite resistance when they are open. They can handle a range of many orders of magnitude of voltage and current of either polarity. They also have, as might be expected, some undesirable characteristics. First, they have contact bounce whenever the contacts are opened or closed. This may result in significant signal noise that must be filtered out. Mercury-wetted contacts often help to lessen bounce noise, but in no way eliminate it. Second, electromechanical switches are rather slow in switching; the faster ones take 1 msec or so to open or close.

Electronic analog switches are usually constructed from *junction field-effect transistors* (JFETs) or *metal-oxide-silicon field-effect transistors* (MOS-FETs). FETs used as switches are usually driven by transistor driving circuits that open or close them. A typical FET analog switch might be closed when a logical 0 from integrated-circuit logic is applied to its driver and open when a logical 1 is applied. FET switches have advantages over electromechanical analog switches in one principal area—speed. They have orders-of-magnitude greater switching speeds with turn-on or turn-off times of less than 1 μsec. Since they have no contacts, they do not exhibit contact bounce noise. They may, however, if not designed into a circuit correctly, generate electronic switching spikes.

Electronic analog switches also have some shortcomings. They have finite on and off resistances. JFETs usually have the smaller on or closed resistance, some types as low as 1 Ω or so. MOSFETs, on the other hand, have the higher off or open resistance, typically on the order of 10^{16} Ω. In other words, FET switches are not ideal switches, but rather can be considered as electronically variable resistors.

Because of their electronic nature, FET switches exhibit voltage and current magnitude and polarity restrictions. Common signal voltage levels are on the order of 10 V or so, but some specialized units go as high as 100 V. Maximum currents are usually no greater than 100 mA. FET switches usually do not work well for signal levels less than a few millivolts because of a small inherent voltage drop in the transistors that may vary with operating conditions.

Generally speaking, one should use electromechanical analog switches for low-level or high-level signals that need not be switched at times less than 1 msec or so and electronic analog switches for moderate signal levels with switching times less than 1 μsec or so. Both electromechanical and electronic types are available in configurations that can be driven directly from integrated-circuit digital logic.

8-3 TIMING AND CONTROL ELEMENTS

In this section, we will discuss the more common digital hardware components used to provide the timing and control-logic functions necessary in an experimental interface. Figure 8-40 illustrates the most important experimental interface function—the generation of a stable time base. The time-base generator, or digital clock, for an experimental system is generally a combination of a crystal-controlled, stable, fixed-frequency oscillator which outputs a pulse train of very accurately known frequency and a counter or scaler logic section used to divide this frequency into a variety of output frequencies. The output frequency can be incrementally variable through the use of counters and can be controlled by enabling or disabling the counter system. We will generally use in future discussions the block diagram presented in Fig. 8-40b, which shows only an enable/disable input and a series of outputs.

Another useful control element is the *switch-bounce filter* presented in Fig. 8-41. We normally cannot connect mechanical switches directly to digital inputs because the contacts within the switch bounce and produce noise signals

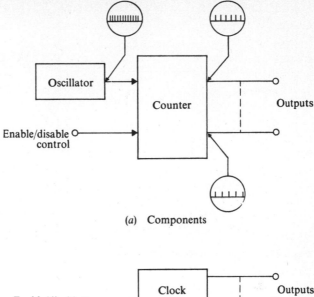

(a) Components

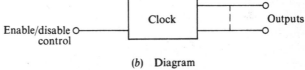

(b) Diagram

Fig. 8-40 Digital clock.

that interfere with correct operation of the logic. This problem can be eliminated with the simple circuit presented in Fig. 8-41, which consists of two cross-coupled NAND gates forming an RS flip-flop. The RS flip-flop is set and reset with a single pole double throw (SPDT) switch.

A useful timing function often required in an interface system is the generation of accurately known and reproducible adjustable time delays. The *monostable multivibrator* or *one-shot* presented in Fig. 8-42 is often used for generating this **type** of a delay function. The diagram shown in Fig. 8-42 is for a gated one-shot in that there is a series of gates which allow logic to be performed

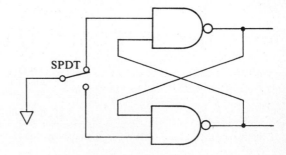

Fig. 8-41 Switch-bounce fil-ter.

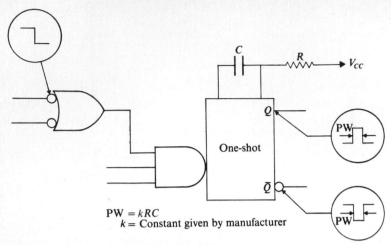

$$PW = kRC$$
$$k = \text{Constant given by manufacturer}$$

Fig. 8-42   One-shot or monostable multivibrator.

at the input of the device. The one-shot has the characteristic of remaining in a stable state until a trigger pulse is applied to the input, at which time the output changes state for a period of time called the *pulse width* (PW), which is dependent on the value of the resistance-capacitance (RC) timing network employed in the circuit. Generally, the one-shot device has a provision for attaching different resistances and capacitances. Delay times can generally be varied over wide ranges, typically from nanoseconds to many seconds or even minutes. Many integrated-circuit one-shots are currently available.

Another way of generating signal delays, generally short delays, is to

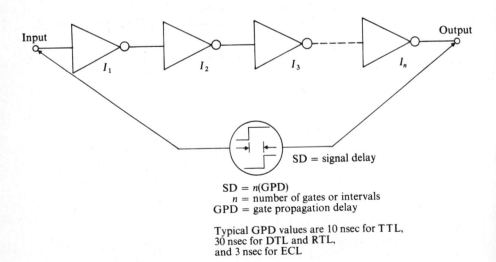

$$SD = n(GPD)$$
$$n = \text{number of gates or intervals}$$
$$GPD = \text{gate propagation delay}$$

Typical GPD values are 10 nsec for TTL,
30 nsec for DTL and RTL,
and 3 nsec for ECL

Fig. 8-43   Gate propagation signal delay.

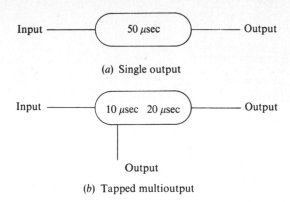

(a) Single output

(b) Tapped multioutput

Fig. 8-44   Delay elements.

connect a series of inverters or gates in the manner shown in Fig. 8-43. Since each gate requires a finite time or propagation delay for a signal to pass through it, the total signal delay will be equal to the individual gate propagation delay multiplied by the number of gates. This technique is useful for generating short time delays in order to get proper logic synchronization.

There are other signal delay elements including both RC and LC (inductance-capacitance) networks. They are often given the symbols presented in Fig. 8-44. They are characterized by an input and an output with the signal delay time written in the diagram. In Fig. 8-44b, a tapped multioutput signal delay is illustrated. RC and LC signal delay networks are often used for short signal delays up to 1 msec or so.

In addition, there are still other methods for generating signal delays. Synchronous signal delays can be performed using the serial shift registers presented in Chap. 7. The signal delay for an n-length shift register is equal to n clock pulses.

8-4   SIGNAL TRANSMISSION ELEMENTS

An important aspect of analog signal handling which often must be considered is the distance over which a signal must be transmitted. Oftentimes, the ADC is located at some distance from the experimental voltage source. Analog signals must then be sent over some kind of transmission line. Usually there is a line driver and receiver between the experimental transducer and the converter, as is illustrated in Fig. 8-45. The line-driving section might contain an amplifier, an

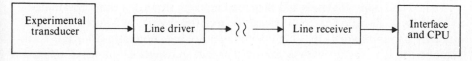

Fig. 8-45   Signal transmission elements.

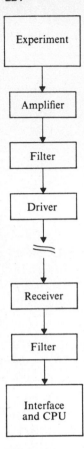

Fig. 8-46   Analog data transmission system.

active filter, and a high-current driver. The receiving section might contain an amplifier, active filter, etc. These are illustrated in Fig. 8-46.

A transmission line consisting of two-conductor, twisted, shielded cables illustrated in Fig. 8-47 can be used to transmit satisfactorily analog signals up to 1 kHz. For signal frequencies in the range of 1 Hz to 1 kHz, the signal should be amplified up to the level of several volts before transmission. This is because the 60-Hz line and other frequency-noise pickup is difficult to filter out of low-level signals. For signal frequencies less than 1 Hz, it is possible to transmit millivolt signal levels provided the output is well filtered. All these characteristics are illustrated in Fig. 8-47.

Figure 8-48 illustrates another method of sending analog signals over long distances. The basic idea is to transmit the experimental signal by splitting it into two equal and opposite levels and then sending them through a transmission line as described above. The split signal is received by an *absolute-difference amplifier* or *subtractor* with a high common-mode rejection ratio (CMRR). This amplifier, if it is set up with a gain of one-half, provides a ground-referenced

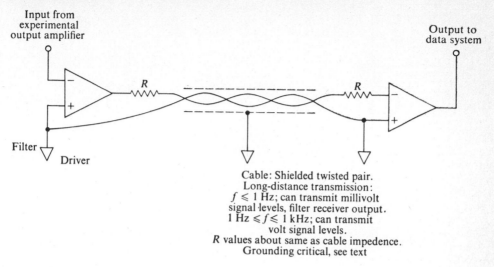

Cable: Shielded twisted pair.
Long-distance transmission:
$f \leqslant 1$ Hz; can transmit millivolt
signal levels, filter receiver output.
1 Hz $\leqslant f \leqslant 1$ kHz; can transmit
volt signal levels.
R values about same as cable impedence.
Grounding critical, see text

Fig. 8-47   Grounded analog signal transmission.

output signal equal to the original experimental signal. The use of the split input and the high common-mode rejection differential amplifier reduces the effect of noise pickup dramatically because the noise is seen by the subtractor as a common-mode signal. In addition, it eliminates problems that often occur when connecting grounds together between two electronic devices a long distance apart, thus reducing ground loop errors.

When experimental signal sources and data-handling systems are far removed from each other, one may also be required to send digital signals over

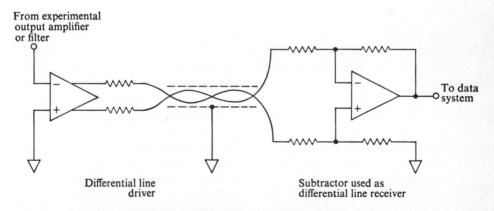

Cable: Long-distance transmission type.
Receiver must have high common-mode rejection

Fig. 8-48   Differential analog signal transmission.

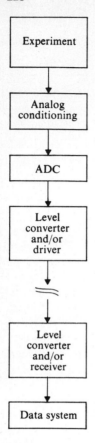

Fig. 8-49   Digital data transmission system.

long distances. In some situations, for example, an ADC is present at an experiment, and digital signals are sent from it to a computer or data system some distance away. This is illustrated in Fig. 8-49. A system such as this might consist of the experiment, analog conditioning elements such as operational amplifiers, an ADC, a level converter and driver, and a long-distance connection to another level converter or receiver, the output of which finally is connected to the data system. The level conversion functions necessary might consist of changing negative logic voltage levels at the output of the ADC to positive voltage levels that can be received by the data system or of changing lower-voltage digital output signals from the ADC to higher-voltage digital output signals required by the data system. For example, suppose the ADC puts out logic levels of 0 and 5 V and the data system requires output levels of 0 and 12 V. Obviously, some device is needed to change from the 5-V logical 1 level of the ADC to the 12-V logical 1 level required by the data system. These functions are generally performed by the level converters presented in Fig. 8-50. In Fig.

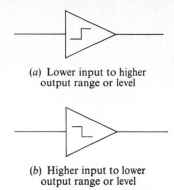

(*a*) Lower input to higher
output range or level

(*b*) Higher input to lower
output range or level

Higher: more positive
Lower: less positive

Fig. 8-50   Digital voltage level converters.

8-50*a*, a lower-input-to-high-output level converter is presented. Thus, in Fig.
8-50, the logical 1 input signal is lower or more negative than the logical 1
output signal. By contrast, in Fig. 8-50*b* the logical 1 input is more positive than
the corresponding output.

The digital signals must be transmitted in some manner to the computer or
data system from the experiment. This can be accomplished in much the same
way as for analog signals. The simplest methods are those presented in Fig. 8-51.

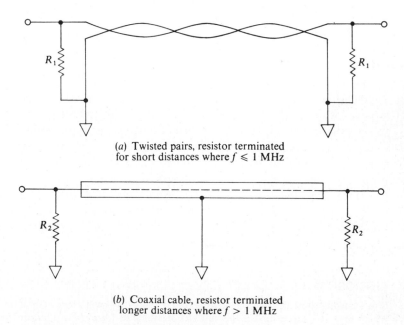

(*a*) Twisted pairs, resistor terminated
for short distances where $f \leq 1$ MHz

(*b*) Coaxial cable, resistor terminated
longer distances where $f > 1$ MHz

Fig. 8-51   Digital data transmission lines.

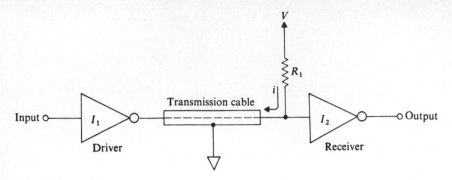

Fig. 8-52 Grounded-output digital data transmission.

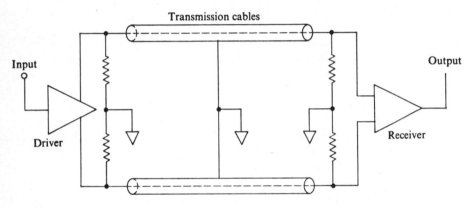

Fig. 8-53 Differential-output digital data transmission.

These include twisted wire pairs for signals with frequencies less than 1 mHz and low-capacity coaxial cable for frequencies greater than 1 MHz. A special case of this is presented in Fig. 8-52. For current-sinking logic types (refer to Appendix B for a description), a pullup resistor is placed at the receiving end of the transmission cable to provide significant current drive down the cable. This keeps the cable impedance low and keeps its capacitance charged up to reduce the tendency of noise pickup. Notice in Fig. 8-53 that differential digital data transmission can be performed in a manner similar to the analog signal transmission presented in Fig. 8-48. The drivers and receivers necessary are available in many integrated-circuit forms.

BIBLIOGRAPHY

The following short list contains a group of references that will prove helpful for the reader who wishes to pursue further the material presented in Chap. 8. In addition, much useful information is provided in manufacturers' catalogs, such as those published by Digital

Equipment Corp., Maynard, Mass. The references presented below should, in many cases, be used with those presented at the end of Chap. 7.

Much of the information presented in this chapter has been presented in more detail in Ref. 1. Reference 2 provides a great deal of insight into the use of operational-amplifier systems. Reference 3 discusses many of the considerations necessary for data transmission. Finally, mention must be made that much of the information used to describe low-level analog-to-digital conversion came from Hewlett-Packard's "Electronic Instruments Catalog." It is, in itself, a fine reference text for many electronic instruments.

1. Hoeschele, David F., Jr.: "Analog-to-Digital/Digital-to-Analog Conversion Techniques," Wiley, New York, 1968.
2. Korn, G. A., and T. M. Korn: "Electronic Analog and Hybrid Computers," McGraw-Hill, New York, 1964.
3. Morris, R. L., and J. R. Miller: "Designing with TTL Integrated Circuits," McGraw-Hill, New York, 1971.

9

Hardware and Software Fundamentals for On-Line Computer Experimentation

A digital computer is considered on-line to an experimental system when a direct electronic communication link exists between the experiment and the computer. In Chap. 6, we considered in some detail the hardware characteristics of the communication link—or interface—between the computer and standard peripheral devices. In Chaps. 7 and 8, we defined the characteristics of analog and digital devices important for instrumental interfacing. In this chapter, we will consider the general design characteristics for interfaces between experimental systems and the digital computer, as well as software approaches for communicating with experimental systems.

9-1 EXPERIMENTAL INTERFACE DESIGN

Using the analog and digital devices described in Chaps. 7 and 8, one can begin to design the specific communication link—or interface—between the digital computer and experimental systems to be operated on-line. Several fundamental considerations must be kept in mind: How does the particular computer use, recognize, and interpret information from the outside world? How does the computer transmit information? What are the computer machine-language instructions available for input/output (I/O) functions? These considerations were discussed in Chap. 6 in connection with standard peripheral devices. Now it is appropriate to consider the computer I/O hardware in more detail for applications with nonstandard experimental systems.

One fundamental principle that should be emphasized here is that *the computer communicates with the outside world by recognizing or causing binary voltage level changes at particular terminals at particular times.* Thus, the interface design is reduced to the problem of monitoring and interpreting voltage level changes from the computer and/or causing voltage level changes to be detected and properly interpreted by the computer. Thus, interface design can be completed only if the I/O hardware *and* software of the computer are well understood. The experimenter must also have a good appreciation for the availability and characteristics of the hardware interface components described above.

The general characteristics of the I/O hardware on a single I/O channel of an H.P. 2100 family computer are illustrated in Fig. 9-1. The illustration is in the context of communication with a standard peripheral device. However, it is pertinent to consider only the general-purpose characteristics of the standard interface hardware. For this particular computer, I/O functions are generated through a plug-in, 16-bit duplex register card which occupies a single I/O channel (H.P. 02116-6195). This card has the following characteristics:

(1) Two 16-bit flip-flop registers, one for input and one for output.
(2) A flip-flop flag bit which can be set externally and set or cleared internally. The status of the flag bit can be sensed by the computer.
(3) A flip-flop encode bit which can be set or cleared by the computer and the status of which is available externally.
(4) A flip-flop control bit, the status of which is not directly available externally. The control bit must be set (with an STC instruction) to allow interrupt requests on the selected channel. (See Chap. 6.) The encode bit is also set when an STC instruction is executed; when the control bit is cleared (CLC), the encode bit is also cleared. (Normally, the encode bit will *follow* the control bit at all times. However, the encode bit can also be wired to be cleared whenever an external signal sets the flag bit. This option is selected by the user with a jumper connection on the interface card.)

The assembly-language instructions which allow the use of the hardware characteristics of these general-purpose interface cards have been summarized in Table 6-1. In addition, the general approach to I/O programming used for standard peripheral devices is completely applicable to programming for nonstandard experimental systems. Thus, the reader is urged to review these discussions in Chap. 6. (See also Refs. 1 and 2.)

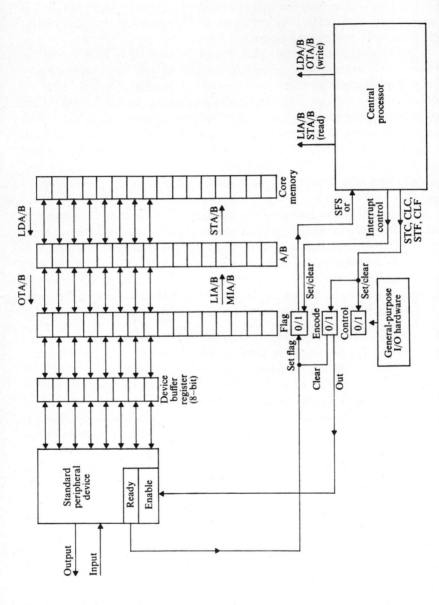

Fig. 9-1 Communication with standard I/O peripheral device (Teletype printer, high-speed reader, magnetic tape or disc, etc.).

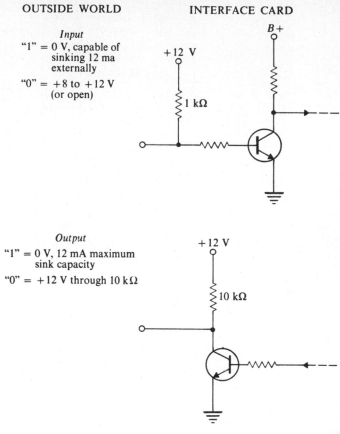

OUTSIDE WORLD   INTERFACE CARD

Input
"1" = 0 V, capable of
sinking 12 ma
externally

"0" = +8 to +12 V
(or open)

Output
"1" = 0 V, 12 mA maximum
sink capacity

"0" = +12 V through 10 kΩ

Fig. 9-2 Digital I/O characteristics for Hewlett-Packard general-purpose interface card.

Figure 9-2 illustrates the specific electrical characteristics associated with the input or output of a single bit of information through the Hewlett-Packard general-purpose interface card, 12-V negative-logic version (H.P. 02116-6195). From inspection of the diagrams, it can be seen that on each input +12 V is seen through a 1K Ω resistance. A logical 1 input is generated when the input voltage level goes to 0 V (or ground); the input device must be capable of sinking 12 mA externally. A logical 0 input is generated when the input terminal goes to +8 to +12 V, or open circuit. (At open circuit, the interface card supplies its own logical 0 reference voltage, +12 V.) At each output, a ground voltage is considered a logical 1 where the card terminal is capable of sinking 12 mA maximum. A logical 0 is seen when the output transistor is biased off, and a +12-V level is seen at the output terminal through a 10K Ω resistance. [Alternatively, the interface card may be configured with an *open-collector*

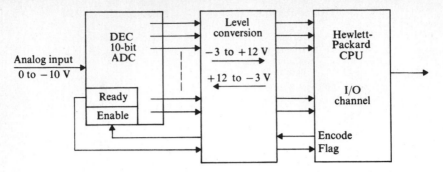

Fig. 9-3   Requirements for DEC/H.P. interface (DEC R-series positive logic to H.P. 12-V negative logic).

output transistor; here, the reference voltage and resistor can be supplied externally, allowing more latitude in logic levels (+5 to +12 V).]

When the external hardware generates logic levels which are not directly compatible with the electrical characteristics of the digital computer being used, level conversion hardware must be incorporated into the interface (Chap. 8). For example, interfacing a DEC C002 R-series logic ADC to a Hewlett-Packard computer requires level conversion as shown in Fig. 9-3. DEC level conversion cards W-601 and W-510 are used. Similar considerations are required in almost every interface design problem. Even when logic levels are identical for interface elements and the computer, the load capabilities and requirements of the two may not be identical, and buffer driver elements are required to match load requirements without changing logic levels.

Program-controlled data acquisition   As stated in Chap. 6, two kinds of data-acquisition approaches can be defined. One is to operate the data-acquisition programming under *interrupt* control. That is, the computer is operated in a mode where data-acquisition devices are serviced upon demand. The other approach is called *program-controlled* data acquisition. This approach involves programming the computer to look for service requests from specific devices and wait for these requests if necessary. The latter approach will be discussed here; whereas interrupt servicing will be discussed in a later section.

Perhaps the best way to discuss program-controlled data acquisition is to consider a specific problem. Consider the case where an experiment of a transient nature is to be conducted and data acquired during the lifetime of the experiment. It is desired to use the computer to initiate the experiment and simultaneously initiate data acquisition. Data acquisition is desired at a constant rate of 10 Hz. When a specified total number of data has been taken, it is desired to have the computer terminate data acquisition and reset the experimental instrumentation to original conditions. Each data point is to be taken in from the ADC and stored in memory for later processing.

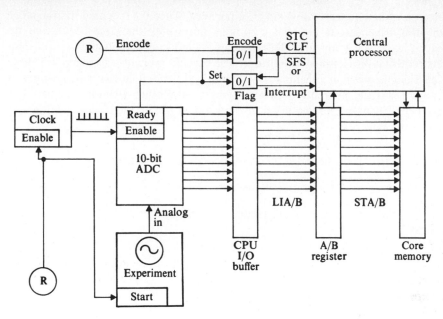

Fig. 9-4   Data acquisition with external clock controlling data rate. *Note*: (1) CPU initiates data-acquistion cycle with STC, CLF; (2) data taken by servicing interrupt *or* by waiting for flag with SFS instruction (programmers option); (3) CPU counts data points.

A schematic diagram of the computerized data-acquisition system is illustrated in Fig. 9-4. The computer initiates the data-acquisition cycle by executing an STC instruction, which sets the encode bit to a 1 state. This terminal is connected to the enable input of a 10-Hz clock, the output of which enables a 10-bit ADC every 0.1 sec. Simultaneous with enabling the clock and the data-acquisition process, the encode output initiates the experiment. The experimental output is continuously available at the input to the ADC. Every time a conversion is completed, the ADC sets a ready flip-flop which is connected to the flag flip-flop on the general-purpose interface card. When that flag goes to a 1 state, the computer can determine that a conversion has been completed and that the digitized datum has been strobed into the input buffer register.

Under program-controlled data acquisition the computer will be programmed to test the status of the flag bit (with an SFS instruction) to determine when each conversion has been completed. When the computer gets a "true" answer in querying the flag flip-flop, it goes to a data input routine which loads the contents of the buffer register into the A or B accumulator, clears the flag bit on the interface card, and then stores the datum in core memory. This routine must keep track of the total number of data taken and handle sequential storage of data in a specified block of memory. When the specified total number of data has been taken, the computer terminates the data acquisition by executing a CLC command. This disables the clock and resets the experimental instrumentation to initial conditions. The specific program to accomplish these functions is given in Example 9-1. (Note in this example that it is possible to execute a CLF instruction simultaneously with other I/O instructions. The assembly-language representation involves a ,C suffix on the I/O instruction. See Appendix A.)

Example 9-1   Data acquisition with external clock-controlled ADC

> *Assume:*   1. ADC is on channel 10_8.
>   2. Hardware as in Fig. 9-4.
>   3. External logic initialized manually.
>   .
>   .
>   .

```
              CLF   0          /TURN OFF INTERRUPT SYSTEM
              STC   ADC,C      /START CLOCK & EXPERIMENT; CLR FLG
      CNVRT   SFS   ADC        /WAIT FOR ADC FLAG
              JMP   *-1
              LIA   ADC,C      /GET DATUM FROM ADC & CLR FLG
              STA   PBFFR,I    /SAVE DATUM
              ISZ   PBFFR
              ISZ   DCNTR      /GOT ALL DATA?
              JMP   CONVRT     /NO, GET NXT DATUM
              CLC   ADC        /YES, TURN OFF CLOCK WITH ENCODE BIT.
              .                /CLR I/O CHANNL
              .
              .

      ADC     EQU   10B        /SET "ADC" EQUAL TO 10B, CHANNL # FOR ADC
              .
              .
              .
```

It is useful here to consider the case where the transient experimental output presents a voltage waveform which changes very rapidly at first but only gradually at later times. (A good example is the exponential decay curve followed during the discharge of a capacitor.) In such a case, a constant data-acquisition rate will be inappropriate. A slow rate which was compatible with the data generated at long times will not be adequate for short-time data. On the other hand, if a data-acquisition rate is employed which is adequate for the rapidly changing initial portion of the experimental output, more data points than necessary will be acquired during the latter part of the experiment. If the experiment is to be conducted over a long time interval, it is conceivable that core memory space might be exceeded. Therefore, it is sometimes useful to build into the data-acquisition program a variable data-acquisition rate. One way to accomplish this is to use a clock rate higher than the data-acquisition frequency but incorporate into the data-acquisition loop a software counter. Example 9-2 presents a program which illustrates how one generates a data-acquisition frequency less than the clock frequency by using a software counter. It is left as an exercise for the reader to design a program which will take data at one rate for a specified number of initial data points and then at another rate during the rest of the experiment.

Example 9-2   Data acquisition with an external clock and program-scaling of the data rate

```
            ·
            ·
            ·
            CLF   0           /DISABLE INTERRUPT SYSTEM
            STC   ADC,C       /STRT FIXED FREQ CLOCK & EXPERIMENT
    CNVRT   SFS   ADC         /WAIT FOR ADC FLAG
            JMP   *–1
            CLF   ADC         /CLR ADC FLG AFTER EACH CONVRSN
            ISZ   RCNTR       /TAKE THIS DATUM?
            JMP   CNVRT       /NO
            LIA   ADC         /YES
            STA   PBFFR,I     /SAVE DATUM
            LDA   RCTST       /RESET DATA RATE COUNTER
            STA   RCNTR
            ISZ   PBFFR
            ISZ   DCNTR       /ALL-DATA IN?
            JMP   CNVRT       /NO–GET NXT DATUM
            CLC   ADC         /YES, DISABLE CLOCK, STOP EXP'T.
            ·                 /CLR I/O CHANNL
            ·
            ·
    ADC     EQU   10B         /SET "ADC" EQUAL TO 10B
            ·
            ·
            ·
```

In the development of on-line experimentation it is often useful to have available a general-purpose digital I/O facility. That is, it is necessary not only to provide for analog data acquisition but also to provide for a wide variety of digital I/O communication. One way to accomplish this is by providing a digital I/O module with external terminals connected to the respective contacts of the general-purpose interface card of the computer. In this fashion, it is possible to bring 16 bits of information to or from the computer. It is also possible to use the encode and flag bits for general interfacing functions. (It should be recalled: *The transfer of external digital information to the buffer register on the general-purpose interface card is accomplished when the flag bit is set externally*.) For example, one or more of the input bits could be used to provide for *status checks*. That is, the computer could be programmed to determine the status of any one of these bits at any given time and use the status of any one of these bits to determine programmatically if and when external events had occurred. Thus, these external input terminals could be used as sense switches for general-purpose communication functions. They could be set manually or electronically to appropriate logical levels. Output bits could be used for experimental control functions and for outputting coded binary commands to be decoded and implemented at external devices (such as a multiplexer or ranging amplifier). (Specific examples will be given in later illustrations.)

Timing and synchronization in data acquisition   Although the need for synchronization between data-acquisition operations and experimental events has been mentioned above, the importance of this has not been illustrated. Figure 9-5 describes what can happen when synchronization error occurs. The trace representing the real data which starts at time zero t_0 is *presumably* sampled by the data-acquisition system at time points indicated by the clock pulses on the x axis. Because the clock can generate pulses at an accurately known frequency f, the time between pulses $1/f$ is precisely known. The program generally will assume that the first clock pulse is seen at exactly the fundamental time interval $1/f$ after t_0. However, this will only be true if data acquisition has been synchronized exactly with the start of the experiment. If synchronization has not occurred, the first clock pulse can come anywhere during that first time interval as shown in Fig. 9-5. If the first clock pulse occurs early, as shown in Fig. 9-5, and the program is not aware of the synchronization error, the program may assume that the first datum obtained really corresponds to the time assigned to the first clock pulse on the diagram. Thus, the data points taken at the x's on the real data trace are effectively displaced along the time axis to the points indicated by the squares on the diagram, and the digitized waveform seen by the computer has the appearance of having been translated on the time axis as shown in Fig. 9-5. For experiments where the data density is

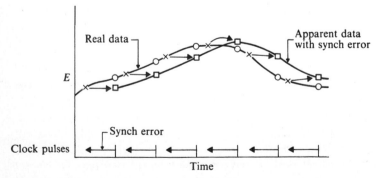

Fig. 9-5   Effects of synchronization error in data acquisition.

great, an error of this sort may be insignificant; however, for most experiments, this type error may cause severe difficulties in data processing. (An example of such an experiment is one involving ensemble averaging, described below.)

Generally, a crystal oscillator is used for precise timing. However, this type clock provides a continuously available pulse train at a fixed frequency. Thus, there is no way to determine when a given clock pulse occurs in real time. The uncertainty can be minimized if one selects a clock with a very high frequency value and scales this down to the desired frequency range. (See Fig. 8-40.) The countdown logic can be initialized, enabled, or disabled. Thus, the output of the scaler provides a pulse train where the uncertainty in the duration of the first time interval is no greater than the time interval $1/f_0$ of the crystal oscillator. For example, if the basic clock rate f_0 is scaled by a factor of 100, then the uncertainty in the initial scaled time interval will be no greater than 1 percent.

The scaled clock is most valuable for establishing a time base for experiments which cannot be started at a precisely known moment by external control. If the start of a spontaneously initiated experiment can be detected electronically, this signal can be used to enable the scaler logic of the clock. Thus, a time base precisely synchronized with the experiment is obtained.

If the exact frequency of a free-running clock is to be used for data-acquisition timing, synchronization can be achieved by simple gating. This is demonstrated in Fig. 9-6. The output of the free-running clock is brought to

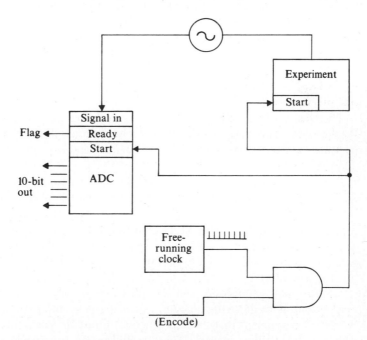

Fig. 9-6   Synchronization by gating a clock.

one input of an AND gate. The other input can be enabled by the encode output of the computer or some other source. When this input of the AND gate is conditioned true, clock pulses can get through the gate and are seen at the output. The first clock pulse that gets through the AND gate is used to initiate data acquisition and simultaneously start the experiment. Subsequent pulses are seen at exact multiples of the fundamental time interval $1/f_0$ of the free-running clock. Thus, data acquisition is exactly synchronized with the start of the experiment because the first available clock pulse was used to initiate the experiment. *Note*: This synchronization approach is applicable only if the experiment can be initiated externally. For an experiment which initiates spontaneously, the alternative approach of enabling scaler logic on a high-frequency clock should be used. The scaled clock is the most generally applicable and is the type implied in most illustrations here. *The laboratory computer user should be keenly aware of the synchronization limits and capabilities of the time-base generator (clock) used in his system.*

Ensemble-averaging application   A good example of many of the principles we have just discussed, and a useful application of the digital computer for enhancing experimental measurements, is the technique of *ensemble averaging*. This technique can be applied in cases where experimental data are obtained with large amounts of superimposed background noise. Although many approaches can be taken to handle instrumental problems leading to noisy data, it is not always possible to eliminate noise. (For example, a situation where standard noise elimination procedures may be inadequate is the case where the source of the noise is not in the electronics but rather an inherent part of the experimental system.) When the frequency of the noise is similar to the frequency of the fundamental waveform of interest, conventional filtering techniques are not adequate. In such cases, some sort of signal-averaging approach must be used in order to extract the fundamental signal from the noise background. However, two conditions must be met: First, the signal must be repeatable; second, the noise must be random and not synchronized with the experimental output.

The ensemble-averaging approach involves running repetitive experiments and using the computer to acquire the digitized waveform from each experiment and sum the repetitive waveforms. When many such experimental outputs have been summed in this coherent fashion, the random noise fluctuations in the individual waveforms will begin to cancel. The signal-to-noise ratio, in fact, should increase proportionally to the square root of the number of averaging cycles. This approach is illustrated in Fig. 9-7.

The data-acquisition and interface design for an ensemble-averaging experiment is outlined in Fig. 9-8. It is extremely important in an ensemble-averaging experiment that the experimental output be synchronized exactly with the data-acquisition process. If any significant fluctuation in synchronization

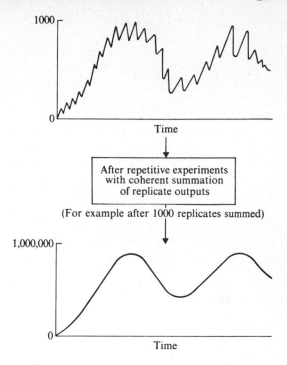

Fig. 9-7   Ensemble averaging

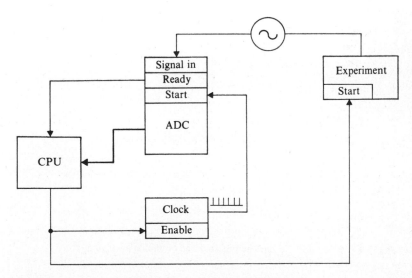

Fig. 9-8   Interface design, ensemble-averaging experiment.

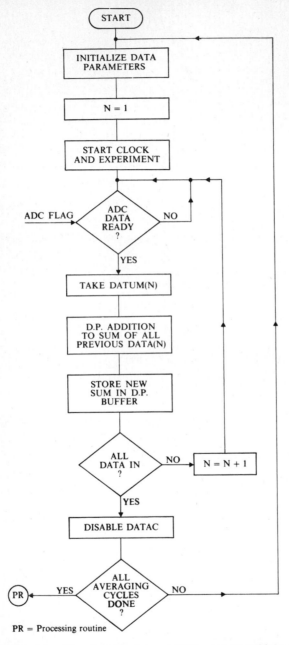

Fig. 9-9   Flowchart for data acquisition and ensemble averaging.

with the time base occurs, a distortion in the extracted fundamental waveform will be observed.

A flowchart of a program which will generate an ensemble-averaging experiment is shown in Fig. 9-9. One feature of the algorithm that should be pointed out is that the summation of replicate data points is carried out in double precision. Example 9-3 presents a program corresponding to the flowchart of Fig. 9-9. Note that because of the amount of data processing done between acquired data points, the maximum data-acquisition rate is limited to about 25 kHz with an H.P. 2116 computer (1.6-μsec cycle time). Normalization of the double-precision data is simplified if the number of averaging cycles is

Example 9-3   Data-acquisition routine for an ensemble-averaging program

```
              .
              .
              .
        CLF  0          /TURN OFF INTERRUPT SYSTEM
  CLOCK STC  ADC,C      /START CLOCK
        SFS  ADC        /WAIT LOOP. WAIT FOR ADC FLAG
        JMP  *-1
        LIA  ADC,C      /GET ADC DATUM. CLEAR FLAG
        CLE             /START DOUBLE-PRECISION ADDITION ROUTINE
        ADA  .LSBF,I
        SEZ             /CHECK FOR CARRY TO MSH
        ISZ  .MSBF,I    /CARRY "1" TO MSH, IF "E" WAS "1"
        STA  .LSBF,I    /STORE "A" IN LSH BFFR
        ISZ  DTCNT      /ALL DATA PTS. IN?
        RSS
        JMP  DSABL      /YES--DISABLE CLOCK
        ISZ  .LSBF      /NO--INCRMNT DATA STORAGE ADRSS
        ISZ  .MSBF
        JMP  CLOCK+1    /GO WAIT FOR NEXT FLAG
  DSABL CLC  ADC        /CLEAR ENCODE COMMAND. DISABLE CLOCK
        LDA  .MSST      /REINITIALIZE DATA BUFFR ADRSS'S
        STA  .MSBF
        LDA  .LSST
        STA  .LSBF
        ISZ  AVGCT      /AVGNG CYCLES FINISHED?
        JMP  TCLCK      /NO--REENABLE CLOCK
        JMP  NRMLZ      /YES--NORMALIZE AND PROCESS DATA
  TCLCK LDA  DCTST      /RESET DATA COUNTER
        STA  DTCNT
        JMP  CLOCK      /RE-ENABLE CLOCK FOR TAKING NXT DATA SET
              .
              .
              .
  ADC   EQU  XXB
  NRMLZ .              /NORMALIZATION ROUTINE, DATA BUFFR, ETC.
              .
```

always selected to be some integral power of 2. Normalization of the double-precision data can be achieved quite simply, then, by shifting the double-precision sum for each data point to the right. (See Chap. 5.)

9-2   SAMPLING OF EXPERIMENTAL DATA

Signal tracking   In the preceding discussions of digital data acquisition, little consideration was given to the signal *tracking* capabilities of the ADC. That is, what are the uncertainties in the time base due to the finite aperture time of the ADC? If the sampled waveform is changing during the sampling interval (as shown in Fig. 9-10), there may be some uncertainty in the time assigned to the digitized datum. That is, the final digital value may not represent the analog value of time t_0, but will likely represent a value reached at some time between t_0 and t'. If the *real time* assigned to the datum is t_0, the datum is inaccurate. The limits of uncertainty can be estimated from the ADC characteristics— namely the *aperture time, Δt* (referred to also as the *conversion time*), the number of bits per word, and the analog input range.

Ideally, the change in the analog input level ΔE during the aperture time should be less than the equivalent of the least significant bit (LSB) of the ADC. For a 10-bit 10-V ADC, the voltage equivalent of the LSB is $(1/1{,}024) \times 10$ V = 9.76 mV. Thus, if ΔE exceeds 9.76 mV during the ADC aperture time Δt, the digitized datum *assigned to time t_0 contains less than 10 significant bits*. The number of significant bits retained will depend upon the actual magnitude of $\Delta E/\Delta t$.

If the sampled waveform is a pure sine wave of amplitude $E_{\max}$, the maximum voltage error ΔE due to time uncertainty Δt is given by

$$\Delta E = E_{\max} 2\pi f\, \Delta t \tag{9-1}$$

where f is the sine-wave frequency. Thus, to have ΔE less than 9.76 mV for a sine waveform with $E_{\max} = 10$ V, the frequency must be less than $155/\Delta t$, where Δt is in microseconds. (For example, if Δt is 10 μsec, f must be less than 15.5 Hz to retain 10-bit significance in digitization.)

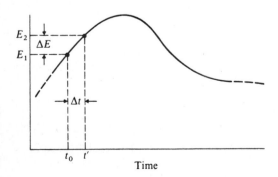

Fig. 9-10   Tracking error for ADC aperture time Δt.

If the frequency and amplitude of the sample waveform are known, the significance of digitized values can be estimated. For example, we calculated above that for a 10-bit 10-V 10-μsec ADC, a 10-V sine wave of frequency less than 15.5 Hz will be sampled with 10-bit resolution. To retain at least 9 bits of significance, the frequency must be below 31 Hz. Moreover, if the maximum allowable frequency for n bits of significance is designated f_n, the value of f_{n-1} is obtained by the relationship

$$f_{n-1} = 2f_n \tag{9-2}$$

or, more generally,

$$f_{n-k} = 2^k f_n \tag{9-3}$$

Thus, for the example here, a 62-Hz signal could be sampled with 8 bits of significance, a 124-Hz signal with 7 bits, etc.

Using track-and-hold amplifiers   The preceding discussion should illustrate that the *tracking* capabilities of even a high-speed ADC are relatively limited. To substantially improve the tracking features of a data-acquisition system, a track-and-hold (T/H) amplifier (see Chap. 8) should precede the ADC.

During each hold period, the voltage output of the T/H amplifier remains constant so that digitization may take place. (See Fig. 8-33.) The digital output of the ADC will reflect the voltage level at the specific time corresponding to the beginning of each hold period, despite the fact that the conversion is completed at some finite time later. Thus, the T/H amplifier allows waveform sampling with time-base precision independent of the conversion rate of the ADC. The uncertainty in the timing of the sampling is dependent on the switch opening time and is a characteristic of the T/H amplifier—the aperture time. T/H amplifiers are available commercially with aperture times on the order of 10 to 100 nsec. Other important characteristics include *response* and *settling* times, which refer to the amplifier's ability to follow rapidly changing signals. These features actually limit the overall acquisition or sampling rate. Sampling intervals (the time from the end of one hold period to the beginning of the next) on the order of one to several microseconds can be attained with currently available devices.

Sampling frequency   The sampling frequency selected for data acquisition obviously should be related to the bandwidth of the sampled waveform. From information theory [3, 4], the criterion for adequate sampling states that the minimum sampling frequency (*Nyquist frequency*) must be twice the bandwidth of the sampled waveform. Thus, for a 100-Hz signal, the sampling frequency must be at least 200 Hz to retain all the information inherent in the waveform.

This criterion is strictly applicable in such applications as the sampling of

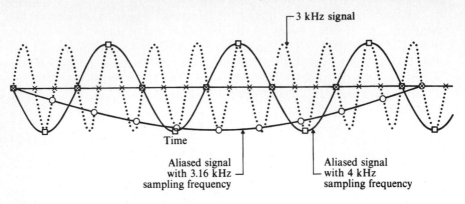

Fig. 9-11   Aliasing errors due to undersampling.

interferograms for Fourier transform analysis [5]. However, sampling frequencies considerably greater than the Nyquist frequency should be used to allow faithful reproduction of the signal for straightforward data processing algorithms. A rule-of-thumb criterion is that the sampling frequency should be at least 10 times the bandwidth of the waveform. (Of course, the previously discussed limits imposed by ADC or T/H aperture times or amplifier response place an effective upper limit on sampling frequencies with a specified resolution.)

Oversampling of an experimental signal can cause problems mainly because of excessive memory requirements. On the other hand, undersampling can cause more serious problems. One of these is the possibility of producing signal artifacts by *aliasing*. This phenomenon can occur when sampling frequencies lower than the Nyquist frequency are used. Figure 9-11 illustrates how a 3-kHz sine wave can be aliased to a 1-kHz signal or a 158-Hz signal by using sampling frequencies of 4 kHz and 3.16 kHz, respectively.

Sampled noise   The relative merits of integrating versus nonintegrating types of ADCs for noise rejection in data acquisition have been discussed in Chap. 8. It should be pointed out here, however, that when the ADC is preceded by a T/H amplifier, the high-frequency noise contribution to the data will be retained, although it may show up at aliased frequencies. Preceding an integrating ADC with a T/H amplifier would obviously negate its noise rejection characteristics, and this practice obviously would not be recommended where noisy data are observed.

Multiplexing   It is sometimes necessary to sample more than one experimental waveform simultaneously during a single experiment. To accomplish this, an analog multiplexer device can be used. The general characteristics of such a device have been discussed in Chap. 8. We will discuss here the configurational and sampling considerations that must be made when multiplexing analog signals.

The primary characteristic of an analog multiplexer, of course, is the fact that it can accommodate multiple analog inputs, any one of which can be sampled through a single output channel. The selection of the input channel to be transmitted to the digitization hardware can be accomplished by the output of a binary-coded command from the computer or by generating an appropriate external sequencing code. The critical characteristic that we must consider here is the time required to switch between input channels. With solid-state analog switches, the sequential selection of analog inputs can proceed with time intervals on the order of a few microseconds or less between channels.

The reader should recognize immediately that the analog multiplexer need not provide the slow step in an overall data-acquisition process. Indeed, the slower processes will be associated with the data-acquisition hardware and software which follow the analog multiplexer. Another point which should be emphasized here is that it is impossible for the multiplexer to sample independent waveforms in a truly simultaneous fashion. Some finite time interval must exist between samplings of different channels. The manner in which this problem is handled will depend on the need for acquiring truly simultaneous data from different channels.

Figure 9-12 illustrates two alternative configurations for multiplexing analog signals from four independent sources. In Fig. 9-12a, the multiplexer is followed by a T/H amplifier and ADC. This configuration is used if there is no need to achieve simultaneous sampling of the four input channels. Moreover, if the total time required to complete data acquisition from the four channels is small compared to the time interval between samplings, the configuration of Fig. 9-12a can be used. For example, if an overall data-acquisition frequency for all four channels of 10 Hz is employed and the total sampling time per channel is 100 μsec, the maximum *skew* of the sampled data will be 300 μsec. That is, the fourth channel will be sampled 300 μsec after the first channel. This amount of skew can be considered negligible compared to the 100-msec time interval between samplings.

For the case where simultaneous sampling is required and the time required to sequentially sample each of the input channels is long compared to the overall data-acquisition time base, the configuration shown in Fig. 9-12b is recommended. In this alternative case, each analog signal is funneled through a T/H amplifier. Because all the T/H amplifiers can be gated to the *hold* mode simultaneously, the time required to sample all channels through the multiplexer will be inconsequential provided that it does not exceed the overall data-acquisition time-base interval. Another consideration, of course, is the *droop* specification on the T/H amplifier. That is, the T/H amplifier must be capable of holding the analog signal without significant decay until the particular channel is sampled and the digitization complete.

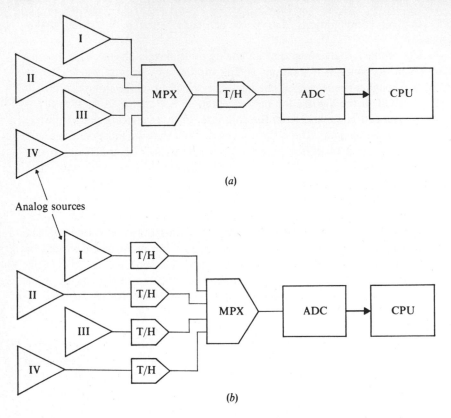

Fig. 9-12   Alternative multiplexing configurations.

Figure 9-13 illustrates how one might interface the multichannel data-acquisition system to the computer. The configuration used is patterned after Fig. 9-12*a*. Let us assume that the particular channel to be sampled is selected first by outputting a binary-coded select code to the multiplexer and that the selected switch is closed only when the *sequence* input to the multiplexer is true. This control can easily be implemented by a digital output word from the computer. However, other timing characteristics must be considered. For example, it would not be desirable to gate the T/H amplifier to the hold condition until *after* the selected multiplexer switch has had time to settle. In addition, because the T/H amplifier may have to undergo a large potential excursion when the new analog signal is provided, the overall response and settling time of the T/H amplifier must be considered. Thus, it may be necessary to delay the hold command to the T/H amplifier by several microseconds after the multiplexer has been sequenced. To accomplish this, a one-shot delay monostable is used to delay the hold command by a time τ after the sequence command. The settling time required is shown in the timing diagram of

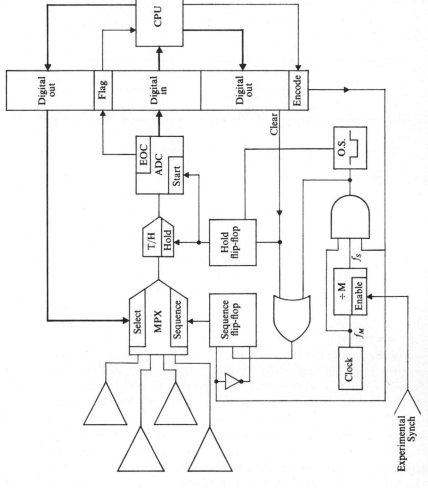

Fig. 9-13 Interface for multichannel input.

249

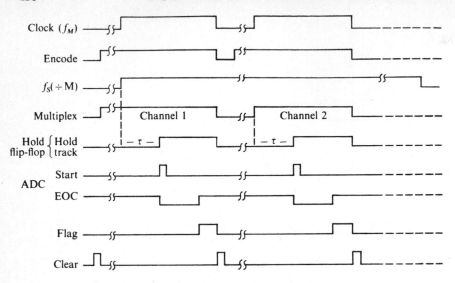

Fig. 9-14　Timing device for sampling of first two channels of multiplexed input

Fig. 9-14. The ADC conversion can be initiated simultaneously with the hold command to the T/H amplifier.

The overall data-acquisition frequency f_s (that is, the clock frequency which determines the time interval between each burst of multichannel samplings) is provided by a divide-by-M counter in the diagram of Fig. 9-13. Although not obvious from Fig. 9-14, f_s is going to be small compared to the fundamental clock frequency f_M, which is used to control the timing for the multiplexer and T/H devices. The overall sequence of events which must occur to achieve multichannel data acquisition is as follows: The initiation of a sampling sequence will take place when the output of the divide-by-M counter goes true. Prior to that point in time, the computer will have output an appropriate digital word to select the appropriate multiplexer channel and to clear the external logic. Following that, the encode bit must be set true in anticipation of the initiation of the sampling sequence from the output of the divide-by-M counter. When the divide-by-M counter (or point f_s in Fig. 9-13) goes true, the high-frequency clock pulses are let through to switch multiplex channels and initiate the digitization process after the appropriate time interval. At the completion of the sampling interval from a single channel, the flag bit is set, and the computer can acquire the digitized value. The flag and encode bits are immediately cleared by the computer. After a time interval sufficient for appropriate data handling and bookkeeping, etc., the computer outputs another digital word to select a new input channel and to clear the external logic followed by a command through the encode bit to allow sampling of the next channel at the next f_M pulse. Example 9-4 provides a program segment which might be used for the multichannel data-acquisition process.

Example 9-4   Multichannel data-acquisition program

```
              .
              .
              .
          LDA   DCTST   /SET TOTAL # MULTI-CHNNL SAMPLINGS
          STA   DCNTR
MPX       LDA   CNTST   /SET CNTR TO # MPXR CHNNLS TO SAMPLE
          STA   CHCNT
          LDA   CHNL1   /GET INITIAL CHNNL SELECT
LOOP      STA   CHNNL   /SAVE--CODE
          OTA   ADC     /OUTPUT TO ADC I/O CHNNL (CLRS LOGIC
                          & SELECTS MPXR CHNNL)
          STC   ADC,C   /CONDITION EXTERNL LOGIC FOR STRT OF
                          SEQUENCE
          SFS   ADC     /WAIT FOR DATUM
          JMP   *-1
          LIA   ADC,C   /GET DATUM
          CLC   ADC     /CLR ENCODE
          STA   PBFR,I  /SAVE IN MIXED BFFR
          ISZ   PBFR
          LDA   CHNNL   /GET CURRENT CHNNL CODE
          INA           /GENERATE--NEXT CHNNL CODE
          ISZ   CHCNT   /SAMPLED ALL CHNNLS?
          JMP   LOOP    /NO, GET NEXT CHNNL
          ISZ   DCNTR   /ENUFF DATA ALL CHNNLS?
          JMP   MPX     /NO, RECYCLE
          JMP   PRCSS   /YES--PROCESS DATA
              .
              .
              .
```

9-3   REAL-TIME COMPUTER OPERATIONS

The term "real-time clock" is often used in the literature. This term should be used to refer to a digital clock of the type that we have described earlier, which can provide a time base synchronized with some external experimental event. We have discussed extensively the characteristics of such a clock and synchronization hardware to achieve this objective. Figure 9-15 provides a typical timing diagram which represents the real-time and non-real-time segments during data acquisition from a repeatable waveform. Any computer operations executed during the real-time segment can be considered real-time operations. However, in the most demanding situation a specified sequence of operations must be completed within the time interval between real-time clock pulses. Thus, for the

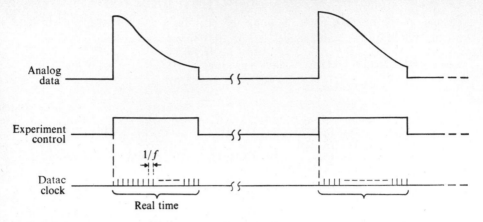

Fig. 9-15   Real-time and non-real-time segments for repeatable experiment.

rest of our discussion here, we will consider real-time computer functions as those which must be executed between clock pulses, assuming that each clock pulse represents a point in time where the computer must devote its attention to the experiment, either for data acquisition or control or both. Also, this discussion will be oriented toward a single-channel data-acquisition system; although extension to multichannel data acquisition will be obvious.

There are many reasons for executing real-time computer operations. For example, the simplest and most fundamental task of the computer in real time must be to handle acquired data. This involves storing the data as well as bookkeeping operations to keep track of where the data go and how much has been acquired. In addition, the experiment may require that some computational tasks be executed in real time. For example, in the ensemble-averaging experiment described previously, it was necessary to execute a double-precision summation operation after each data point was acquired. Other types of real-time computations are common. For example, real-time smoothing of acquired data can be carried out. One approach is to apply an n-point moving average to the data as they are acquired. The algorithm for this is represented in Fig. 9-16. More sophisticated smoothing algorithms can also be carried out in real time, such as digital filtering or least-squares smoothing techniques. Digital filtering might involve an algorithm as simple as looking for and suppressing data excursions which exceed a specified bandwidth. For more detailed discussions of data-smoothing methods, the reader is referred to Refs. 6 and 7.

Another very useful real-time computational function is that of data compression. A simple approach to data compression was discussed in Chap. 4 when a real-time thresholding algorithm was proposed to confine data acquisition in gas chromatography to the region of elution peaks. A more sophisticated data compression algorithm is illustrated in Fig. 9-17. In this algorithm, the computer first establishes a background noise level for the

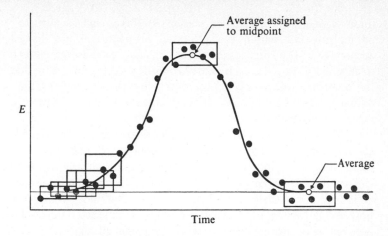

Fig. 9-16 Schematic algorithm for n-point moving average.

experimental data source. This might be established before the start of the experiment by sampling the quiescent analog output. When the experiment is initiated, the computer acquires the first few data points, and an imaginary straight-line segment is projected with an imaginary window on either side of the line equivalent to $\pm S$, the standard deviation of the background signal. If the third data point falls within the window, it need not be saved because its deviation from the projected linear segment does not exceed the noise level. Thus, in the reconstruction of the original waveform, a datum could be inserted at that point which would lie on the projected linear segment. Moreover, the reconstructed data point would have no less significance than the original data point because the difference between the two would never exceed the fundamental noise level. Continuing on in real time, the computer then picks up the fourth data point and may find that it does fall outside the projected linear window. When this happens, the data point is acquired and stored and becomes the new second point of a two-point sequence determining a new projected

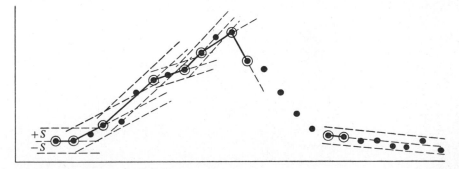

Fig. 9-17 Real-time data compression algorithm.

linear segment. Subsequent data points are taken and checked to see if they do or do not fall within the projected segment and are discarded or saved accordingly. The procedure of continually establishing new projections is maintained during the entire data-acquisition cycle. To ensure that large gaps in the waveform will not occur even during flat base-line regions, the algorithm can be modified to continuously diminish the size of the window in some arbitrary fashion as successive data points are acquired.

The compression ratio obtained by the procedure described above depends on the nature of the waveform, of course, as well as on the method of establishing the window. Obviously, one could arbitrarily fix the magnitude of the window to be larger or smaller than the background noise level to achieve higher or lower data compression ratios. The most important point here, however, is that by such an algorithm it is possible to compress data and save on required core memory space while retaining all significant data excursions. Moreover, the original waveform can be accurately reconstructed from the compressed data without any loss of significant information.

We should now consider the characteristics of *real-time computer control* of experimental systems. Figure 9-18 represents schematically the configuration of an experimental laboratory computer system. Four of the elements in the diagram are typical of conventional laboratory experimentation. These are the fundamental aspects of the experiment itself, the monitoring instrumentation, the control instrumentation, and the scientist who supervises the implementation of these other elements. For example, if we are considering instrumentation for gas chromatography, the fundamental elements of the experiment will include the chromatographic column and the distribution equilibria associated with the passage of a mixture of components through that column, which

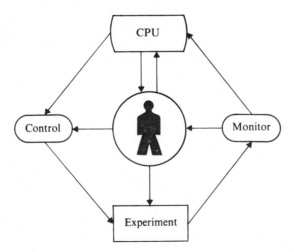

Fig. 9-18   Laboratory computer system.

results in the eventual separation of those components and their sequential elution from the column. The control elements include devices maintaining the column temperature and flow rate; monitor elements include the transducer, which translates some property of the separated eluents into electrical signals, any signal-conditioning electronics, such as current or voltage amplifiers and filters, and the final output device such as a potentiometric recorder. The scientist supervises the overall operation of the experiment by determining the control conditions as well as the column composition and nature of the carrier gas. He can also interact with the experiment to a certain extent by observing the output of one experiment and modifying conditions for the next so that more nearly optimum experimental conditions will exist. Moreover, the scientist can interact with any given experiment in real time if the time scale of the experiment is long enough and the monitor output provides the appropriate correlations to allow intelligent decisions regarding the modification of experimental conditions to improve the experiment. The fundamental limitation here, of course, is the response time of the human as a feedback element.

Figure 9-18 shows an alternative feedback path through the digital computer. Because the computer has a memory, because it can execute computational and logical operations, and because it can execute data-acquisition and control functions, it can act as a pseudointelligent feedback element in chemical instrumentation. Moreover, because of its speed, computerized feedback control can be accomplished on a microsecond or millisecond time scale. Thus, an advantage of several orders of magnitude in the time scale for real-time experiment interaction over human real-time functions can be gained with a computerized feedback loop.

Response characteristics, transfer functions, and stability requirements for instrumentation incorporating computerized feedback loops can be defined by the application of conventional Laplace transform and z-transform analysis [8, 9]. Because sampled-data systems provide discontinuities in the monitoring and/or control loops, the response and stability characteristics of these systems can be quite different than for pure analog systems. Although a detailed discussion of system analysis for sampled-data systems and periodic controllers is beyond the scope of this text, a discussion of response limitations for computerized feedback can be presented in terms consistent with the background presented thus far.

The response limitations for computerized real-time interaction cannot be analyzed similarly to the analysis of analog electronic devices. The digital computer can be considered as an active electronic element in instrumentation, but the electronic characteristics which it displays as a feedback element in instrumentation are unique. For example, the operational amplifier can be programmed to differentiate, integrate, add, subtract, or generate a combination of mathematical functions by providing the appropriate input and feedback elements such as resistors, capacitors, inductors, transistors, etc. Likewise, the digital computer can be programmed to accomplish these mathematical functions. The fundamental difference, however, is that the analog devices generate essentially instantaneous functions whereas the computer must execute a sequence of programmed steps in order to provide the mathematical results. Whereas with an analog operational amplifier, the "power" of that device is related to the voltage *gain*, the "power" of the digital computer as an instrumental element is directly related to the *computational time available*. That is, because the computer can function only by executing a programmed sequence of discrete steps, the computer can accomplish a given computational or control objective only if it has sufficient time available. The greater the amount of time available, the greater the power of the computer to execute computational or intelligent feedback operations.

The amount of computer power available in real time will depend on the *stimulus frequency f*; this is the frequency at which the computer is "poked" by the experiment requesting some external service. In the simplest situation, f is the data-acquisition frequency—the rate at which digitized data are made available to the computer from the experiment. At a minimum, the computer's service, as each datum is made available from the digital data-acquisition system, involves inputting the datum, saving it in memory, and some bookkeeping. Typical *minimum service time τ_M* may be 20 or 30 μsec. The *real-time computer power* can be equated to the *available computational time τ_A* between stimuli as given in Eq. (9-4):

$$\tau_A = \frac{1}{f} - \tau_M \tag{9-4}$$

To provide any useful feedback functions—where some computational evaluation of the data, logical decisions, and possible experimental control operations may be carried out in real time—the value for τ_A must be large enough to accommodate the program execution required.

Certain additional factors related to the system response must be considered. First of all, the discussions here are presented in the context of *dedicated* systems, where the computer is interfaced to a single instrument. This implies that response to stimuli from the particular experiment is the highest-priority activity of the computer. Thus, it is assumed that the execution of the real-time service program begins immediately upon request from the interfaced instrument. In fact, there is some *minimum response time τ_0* required. The magnitude of τ_0 depends on the computer hardware features, particularly the cycle time, and also depends on the specific instruction sequence used to identify and/or respond to a stimulus. However, regardless of whether the computer's interrupt system or program-controlled response is used, τ_0 may be several microseconds. (Moreover, for time-sharing systems, τ_0 may be many milliseconds.) This must be included in estimating τ_M.

Another relevant consideration is that the values for τ_M or τ_A used in any calculations should be the *worst-case* values. That is, where alternative program pathways exist, assume that conditions will always require the longest path.

Some sample calculations　　Consider now how one might use the system response analysis described above for the design of a particular experimental application involving real-time computer interaction. First, the essential elements of the minimal experimental service program—including the required I/O and bookkeeping instructions to be executed for each stimulus—must be established. The time required for these operations and τ_0 correspond to τ_M. The programmer can then calculate τ_A for the specific data-acquisition or service frequency f required in his application. Then, he must establish τ_R, the *real-time computational time* required to provide the required calculations, decisions, and experimental control operations to allow computer interaction with the experiment. The linear combination of the two programming segments results in a *total real-time service time* τ_T, where

$$\tau_T = \tau_M + \tau_R \tag{9-5}$$

If the programming is such that $\tau_R \leqslant \tau_A$, the proposed application is feasible. Alternatively, for a given value of τ_T, one can calculate the maximum data-acquisition frequency allowed. This can be obtained by writing

$$\frac{1}{f} = \frac{1}{f_{max}} = \tau_T \tag{9-6}$$

A specific example should illustrate the above discussion. Assume the computer system has a basic machine cycle time of $2.0\,\mu$sec and most instructions are some integral multiple of this value. For a particular application, program-controlled service is used requiring a worst-case response τ_0 of $6.0\,\mu$sec. The additional minimal service programming requires eight cycles, i.e., $16.0\,\mu$sec. Thus, τ_M is $22.0\,\mu$sec. The desired data-acquisition frequency f is 10 kHz. Therefore, the value of τ_A is computed from Eq. (9-4) to be $78.0\,\mu$sec. Thus, the real-time interactive programming has to have a worst-case execution time τ_R less than $78.0\,\mu$sec. This allows real-time programming requiring no more than 39 machine cycles. Alternatively, if the programmer determines τ_R and τ_T, the limiting data-acquisition frequency can be determined from Eq. (9-6). Thus, for $\tau_R = 140\,\mu$sec and $\tau_T = 162\,\mu$sec, the limiting frequency is about $6,173$ Hz.

It should be noted here that as a feedback element, the digital computer is a relatively *slow* device. In fact, it can often be assumed that analog instrumentation controlled or measured by the computer does not limit the overall system response. Thus, the overall system response and transfer function is often limited by the computer response characteristics discussed here. However, because the computer can provide intelligent feedback, the experimenter can trade speed for sophistication of interaction.

9-4   INTERRUPT-CONTROLLED DATA ACQUISITION

Considerable discussion was presented in Chap. 6 regarding characteristics of interrupt-controlled data transfers in the context of servicing standard peripheral devices. These same general considerations can be applied in the case of interrupt-controlled servicing of experimental systems. In this section, we will consider some of the hardware and software features of interrupt-controlled data-acquisition systems.

Figure 9-19 shows the hardware configuration for a multichannel I/O priority interrupt system. Each external experimental system is connected to the computer through a particular I/O channel. When the computer's interrupt system is activated (by an STF 0 instruction), then any external device can cause an interrupt to occur by setting a flag bit on the particular channel. An interrupt occurs, however, *only* if the control bit has been set previously (with an STC) on that channel. When the computer recognizes an interrupt on a given channel, it subsequently executes the instruction stored in the location whose address is equal numerically to the I/O channel on which the interrupt occurred. (This is inherent in the hardware of the H.P. 2100 family of computers. Details were discussed in Chap. 6.)

The most important advantage of using the interrupt system for data-acquisition applications is that the computer does not have to wait for service requests from the data-acquisition devices. The computer can be

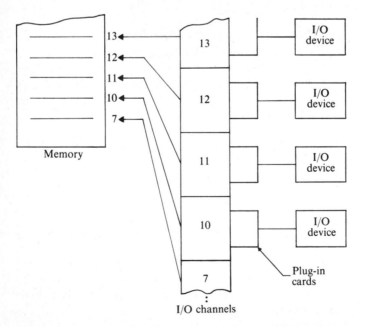

Fig. 9-19   Multichannel hardware priority interrupt structure.

programmed to be executing computational, compiling, printing, or data processing functions most of the time and be able to acquire data upon interrupt request when the data are actually available. Thus, the computer can use its time most efficiently. (See Chap. 10 for more extensive discussions of these points.)

Single-channel data acquisition　　Figure 9-20 presents a schematic description of a data-acquisition system operating under interrupt control. In this example, the data-acquisition hardware provides many of the control and bookkeeping functions previously handled by computer software. This simply illustrates that it is sometimes possible, necessary, or advantageous to handle these functions by external hardware. The experiment and data acquisition here are initiated externally, and the number of data points taken is counted by an external hardware counter. Whenever a conversion is completed, the ADC sets the flag on the I/O channel, causing an interrupt to occur. The computer immediately services the interrupt and takes the digitized data point and stores it in the appropriate memory block. In addition, the computer checks the status of bit 15 on the interface buffer register to see if all data have been taken for this experiment.

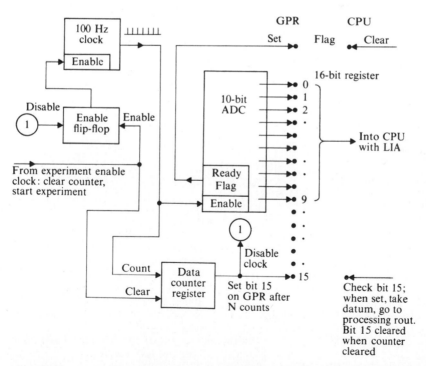

Fig. 9-20　　Interface hardware for typical system with interrupt-controlled data acquisition.

The software required to handle the data-acquisition system described in Fig. 9-20 is given in Example 9-5. The interrupt service instruction is stored in location 10_8, with the assumption that the data-acquisition channel is number 10_8. The interrupt instruction is a JSB *m*,I which provides for entering the ADC service routine. As described in Chap. 6, when the interrupt service instruction is a JSB, the *address* of the next instruction in the main program, which would have been executed had the interrupt not occurred, is stored in the first location of the service subroutine.

Example 9-5   Interrupt-controlled single-channel data-acquisition program

```
              .
              .
              .
         ORG  10B         /ASSUME ADC ON CHANNL 10B
         JSB  .ADC,I      /PUT INTERRUPT SERVICE INSTRUCTION
              .             IN LOC'N 10B
              .
              .
.ADC     DEF  ADSVC       /POINTR TO INTRRPT SERVICE ROUTINE
              .
              .
              .
START    NOP              /START OF MAIN PROGRM
              .
              .
              .
         CLF  ADC         /CLR FLG ON CHNNL 10B BEFORE TURN
                            INTRRPT ON
         STF  0           /TURN ON INTRRPT SYSTEM
         STC  ADC         /PUT CHNNL 10B ON THE INTRRPT LINE
              .           /(BACKGROUND PROGRAMMING)
              .
              .
ADSVC    NOP              /ADC SERVICE ROUTINE
         STA  ASAVE       /SAVE "A"
         STB  BSAVE       /SAVE "B"
         ELA              /GET "E" INTO LSB OF "A"
         STA  ESAVE       /SAVE "E"
         CLA
         SOC              /CHECK "O" REGISTER
         INA
         STA  OSAVE       /SAVE "O"
         LIA  ADC         /GET DATUM FROM ADC
         STA  PBFFR,I
         ISZ  PBFFR
         SSA,RSS          /CHECK MSB OF INPUT FROM 16-BIT REGISTER
                            TO SEE IF THIS IS LAST DATA PT.
         JMP  OUT         /BIT 15 ≠ 1, EXIT NORMALLY
```

Example 9-5 (*continued*)

```
            CLC   ADC      /REMOVE CHNNL 10B FROM INTRRPT SYSTEM,
                             BIT 15 = 1
            (CLF 0)        /(DISABLE INTERRUPT SYSTEM)
    OUT     LDA   ESAVE    /EXIT AFTER RESTORING WRKING REGISTERS
            ERA            /RESTORE "E"
            CLO
            LDA   OSAVE
            SZA
            STO            /RESTORE "O"
            LDA   ASAVE    /RESTORE "A"
            LDB   BSAVE    /RESTORE "B"
            CLF   ADC      /CLR FLG CHNNL 10B LASTLY TO PRESERVE
                             PRIORITY
            JMP   ADSVC,I  /RETRN MAIN PROGRM WITH REGISTERS RESTORED
             .
             .
             .

    ADC     EQU   10B      /DEFINE ADC ON CHNNL #10B
             .
             .
             .
```

The first steps in the ADC service routine involve saving the contents of all the working registers so that whatever computations might have been interrupted in the main program can be continued after the interrupt request has been serviced. After the working-register contents are saved, the digitized datum is loaded into the A register and stored in memory. The status of bit 15 is then checked to see if this is the last data point. If it is, a CLC instruction can be executed to remove that channel from the interrupt system. A CLF 0 instruction may be executed also if the interrupt system is to be disabled at this point. (Neither of these instructions need be executed if the data-acquisition channel is the only one on the interrupt system and the external hardware disables itself from setting flags after the final data point is taken.) Regardless of whether the last datum has been taken, the next steps are to restore the contents of all the working registers, clear the flag on the I/O channel, and then exit to the main program where the interrupt occurred, continuing as if nothing had happened.

If one compares this data-acquisition subroutine with the program-controlled data-acquisition software given in Example 9-1, it becomes obvious that the amount of program execution required for each data point acquired is considerably greater under interrupt control. This is because of the bookkeeping overhead required of the computer in servicing each interrupt. It is interesting to compare quantitatively the two data-acquisition approaches by calculating the maximum data-acquisition rate that could be allowed for each case. This is left as an exercise for the reader, using the instruction execution times listed in Sec. A-1.

One advantage of using interrupt-controlled data acquisition is that the computer can be executing background programs without any concern of missing experimental data. Thus, the computer's time is used more efficiently, but there is some sacrifice in the maximum data-acquisition rate allowed because of interrupt service overhead required.

Direct memory access   It should be obvious by now that on-line data-acquisition rates, whether under program or interrupt control, cannot easily exceed 50 to 100 kHz, assuming a memory cycle time on the order of 1 μsec. However, it is possible to transfer data into or out of memory at rates limited by the cycle time. To take advantage of this speed, computer hardware is required which provides a more direct line of communication between memory and external devices. Such hardware describes a *direct-memory-access* (DMA) channel or *memory port*.

The requirements of DMA hardware include the ability to address memory, maintain a word count, and execute normal I/O operations such as command outputs, buffering, interrupt acknowledgments, and digital word transfer between memory and external devices. These functions must be executed independently of the CPU and bypass the arithmetic registers. Thus, the DMA device could effect I/O transfers by "stealing" memory cycles from the CPU. That is, whenever a DMA transfer takes place, the normal program functions of the CPU are suspended while DMA utilizes a memory cycle for a read or write operation. One or more machine cycles may be required to complete the data transfer to or from the external device.

For the computer system considered here, the DMA hardware can be initialized under program control prior to the initiation of the I/O operation. Initialization includes specifying the particular I/O channel to which the DMA is to be assigned, setting the word count (up to 16K), identifying the first address of the block of memory to be accessed (0 to 32K), and identifying which logical I/O operations are executed with each data transfer (e.g., CLC, CLF, STC, and packing or unpacking of 8-bit bytes). (See Ref. 1 for additional details.) Once the DMA hardware is initialized and the external device activated, no further program attention to DMA is required. The data transfers proceed to completion without any interference to executed programs, except for slowing them down.

Thus, DMA can be used to accomplish very high-speed transfers of data *blocks* between memory and external devices. Such hardware is often required to utilize high-speed peripheral memory devices, such as a magnetic disc. (See Chap. 10.) Moreover, DMA can be used to take on-line data from experiments at high rates. For example, with DMA requiring two machine cycles per data transfer, 16-bit data can be dumped into memory at a 312.5-kHz rate with an H.P. 2116 computer. This rate can be doubled when two DMA channels are used. Other computer systems are available with shorter cycle times providing DMA word transfer rates in excess of 1 MHz.

Time-sharing servicing of multiple experimental systems   The primary utilization of the interrupt system is in developing time-sharing data-acquisition systems. Figure 9-21 provides the schematic diagram of a time-sharing system for data acquisition from four experimental systems. System A is the one requiring the highest service rate and is assigned to the highest-priority channel. System D requires servicing at the lowest rate and is assigned the lowest-priority channel of the four. The software to provide time-sharing service to these four systems is given in Example 9-6. The interrupt service instructions for channels 10_8 to 13_8 are stored in memory locations 10_8 to 13_8. The service routine for system A cannot be interrupted by any of the other experimental systems; whereas the service routine for system D can be interrupted by systems A, B, or C. It is necessary that the CLF instruction be the last instruction executed before

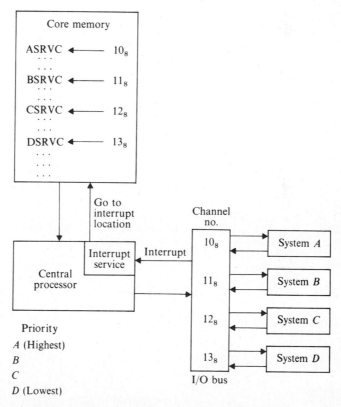

Fig. 9-21   Schematic diagram of time-sharing linkage to four experimental systems.

Example 9-6   Time-sharing program (with priority interrupt)

Four Systems—A, B, C, D. A = highest priority; D = lowest
(using I/O channels 10_8 to 13_8).

Location		Contents	Comments
10_8		JSB .ASVC,I	STORAGE OF INTERRUPT–
11_8		JSB .BSVC,I	FETCHED INSTRUCTIONS FOR
12_8		JSB .CSVC,I	ALL FOUR CHNNLS IN SYSTEM
13_8		JSB .DSVC,I	
•		•	
•		•	
•		•	
50_8	.ASVC	DEF ASVC	/POINTRS TO INTRRPT SERVICE
•	.BSVC	DEF BSVC	ROUTINES FOR EACH DEVICE
	.CSVC	DEF CSVC	
•	.DSVC	DEF DSVC	
		•	
•		•	
		•	
2000_8	START	NOP	/START OF MAIN PROGRM
		•	
		•	/INITIALIZE PROGRM
		•	
		STC 10B,C	/ENABLE ALL SYSTEMS
•		STC 11B,C	
		STC 12B,C	
		STC 13B,C	
		STF 0	/ENABLE INTRRPT SYSTEM
•		•	
		•	
		•	/EXECUTE MAIN PROGRM
	ASVC	NOP	/INTRRPT SERVICE ROUT FOR
•		•	DEVICE A /STORE CONTENTS OF
		•	ALL WRKING REGISTERS
		•	
		LIA 10B	/GET DATUM FROM DEVICE A
(ASVC cannot		STA ASTR,I	
be interrupted		ISZ ATOTL	
by any other		STC 10B	/RE-ENABLE DEVICE IF NOT
device)			THRU
		LDA ESAV1	/RESTORE WRKING REGISTERS
		•	BEFORE EXIT
		•	
		•	
		LDA OSAV1	
		•	
		•	
		•	
		LDA ASAV1	
		LDB BSAV1	
		CLF 10B	/RELEASE PRIORITY ON CHNNL
			10B

Example 9-6 *(continued)*

```
                     JMP ASVC,I    /RETURN MAIN PROGRM
            BSVC     NOP           /INTERRUPT SERVICE FOR DEVICE B
                       .
                       .           /STORE CONTENTS OF ALL
                       .           WRKING REGISTERS
(BSVC can be         LIA 11B       /GET DATUM
interrupted          STA BSTR,I    /HANDLE DATUM
by device A            .
only)                  .
                       .
                     STC 11B       /RE-ENABLE DEVICE IF NECESSARY
                     LDA ESAV2     /RESTORE WRKING REGISTERS
                       .
                       .
                       .
                     LDB BSAV2
                     CLF 11B       /RELEASE PRIORITY ON CHNNL 11B
                     JMP BSVC,I    /RETURN MAIN PROGRM
            CSVC     NOP
(CSVC can be           .           /SERVICE ROUTINE SIMILAR TO A
interrupted by A and B) .
                       .
            DSVC     NOP
(DSVC can be           .
interrupted by         .           /SERVICE ROUTINE SIMILAR TO A
A, B, or C).           .
                     JMP DSVC,I
```

exiting the particular service routine in order to preserve the priority of that particular channel. That is, the priority of the servicing of a given channel is maintained only as long as the flag bit of that channel remains set. As soon as the flag bit is cleared, the computer hardware interprets this as an indication that the servicing of that channel has been completed, and interrupt requests from lower-priority channels can be serviced. If a flag on a higher-priority channel is set while a lower-priority channel is being serviced, that service routine will be interrupted even though the flag bit is still set on that channel. Thus, the service routines of lower-priority channels may be segmented or delayed as illustrated in Chap. 6.

Another feature of the interrupt service routines that should be noted is that each service routine has its own locations allocated to save the contents of the working registers. This is because service routines for lower-priority systems can be interrupted by higher-priority devices.

More extensive discussions of time-sharing experimental systems for laboratory automation are presented in the following chapter. Also, for a comparison with characteristics of interrupt systems using a party-line hardware structure, see Chap. 13 and sec. D-2.

EXERCISES

9-1. (*a*) For the programming of Examples 9-1 and 9-2, calculate the maximum data-acquisition frequencies allowed in each case. (Use the instruction execution times provided in Appendix A, and calculate for H.P. 2116, 2115, and 2100 computers.)

(*b*) With a fixed clock frequency of 10 kHz, how would you modify Example 9-2 to obtain a 2 kHz data-acquisition rate?

9-2. Write a program which will acquire data at a rate of 1 kHz for the first 10 data points, 100 Hz for next 10 data points, and 10 Hz for the final 100 data points.

9-3. For Example 9-5, calculate the maximum data-acquisition rate tolerated by the software. (Assume that an interrupt request is recognized by the CPU and control transferred to the appropriate memory location within two memory cycles. Use the instruction execution times provided in Appendix A.)

9-4. (*a*) Prepare a data-acquisition program which incorporates the following real-time functions:

1. Real-time thresholding
2. Scaling each 10-bit datum by 10_{10}
3. Dividing each scaled datum by 75_{10}
4. Normal bookkeeping, data storage, etc.

(*b*) Using integer arithmetic operations, compute the maximum data-acquisition frequency allowed with an H.P. 2116 computer for the following conditions:

1. Using integer arithmetic subroutines (see Chap. 5)
2. Using whatever shortcut arithmetic methods you can devise (see Chap. 5)

(*c*) What will the maximum data-acquisition frequency be if each datum is first converted to floating point, not scaled, and then divided by 7.5_{10}? (The Float subroutine requires 0.1 msec minimum, 0.3 msec maximum; floating divide requires 0.3 msec minimum, 0.9 msec maximum; values pertain to an H.P. 2116 computer with a 1.6-μsec cycle time.)

BIBLIOGRAPHY

1. "A Pocket Guide to Interfacing Hewlett-Packard Computers," Hewlett-Packard, Cupertino, Calif.
2. Ramaley, L., and G. S. Wilson: *Anal. Chem., 42*: 606 (1970).
3. Bracewell, R.: "The Fourier Transform and Its Applications," McGraw-Hill, New York, 1965.
4. Kobylarz, T.: *Electronics*, 41(8):124 (1968).
5. Horlick, G., and H. V. Malmstadt: *Anal. Chem., 42*:1361 (1970).
6. Savitzky, A., and M. J. E. Golay: *Anal. Chem.*, 36:1627 (1964).
7. Bevington, P. R.: "Data Reduction and Error Analysis for the Physical Sciences," McGraw-Hill, New York, 1969.
8. Barnes, J. E., Jr.: Sampled Data Systems and Periodic Controllers, in E. M. Grabbe, S. Ramo, and D. E. Wooldridge (eds.), "Handbook of Automation, Computation and Control," vol. 1, Wiley, New York, 1958.
9. Luyben, W. L.: *Instrum. Technol.,* 18:58 (1971).

10
Time Sharing—Hardware and Software[1]

10-1 INTRODUCTION

Up to this point, we have confined discussions to applications of dedicated laboratory minicomputer systems. In this chapter, the discussion will be broadened to include larger time-sharing computer systems. Therefore, an introductory comparison between dedicated laboratory computer systems and time-sharing systems is in order. Figure 10-1 illustrates two possible methods of servicing a number of experiments with a computer system. The first is by servicing each experiment with a small dedicated computer with a limited amount of core and a limited number of peripherals per system. In the time-sharing computer system, one processor will service the four experiments. In this system, a computer (which might be classified as medium or large by the scale introduced in Chap. 1) is controlled by some type of monitor program to take data and control the experiments on-line.

[1] This chapter was written by John I. H. Patterson, Department of Biological Chemistry, Hershey Medical Center, Pennsylvania State University, Hershey, Pa., 17033.

Dedicated computer systems

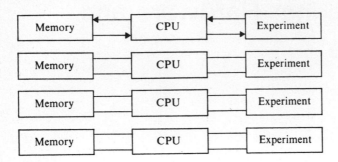

Time-shared computer system

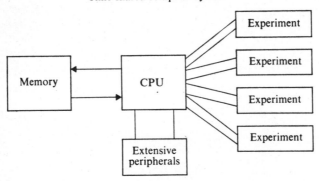

Fig. 10-1   Dedicated and time-sharing computer systems.

The dedicated system has the advantage that the initial expense of placing one experiment on-line is relatively low. However, as the number of experiments placed on-line increases, the expense increases linearly. In the case of the time-sharing system, the initial expense is high, while the incremental cost of adding experiments to the system is fairly small. Therefore, all other things being equal, there is a point at which the cost per experiment on-line will become less for a time-sharing system than for dedicated systems.

The computer which is used for time-sharing types of applications is usually one of greater power than the dedicated system. That is, it will normally have more extensive peripherals and more core memory. This means that more sophisticated programming can be carried out, and the standard I/O can be accomplished faster and easier. However, the fact that the system is shared means that all this power will not be available to one user at all times and it will not be possible to have the very rapid (submillisecond) response of which a dedicated system is capable. In addition, it is possible to have user-user interactions in time-sharing systems which will not occur in dedicated systems.

This interaction can range from two users requesting an I/O device at the same time, thus delaying one of them, to one of the users destroying the system completely, thus necessitating restarting.

In Chaps. 6 and 9, the concepts of interrupt processing and time sharing were discussed. It was pointed out that one of the advantages of using interrupt control is that the computer is able to do useful processing while waiting for an input or output operation to be completed. Time sharing was discussed only briefly in these chapters; this discussion outlined the basic system necessary to service several I/O devices using the interrupt system. This chapter will discuss *monitors*, which are programs that control the flow and/or execution of a series of programs, subprograms, and functions of a computer system, quite often using interrupts and more sophisticated I/O hardware. Monitors include programs such as *schedulers, operating systems, real-time executive*, etc., which are normally supplied by the computer manufacturer and are capable of allocating the resources of the computer system to a number of separate tasks.

Before we discuss the types of monitors available, we should briefly outline the types of environments in which these monitors are used. The most common type of environment is a multiprogramming environment. In the *multiprogramming* environment, a number of programs is resident in the core memory of a computer and/or on a disc memory connected to that computer. The monitor for that unit then schedules the execution of these programs, allowing one to operate until it is superseded by one of higher priority or until it is suspended for an I/O operation or for other reasons. However, at any given instant in time, only one program is operating. The second type of environment which is becoming more common is the *multiprocessing* environment. In this environment, several central processors are connected to the same resources, viz., the memory and I/O devices. In this case, programs may be resident in several processors, and more than one program can be operating at any given time. For example, one program may be handling I/O devices, while calculations are being done in a second processor; or we could have several processors doing calculations or several processors doing I/O at the same time. An example of a multiprocessing computer is the DEC PDP-10. It is possible to configure this computer with two completely equivalent processors which share the memory of the computer and the I/O devices. Multiprocessing has not yet become an important type of environment for the operation of laboratory computers. However, in the future, we will probably see a great deal more of this type of system as the cost of the hardware decreases.

10-2   HARDWARE FOR TIME-SHARING SYSTEMS

In most cases the systems which are to be operated by a monitor, especially a real-time executive, are configured to include a number of I/O peripherals not normally used in a dedicated configuration. Among the most common peripherals added to computer systems for the operation of monitors are disc or drum memories and hardware memory-protect. In addition, it is quite often possible to justify the inclusion of convenient devices such as a card reader and/or a line printer as these devices will keep the system operating efficiently.

Disc memory   A *disc memory* is a device somewhat like a phonograph record and is used to store a large amount of digital data to supplement the core memory storage of the computer. The disc is divided into a number of concentric magnetic tracks, which are further subdivided into sectors. Figure 10-2 shows a very simplified disc, which is divided into three tracks of four sectors each; while real discs contain many tracks each divided into many sectors (50 to 100 per track). It is possible to store a fixed number of bits of information in a given sector, and there are a fixed number of sectors per track. In order to retrieve a block of data, therefore, it is necessary to know the track and sector in which that block is stored.

The disc is rotated at high speed (typically 1,800 rpm) to bring the segment of the disc desired under a *read/write* head. *Moving-head* discs have one read/write head for the disc, and this head is physically moved back and forth to position it over the track desired. The length of time it takes to position the head over the track and sector desired—the *access time*—is determined by the speed at which the head can be moved and the time it takes for the sector to move under the head. (Typical average access times are 50 to 100 msec.) A second type of disc is one which has one head for each track. This is known as a *fixed-head* disc; the access time is determined by the rotational speed of the disc and does not depend on any movement of the head. Thus, the access time is smaller than for moving-head discs, with the average access time being equal to one-half the rotation time (about 10 to 20 msec). Once the datum of interest is in place under the head, it is possible to transfer it at a very high rate—on the

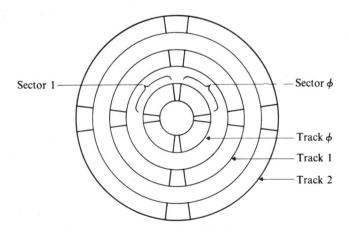

Fig. 10-2   Basic disc layout.

order of several million bits per second. Thus, a disc can be used to store a large amount of data which can be accessed and read at relatively high speeds. The storage capacity of a disc memory is essentially unlimited as multiple units can be implemented. However, a single disc may typically have a maximum capacity of 20 million bits; although discs with greater or considerably lesser capacity are available.

The disc hardware normally includes a write-protect circuit, which, when activated, makes it impossible to write on the area protected. This area is normally the low-numbered tracks, i.e., track 0 to some track determined by the user. The hardware protected area is used to store information and programs which will be *read* only and never changed.

The system is prepared with the write-protect circuit turned off, thus allowing these tracks to be written with the system information. Once the configuration of the system is completed, the write-protect feature is turned on, thus protecting this system information no matter how drastic a mistake is made by the system.

Memory-protect Core memory-protect is very similar in concept to write-protecting an area on a disc. Here, hardware is added to the computer which will protect an area of core memory, thus allowing that area to be read but not written or entered from other areas of core. When an attempt is made to jump into a protected area or write in this area, a memory-protect violation occurs and an interrupt is caused.

There are two types of memory-protect commonly used in laboratory systems. The first protects a contiguous segment of core from the first core location to an address specified by the software. The second protects noncontiguous segments of core. It may protect several areas with the first and final address of each specified by the software; or sometimes core is broken down into blocks, and the software indicates which blocks are protected and which are usable. Memory-protect also provides the ability of determining if an I/O instruction is executed. In some cases, this causes a memory-protect-violation interrupt, or it may generate a special interrupt, which indicates that an I/O instruction was attempted. The system software will process this interrupt to decide if this was an illegal instruction, in which case it will abort the job. In some cases, the software will allow the instruction to be executed. In all cases, the memory-protect feature will generate an interrupt if a program attempts to execute the Halt instruction as the system cannot be allowed to halt when operating in a time-sharing mode.

10-3 TYPES OF MONITORS

There are several types of monitors which are used in multiprogramming environments that will be discussed in this chapter. The first is a simple *scheduler*, a computer program which controls the sequence of execution of a series of other programs, usually core-resident, on a common system time base. The next is the *operating system*, which is a system normally used in conjunction with a mass storage device, such as a magnetic-tape unit or a disc, which controls the flow of a series of jobs through a computer. These systems usually operate until one job is completed or suspended before starting the execution of the next; thus, they are not truly multiprogramming systems. Finally, we will discuss the *real-time executives* in some detail. The real-time executive is a program which, operating in conjunction with a mass storage device, controls the flow and execution of programs on a priority basis, performs all standard I/O operations, and allows all time not used for foreground program execution and upkeep of the executive to be used for background operations, i.e., compilation and execution of programs which do not have a critical time factor.

Monitors for use in laboratory computers are mainly an offshoot of monitors developed in large computing facilities to eliminate the intervention of humans in order to keep the system operating at its highest possible efficiency. For example, if the operator had to take some simple action which took 1 sec, the computer would waste 100,000 to 1 million machine cycles or more. What decision could not be programmed into that many steps? In the smaller laboratory computer, this operator response time is not as critical as in the larger computation-center-type computer. However, as the size of the computer facility becomes larger, the capital investment increases, and this becomes a more important consideration. Also, if several users are operating with one central processor, it may become impossible for the computer to be halted for operator action as some experiment may need essentially instantaneous and/or continuous response. For this type of system, some type of monitor is an absolute necessity.

In many cases, the monitor which is supplied for a computer will be written in such a way as to simplify the I/O operations of the given computer system. These I/O system monitors standardize the format for the I/O operation of the given computer and in most cases provide device-independent I/O operations. In some cases, especially in a real-time executive, the I/O monitor is written such that only the I/O done by the executive is legal and any other attempt at I/O processing will result in an error. It should be noted that in most cases for *standard* I/O, such as Teletype printer, line-printer, disc, etc., this greatly simplifies the I/O operation for the computer programmer as he needs only call the appropriate subroutine to do the I/O operation.

10-4 SCHEDULERS

As defined earlier, a scheduler is a program which controls the execution of a series of subprograms on a common time base. This means that the scheduler is the main program in core and allocates the computer resources, mainly the central processor, to the subprograms on a schedule which the user establishes. The scheduler is run on a system clock. That is, a time-base generator causes interrupts, and the scheduler updates a time list to keep track of the needs of the various subprograms. Usually, a priority is established for the execution of these subprograms; and when it is necessary for a high-priority program to run while a lower-priority program is operating, it is the task of the scheduler to suspend that job and give the central processor to the higher-priority program. This involves saving the status of the various registers and placing the lower-priority program on the suspend list.

While schedulers are the heart of a real-time executive, it is possible to operate a scheduler in a stand-alone environment with a time-base generator being the only special hardware required. It is possible to operate schedulers with no memory-protect option, but when this is done, it becomes the responsibility of the users to ensure that there will be no user-user interactions.

Figure 10-3 is a flowchart for a very simplified scheduler, which will be considered in detail here. First, it is necessary to set up some sort of a table which contains the schedule, that is, the time at which the given programs are to be executed. This schedule will usually indicate in some manner the priority of the programs. For example, if a program appears high in the schedule, it has high priority. Lower programs in this schedule have lower priority. The flowchart shown in Fig. 10-3 starts by entering on an interrupt from the time-base generator. As usual with interrupts, it is necessary to disable the interrupt system and save the working registers.

Once this has been accomplished, it is necessary to check the schedule to find any programs which need to be run. This may be accomplished by updating all the timing counters and checking for zero time remaining, or it is possible to compare the current time with the time at which the program is to execute and check for exact comparison. If no program needs service at this time, the next function is to determine if any program was running when the interrupt occurred. If no program was running, it is necessary merely to enable the interrupt system and go to a wait loop. If, however, a program was running when the time-base generator interrupt occurred, it is necessary to restore the working registers, reenable the interrupt system, and return to the job which was interrupted. If a program is found in the schedule which needs to be serviced and no other program was running, the new job is set up and started. However, if another program was running when the new program needed servicing, it then becomes necessary to determine whether the currently running job is of higher priority. If the currently running job is of higher priority, then the scheduler will merely restore the registers, reenable the interrupt system, and continue working on the present job. If the current job is not of a higher priority than the job which needs to be serviced, it is necessary to suspend the current job. This is accomplished by saving the working registers of the computer and placing the current job on a suspend list.

When a job is completed, it is necessary to return to the monitor to check

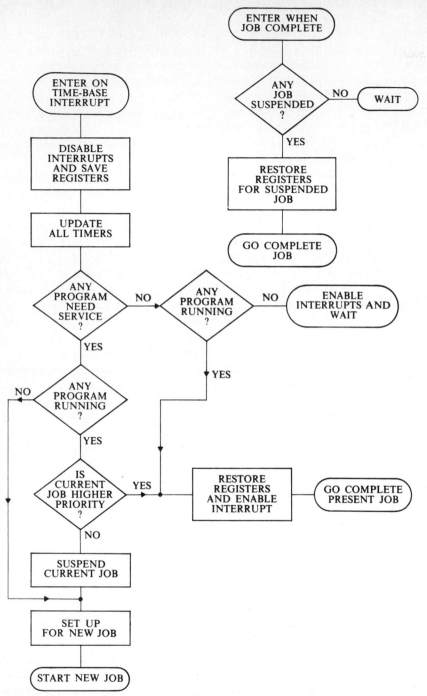

Fig. 10-3 Scheduler flowchart.

274

for jobs in a suspended state. If any job was suspended, it then becomes necessary for the scheduler to restore the working registers for the suspended job and return to that job.

Some simple refinements can be made to this basic scheduler to give the scheduler a great deal more flexibility. For example, an *abort timer* can be added; that is, it is possible to place an additional **time in** the schedule table which indicates how long the program should be allowed to run. If the program runs for a time longer than the allowed time, it is aborted, and other programs of lower priority are allowed to run.

Another possible refinement for a scheduler is the capability of on-line scheduling. Here the operator is allowed to communicate with the monitor to change the time at which any program is to be executed, or a program can decide, possibly on the basis of external events, to change the time at which it or some other program is to be run. Another refinement which is possible is the addition of a *phase time* at the beginning of the operation of this schedule. This concept is illustrated in Fig. 10-4. Here there are three jobs scheduled, each one with a *period*—that is, the time interval between runs—and a *phase*—that is, the time of the first run relative to time 0. For example, job 1, which would be the highest-priority job, is to be run every second. It will be run immediately upon starting of the schedule—that is, its phase time is 0. Job 2 has a period of 0.5 sec; it will be run twice every second. However, it will not be run until 1 sec has elapsed since its phase time is 1 sec. Looking at the time line, we notice that every other time job 2 tries to operate, job 1 will be operating. Because job 1 has a higher priority, job 2 will be delayed in execution for as long as is required for

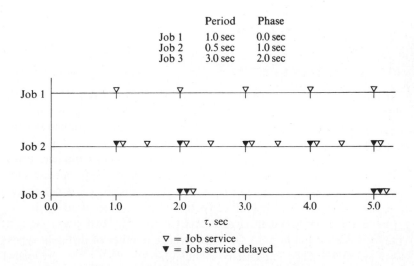

Fig. 10-4   Period and phase timing.

job 1 to be completed. Finally, job 3 has a period of 3 sec and will be executed 2 sec after time 0, or it has a phase time of 2 sec.

Schedulers eliminate the need for the users' programs to handle the timing of the experiments which are on-line. By separating the data-acquisition and the data-reduction routines, it is very easy to control several instruments simultaneously. In this type of system, the data-acquisition programs are given higher priority than the data-reduction programs, allowing the data-acquisition routines to interrupt data processing routines. In this system, it is possible to schedule the data-reduction routines to process a fixed number of data points. For example, a data-acquisition routine can be scheduled 10 times per second, and the data-reduction routine can be scheduled every second, providing 10 data points each cycle. Also, it is possible to schedule the data acquisition for several different experiments on a staggered time base. It should be noted that this type of system is operable only if the data acquisition can be placed on the internal time base of the computer rather than on an external time base.

Hewlett-Packard's DACE (Data Acquisition and Control Executive) is an example of a scheduler which is available commercially. It is especially designed to schedule a number of data-acquisition subroutines. The special hardware which may be included in the system is a time-base generator, a multiplexer, and an integrating, digital Multimeter which is used for digitization. The software includes a scheduler such as described above and the subroutines which will control the special hardware. The scheduler in DACE has a priority system but will not interrupt one program to start another. The priorities are important only when several subroutines need to be executed at the same time. Then, the highest-priority subroutine is executed, and the execution of a lower-priority subroutine is delayed. The DACE system has provision for on-line changing of the time at which the program is to be executed. Also, it has provision for the on-line modification of a table of variables in each subroutine.

10-5   OPERATING SYSTEMS

Operating systems, as defined above, are systems which control the flow of a series of jobs through the computer. These systems are normally used in conjunction with a mass storage device, such as a magnetic-tape unit or a magnetic disc. This definition covers a large number of types of systems. However, the ones with which we are concerned at this time are those which operate in a sequential manner, i.e., systems which will operate one job to completion before starting the operation of another job. These operating systems are the direct descendants of early operating systems used in the computer systems typically found in large computer centers. Most operating systems of this type have a batch mode in which a large number of different types of jobs such as FORTRAN compilations, calculations, assemblies, etc., are loaded into a standard input unit (e.g., a card reader) and then performed in a sequential manner. This type of system usually is used for jobs such as assembling and compiling programs and then for executing the programs, in a *load-and-go* fashion. That is, once the machine language is generated by the Assembler or Compiler, it is combined with the appropriate subroutines, loaded into the computer, and executed. This type of system is very good in cases where a large number of assemblies and compilations have to be made or in systems where a large number of computational programs are to be run. However, it is not usable in cases where the programming is time-critical, such as for data acquisition or real-time control.

Most of the companies which now supply the minicomputers and discs supply some type of disc operating system. For example, Hewlett-Packard supplies a disc operating system which will operate with its discs. This system requires not only the use of the disc but also the use of memory-protect features, memory-parity features, and a time-base generator. Digital Equipment Corp. also supplies disc operating systems with its minicomputers. These systems require a time-base generator. In most cases, the main function of a time-base generator is merely for bookkeeping operations. That is, the system will keep track of the length of time which it takes to compile and/or execute a given program and print an indication of this for the programmer's information. This information is also used in cases where the computer is a service facility and it is necessary to bill the users.

Some computer manufacturers also supply operating systems which will operate in conjunction with a magnetic-tape unit. These systems have many of the same types of features as the disc operating systems. However, in general they are not as complex as the disc operating systems. For example, the Hewlett-Packard magnetic-tape operating system does not require the use of the time-base generator or memory-protect features. However, these systems do not have the same power as the disc operating systems partly because the magnetic-tape units are a slower form of mass storage device. That is, it is not possible to provide random access to information in the magnetic-tape system.

A recent development in operating systems is the use of cassette magnetic-tape storage devices. In these systems, a cassette unit with one or more tape drives is used. If more than one tape drive is used, it is possible to obtain more rapid access to the information desired as all information of one type can be stored on one tape; thus it is not necessary to move back and forth on one tape to access several types of information. For example, if a two-pass compilation is to be performed, the first and second pass of the Compiler could be stored on one tape, the source on a second, and intermediate and final binary on a third. Here, the first drive would read to the end of the first pass and stop at the beginning of the second pass. The computer would then read the source from the second tape and place the intermediate on the third tape rather than reading a section of the source and spacing to the blank tape as is necessary for a one-drive system. Finally, when the second pass is to be started, it is located on the first tape, and it is not necessary to search for it.

The reason for the use of operating systems is twofold: First, when the system is not being used in the batch mode, the system provides ease of operation. For example, once the source program has been entered, it is normally not necessary to handle the source medium (e.g., paper tape or punched cards) at any time during the further compiling and execution of the program. When the operating system is being used in a batch mode, the main advantage is that the system can greatly increase the throughput of the computer. That is, a larger number of jobs can be processed in a given period of time with very little operator intervention. The more powerful operating systems, such as the disc operating systems, have the added advantage that they will simplify the I/O of the computer. In the advanced systems, capabilities are normally included to handle file manipulation by file name rather than worrying about the location of a given segment of information on the disc. This means that when the programmer needs to call a given set of information on the disc, he merely needs to know the name under which that data is filed rather than its physical location on the disc.

In a laboratory environment, the advantages which can be gained by implementing an operating system cannot generally justify the expense of the capital equipment necessary. However, if a mass storage device is attached to the computer, it is often advantageous to be able to implement an operating system. Also, if a heavy load of compiling, assembling, and calculations exists, it may be a good idea to have an operating system which can be used in the batch mode to increase the throughput of the computer. However, the operating-system environment is usually not well suited for any type of on-line operation.

10-6   REAL-TIME EXECUTIVES

The real-time executives, which are programs to control the flow and execution of other programs in a computer on a priority basis, are the most complex form of monitors currently in use in the laboratory environment. They include the scheduler functions as well as a number of utility functions which all operate in real time. The definition of real time can be ambiguous. (See Chap. 9.) For a system to operate in real time, it is necessary for the system to be capable of responding to changes in the environment quickly enough to affect the environment. Thus, real time is relative to the rate of change of the environment. For example, the system must respond in the millisecond range to affect an event which is completed in 1 sec; while a system can respond in the second range and still be considered real-time if the event occurs over several minutes.

All real-time executives are multiprogramming systems and must provide for inputs from several users at the same time. That is, several operators and/or experiments can be inputting information apparently simultaneously on several different input devices using the I/O routines in the executive. It is necessary for the executive to control all I/O so that a program will not input or output information at a given device until any previous program has completed its operation on that device.

The real-time executives which are commercially available require considerably greater investment in hardware than is required in dedicated laboratory systems. Any real-time executive used in conjunction with a laboratory computer requires a disc; some require a fixed-head disc for the rapid access; while other systems will operate using moving-head discs, which have longer access times. In addition, because executive programs can be quite large, these systems generally require a minimum of 16K of core memory. Considerably more memory may be required if the system is to operate effectively. If the computer does not contain a hardware priority-interrupt-type I/O system, it is necessary to add this feature. In addition, a direct-memory-access (DMA) channel is always required to transfer data directly to or from core memory without disturbing the central processor so that the programs and information can be moved in and out of core memory while the processor is operating other programs. Also, a hardware arithmetic unit is generally used to speed up the operation of the system. In all cases, the hardware memory-protect feature is required so that it is not possible for one user's program to destroy other programs or, more importantly, the executive program. Of course, a time-base generator is necessary so that there will be an internal time base on which the scheduler of the executive will operate.

Figure 10-5 shows a typical allocation for core memory of a computer operating in a real-time executive environment. Memory is broken into four main areas. Figure 10-5a is the heart of the executive system, which must always be resident in core to process executive requests and interrupts. It includes the scheduler, I/O routines, and other executive programs. Figure 10-5b is the area in which the real-time core-resident programs reside. The programs which are placed in this area are always core-resident, and the core which they occupy cannot be used for any other purpose. A routine is placed in this area only if it is necessary for that program to respond to an event very rapidly, usually within 50 to 100 msec. Having the program in core at all times means that it will never be necessary to move that program from the disc into core, a process which takes many milliseconds. Normally, the programs which are core-resident are of high priority so that there will be very few events which will prevent the rapid execution of these programs at the time which they are scheduled. Figure 10-5c is the area for the real-time disc-resident programs. A large number of programs can be assigned to this area as these programs reside on the disc and are brought into core (*swapped*) only when **they** are needed. These programs are normally of lower priority than the programs which are core-resident, and it is possible for the execution of these programs to be delayed until they can be moved into core.

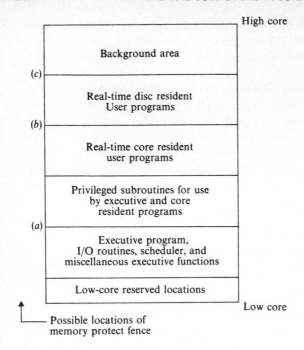

Fig. 10-5   Real-time executive core map.

A logical series of events might be as follows: (1) An experiment causes an interrupt, which is recognized by the executive, and a data point is acquired using executive I/O routines. (2) The data point is transferred to a core-resident program, which is able to make a real-time decision based on that point. (3) Once several data points are collected, the core-resident program calls the disc-resident program to take the data and file them or do other more sophisticated data reduction. Notice that the interrupt is recognized rapidly by the executive program which processes it, and then the data are transferred to the core-resident program after some time delay. The core-resident program then transfers it to a data processing routine when the time becomes available.

The final area of core memory allocated is that used for background processing—such as assembling, compiling, and calculations which can be carried out without concern for the length of time taken to complete the task, i.e., non-real-time tasks. This is the area in which work is done while none of the real-time programs are running. The operations in this area are of lowest priority.

The area of core which is protected by hardware is variable and depends upon the type of program operating. When operations are being done in the background area, all other areas of core are protected. If a real-time disc-resident program is operating, only the areas of the core-resident programs and the executive are protected unless it is possible to protect noncontinuous blocks of core. In the latter case, all areas except the disc-resident area are protected. The memory-protect is disabled only when the executive is in control of the computer. As mentioned earlier, for some types of memory-protect hardware, this is the only time during which an I/O instruction is valid. This means that only routines which are incorporated into the executive can perform I/O functions. If a program attempts an I/O instruction outside the executive area, an interrupt is generated, and the executive interprets the illegal instruction, takes some action to correct this, or aborts the program.

Figure 10-6 shows what might be a typical allocation of the disc space for a real-time executive system. The first area is a *bootstrap* loader, which can be easily loaded and will in turn load the system. The next two areas are core images of the executive programs and the core-resident programs which are used to restart the system when necessary. The real-time disc-resident programs are usually placed on a disc in some relocatable format so that the executive can load them into any free area of core memory within the disc-resident area when they are called. The first four sections of this system are placed on the disc when the system is first configured, and all are protected against writing by the disc hardware. The next area contains background programs which are to be saved, files which are to be saved, and any real-time disc-resident programs which are added after system generation. This area is software-protected by the executive, which keeps track of those areas and will not allow other users to write in that area. The remainder of the disc is used as a *scratch* area, which is allocated by

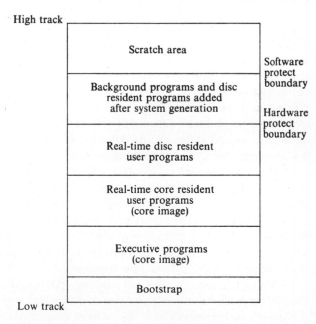

Fig. 10-6   Real-time executive disc allocation.

the executive upon request of any program. This area can be used for intermediate storage of calculations or compiling and assembling.

The software functions which are included in the real-time executives are numerous and complex, and every system has a different set of functions. However, the general concepts are similar and worth discussing. The most important function which is accomplished is the scheduling of programs. The methods used are similar to those discussed in the section on schedulers. However, the algorithms presently being used are much more complex.

Most real-time systems have several lists or *queues* in which a program can be placed depending upon its status. The most important queue is the *execute queue*, in which the programs are placed when they are ready to be executed. A program will be placed in this queue as a result of a request which may be generated when (1) the originally assigned time of execution matches the system time, (2) the operator requests the execution of that program, (3) the program is called by another program, or (4) the hardware interrupt calls for that program. The programs in the execute queue are removed from this queue and allocated the necessary resources on the basis of their priority. However, the priorities in this queue must be dynamic; that is, the longer a program has been in the execute queue, the higher its priority must become, or the low-priority programs might never run.

Other queues include the *I/O queues* and the *dormant queue*. Once a program starts executing, it may request an I/O operation. If the I/O device is free, the transfer is initiated, and the program is moved to the *I/O wait queue*; it remains there until the I/O operation is complete, at which time it will be moved back to the execute queue. If the I/O device is busy when a program requests its use, the program will be placed on an *I/O device wait queue*. It will remain there until that device is free, at which time the highest-priority program in the I/O device wait queue will be assigned to that device. If a program is not in any other queue, i.e., inactive, it will be placed in the dormant queue, which records the location of the dormant programs for later reference. It should be noted that in most cases, actions and events do not result in immediate action but rather result in the program being moved to the appropriate queue.

Real-time executives contain subroutines which will perform all I/O operations for the system and users. These routines are usually *device-independent*; that is, a device is called by a logical unit number. This logical unit number is normally associated with a particular device. However, when that device is "down" or going to be in use for a long time by one user, it is possible for the operator or the system to assign that logical unit number to a different device which can perform the same function.

There are several ways of handling I/O operations for the slow-speed devices and removable media storage devices which require operator attention, and a given executive may have one or more incorporated in its I/O routines. The first is simply to place programs which request a given device in an I/O wait queue and process these requests as the priorities and time available dictate. Second, the executive may assign a particular device to one program and not allow any other user access to that device until the program to which it has been assigned releases it. In this case, a program which requests an unavailable device will be informed that that device is unavailable, and the program can then take any appropriate action, e.g., look for another output device which will meet the needs or suspend itself until that device becomes available. Another method of processing I/O for this type of device is to buffer the output using a rapid access device such as a disc. Here, instead of waiting for the device to become free, the executive will transfer the output file to a buffer on a disc, i.e., *spool* it onto the disc. This buffer is then removed from the disc, *unspooled*, and then output to the device by executive I/O routines as the time becomes available. By using this process, it is possible to respond to output requests rapidly, thus leaving programs in an I/O wait queue a minimum amount of time. It should be noted that spooling is useful only if the output information is not needed immediately; for example, it would be of little value to output alarm messages if they were to sit on a disc for several minutes.

As indicated above, real-time executives normally require DMA channels (see Chap. 9) which allow I/O transfer between an external device and core memory without disturbing the working registers of the central processor. This allows the hardware to set up information for a job which is to be executed when the current job is completed or suspended, thus saving a great deal of time in the setup process. Two basically different types of DMA channels are available. First is a DMA channel which steals cycles from the main processor and uses the computer I/O buses and clock to accomplish the transfer without disturbing the working registers of the computer. This has the advantage of eliminating the need for interrupt-servicing overhead, thus requiring a small amount of computer time to accomplish the transfer. The second type is one which is completely independent of the central processor, in which case no cycles are stolen from the main processor but rather the transfer is made directly to core using special buses. The latter type of DMA channel is preferable as it in no way interferes with or dilutes the power of the central processor.

In some executive systems, the hardware will not allow user programs to perform I/O operations or process interrupts. This hardware places an "authoritarian seal" on the computer, and the programmer cannot manipulate the system or the I/O. The user is forced to allow the executive to handle all I/O operations, even for on-line data acquisition and experimental control. However, when the user sees the computer hardware "through the executive software," the software becomes part of the system; and as long as that software performs the operations the user needs at the speed needed, the programmer should not be concerned that the operations are done by the executive software. The only time a problem arises is when it is necessary to perform I/O operations at speeds greater than the ability of the I/O routines in the executive. In this case, it becomes necessary to add a program to the executive which can perform the desired I/O operation at the speed necessary. These routines must be located in the executive I/O area because this is the only place in which I/O instructions can be executed without causing hardware interrupts which indicate that illegal I/O operations have been performed.

The task of writing an I/O routine which can be added to the executive is a formidable one as it is necessary to conform to the rigid structure which is imposed by the system. There are normally status checks which must be made before any I/O can be executed and status information which must be returned to the executive upon completion. Some systems, on the other hand, allow the user to do I/O and interrupt processing, but, here, it is necessary for the user to check the status of the system and ensure that his I/O does not overlap with other users, thus making the I/O programming more complex and the execution times longer than for dedicated computer systems.

10-7   APPLICATIONS OF REAL-TIME EXECUTIVES

The most common applications of real-time executives in laboratory environments involve cases where a large number of relatively slow instruments are to be controlled. Here, it is possible for the real-time executive to react rapidly enough to acquire data from these slow instruments and, in some cases, control the instrument being operated. It is possible to tie a small number of relatively rapid instruments to a system of this type if the provision is made that only one of these instruments will be on-line at a given time. In this case, most of the computer's resources can be devoted to the data acquisition and processing for one relatively rapid device. An excellent example of this has been reported in the Mulheim computer system.[1] In this system, a PDP-10 computer was used to control a large number of instruments at a slow speed. Inputs from these instruments were handled at speeds less than 20 Hz. Only three fast instruments were attached to the computer. These instruments were allowed to operate at 1.25 to 20 kHz, and only one was allowed to be on-line at a time. However, run times were on the order of 2 to 3 sec; so the computer was not dedicated to any one for a long period of time. On-line control was not attempted for any of these instruments as the response time was fairly slow.

Another application for which a real-time executive is useful is where relatively few instruments are placed on-line but the data reduction for these instruments requires a large amount of complex calculation and a large amount of I/O time. In this case, the resources of the computer are devoted mainly to the data reduction with only a small amount of time devoted to data acquisition.

---

[1] *Anal. Chem.*, **42**(9):51A (1970).

10-8 TIMING CONSIDERATIONS

In order to indicate the complexities of timing considerations in a time-sharing system, we will discuss a specific example which illustrates many of these considerations. Table 10-1 shows a system which includes a mass spectrometer, an electrochemical experiment, a nuclear magnetic resonance (nmr) experiment, 10 gas chromatographs (GCs), and 20 Teletype printers. However, before we can discuss this system, several general timing concepts must be considered.

I/O devices are divided into two basic classes: The first type is the synchronous device, i.e., one which operates such that once it is started it will present data to the computer at fixed intervals in time, regardless of the computer's action. For example, when a magnetic-tape unit is told to read a section of information, it will present data at a fixed rate; if the computer cannot process the data, some of the data will be lost. The second type of device is the asynchronous device, which will process an I/O operation at a rate determined by the computer (up to the device's maximum rate). For example, a Teletype printer will input data one character at a time and will not allow a second character to be read until the computer indicates that it is ready.

Analog experimental inputs should be classified with the synchronous devices as the data are made available at a fixed rate which may not be controlled by the computer. If the data are not acquired at the precise times desired, the stored signal may be a distorted representation of the real signal.

Table 10-1 Timing considerations for time-sharing real-time I/O

Device	Priority	Single service time	Interval[a] between requests	Service time per second
Mass spectrometer	1	100 μsec	200 μsec	500 msec
Electrochemistry	2	150 μsec	500 μsec	300 msec
nmr	3	200 μsec	20 msec	10 msec
10 GCs[b]	4	2.0 msec (250 μsec/GC)[c]	100 msec	20 msec (2.5 msec/GC)
20 Teletype printers (multiplexed)	5	1 msec maximum[d] (50 μsec/unit)	5 msec maximum[d] (100 msec/unit)	10 msec maximum[d] (500 μsec/unit)
				Total 840 msec (i.e., 84% foreground time)

[a]The interval between requests is the reciprocal of the service frequency.
[b]In this system, all GCs will be serviced as the result of a single interrupt. (See text.)
[c]This time represents time which would be necessary to service each GC individually. (See text.)
[d]This figure represents the maximum average rate necessary for servicing the Teletype printers, i.e., the average rate necessary to operate all 20 Teletype printers at 10 characters per second.

When assigning priorities to the devices which are to be time-shared, it is necessary to consider both the *class* of device, i.e., synchronous or asynchronous, and the *rate* at which it is operating. All synchronous devices should be given priority over asynchronous devices so as not to lose data. (This explains the assignment of a higher priority to the GCs than to the Teletype printers in Table 10-1.) In each class, the priority should be assigned on the basis of the transfer rate, with the fastest device given highest priority followed by the slower devices in descending order. Occasionally, exceptions will be made to this priority order for such things as alarm devices, which will be given priority over all types of normal I/O.

Table 10-1 lists the devices which are to be serviced by a hypothetical time-sharing system. The highest-priority device in this system is a mass spectrometer, which requires data acquisition at 5 kHz and 100 μsec to process each data input. Thus, while the mass spectrometer is on-line, the computer spends half the total time available acquiring data from it. The second device is an electrochemical experiment, which requires data acquisition at 2 kHz and 150 μsec to process each data point. A look at the timing diagram in Fig. 10-7 shows that this time is never available as one continuous block if the mass spectrometer is on-line simultaneously, as the mass spectrometer will always interrupt this service. In some cases, it is necessary to wait 100 μsec before the servicing can be started. In this case, it is necessary to buffer the input with a sample-and-hold or a digital buffer following the ADC to avoid acquiring a data point 100 μsec later than it was supposed to be acquired. If this is not done, a

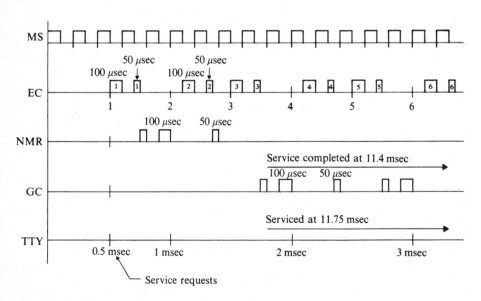

Fig. 10-7   Timing diagram for Table 10-1.

time error of as much as 20 percent ($100\,\mu$sec/$500\,\mu$sec) can occur. Whether a timing error occurs or not depends on the nature of the data-acquisition system. That is, do all experiments share one ADC? How fast is the ADC? Is each experiment sampled precisely with a track-and-hold (T/H) amplifier? Do the high-speed devices have their own ADCs or digital buffers?

The next instrument on-line is an nmr experiment, which requires data acquisition at 50 Hz and $200\,\mu$sec to process each point. For this instrument, the worst-case delay will be $350\,\mu$sec until the beginning of servicing the interrupt; this will result in a 1.75 percent time error if the input is not buffered.

The final instrumental I/O connected to this system is a set of 10 GCs, which are to be serviced sequentially. This means that the data acquisition is placed on the computer time base and every 100 msec a data point is acquired from each GC sequentially via a multiplexed ADC. This has the advantage that all 10 GCs are processed on one interrupt rather than 10 individual interrupts. If each one were treated individually, it would require $250\,\mu$sec per GC; however, in this case, it is possible to perform the overhead operations (e.g., saving the status of the working registers) only once, thus allowing all GCs to be processed in 2.0 msec rather than 2.5 msec. Figure 10-7 shows the delay involved in processing the GC data. The processing will start 1.25 msec (the worst case is 1.35 msec when all interrupts occur simultaneously) after the interrupt occurs, and the processing is completed 10.9 msec after the interrupt, which will result in up to a 10.9 percent error in time. If this amount of error is unacceptable, it is necessary to provide a T/H to buffer the signal from each GC.

Finally, the system is to operate 20 Teletype printers for data I/O. These Teletype printers have an average maximum transfer rate of 200 Hz and burst speeds even larger; however, they are placed lowest in the priority scale because the computer can control the rate of transfer on these devices by not allowing data transfer until it has processed the previous transfer. In the worst case, this means that the Teletype printers will operate at speeds slightly slower than their maximum rate of 10 characters per second, but no data will be lost.

The timing diagram of Fig. 10-7 illustrates the problem of having several high-speed devices on-line at the same time. In the example discussed above, no allowance has been made for the executive functions, executive I/O, and/or background processing; however, we have accounted for 84 percent of the total time available. If one of the two high-speed devices is inactive, this percentage will drop to a much more reasonable value, and the experimental I/O can be accomplished with far fewer problems. If, for example, the electrochemical experiment were inactive, the foreground processing would drop to 54 percent, and the experimental I/O would be handled as shown in Fig. 10-8. This shows that the processing of the GCs will be completed in 4.3 msec (the worst case is 4.4 msec) after the interrupt, an improvement of better than 2.5 msec over the case where two high-speed experiments were on-line simultaneously. If the mass spectrometer is inactive, the time response will be even better as the foreground

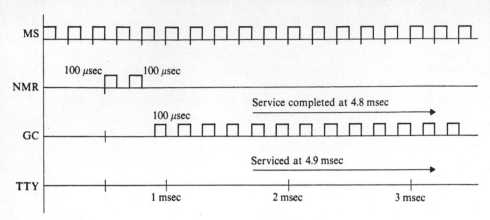

Fig. 10-8   Timing diagram with electrochemical experiment removed.

processing will drop to 34 percent. When neither the mass spectrometer nor the electrochemical experiments are on-line, the foreground load will be only 4 percent. Because these two experiments are designed to operate for only a few seconds at infrequent intervals, the normal foreground load will be 4 percent, and the system will function very effectively for all time-sharing operations, including background processing.

10-9   GENERAL COMMENTS ON THE USE OF MONITORS IN LABORATORY COMPUTERS

The monitors discussed above, especially the real-time executives, have been used with a great deal of success in automating laboratory environments. Most of the systems which have been implemented have been used to acquire data from a number of relatively slow devices such as GCs. (Examples of these are mentioned in Chap. 12.) The usual real-time executive is not useful in cases where high data-acquisition rates (greater than 1 kHz) are necessary because of its slow response to interrupts. However, some systems have been developed which take advantage of the fact that many high-speed analytical instruments have a low duty cycle. A large amount of time is spent in preparing samples, and relatively little time is needed for rapid data acquisition. In a system of this type, it is possible to share the computer among a number of users but allow only one instrument to be acquiring data at a rapid rate. This means that a given user may have to wait from a few seconds to a few minutes before it is possible to acquire data; but when the experiment is on-line, it is given the full capability of a very powerful system. Real-time executive systems are normally not well suited for complex computer *control* operations for even moderately rapid processes. The response time for these systems to provide feedback control is not very fast. This is because the system is required to respond at least three times; once to acquire the data, once to run the program which will perform the control decisions, and finally to output the necessary control information. Executive overhead is dissipated in each step.

Complex monitor systems have been used with success in a research environment where the user takes advantage of the powerful data processing capabilities of the large system. However, when the computer interaction is a basic part of the research, such as when a new computer control system is being developed, the real-time monitor will make the development a much more difficult task because of the restrictions it places on the programmer.

In industrial process control and control of other slow devices, real-time monitors and large computer systems are often the most economic method of computer control. As long as millisecond response is not required, these systems are usually able to control a number of processes. The actual number of processes which are controllable will, of course, depend on the rate of data input and the length of time it takes to digest that data. This type of system has an added advantage in that it is connected with a mass storage device, on which a running record of the processes can be maintained for later comparison and report generation.

In recent years, the cost of computer mainframes has dropped rapidly, and it has become economically feasible to develop systems which contain more than one computer processor. These systems have the advantage that there is more power in the system and it is not necessary to interrupt the various processes as often as with a single processor system. For example, one of the processors could be assigned to handle all I/O operations, while the second handles the executive functions and calculations. In this system, it is not necessary to interrupt the calculations to process an I/O function, and vice versa. This also means that the response to an interrupt can be more rapid as the I/O processor could be waiting for the I/O or at least be able to process it without concern for the calculations being performed or executive functions in progress.

Figure 10-9 shows what might be a basic structure for a multicomputer or multiprocessor system. In this case, the two processors are of approximately the

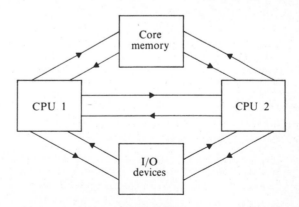

Fig. 10-9   Multiprocessor system.

same power and share at least some segment of core memory. Also, it is possible for both computers to control any or all I/O devices. The first benefit of this system comes in the form of greater reliability. Here, if one of the two processors fails, it is possible for the second to take over all the system functions, although possibly with degraded response. In addition, this type of system is capable of handling an increase in I/O load. Either one computer could be dedicated to the I/O control with little or no computational responsibility, or both computers could spend some time controlling I/O operations and some time calculating; either method would result in increased speed of response. The main drawback of this type of system, aside from the economics, is the complexity of the executive which is required. It is not only necessary to decide when a given function must be done, but it is also necessary to decide which computer will perform that function.

A second type of multicomputer system is one which is known as the *hierarchical* system; it is illustrated in Fig. 10-10. In this system, there is one computer which controls the system and is attached to at least one subordinate *slave* computer. Normally, the central computer is one which is quite powerful and is used for most data processing. The slave computers can be considered high-speed input devices by the central computer; while from the experiments

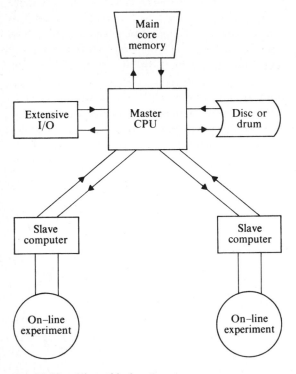

Fig. 10-10   Hierarchical system.

they may be viewed as dedicated computers. There may be one slave computer which is dedicated to the system I/O devices, or the central computer may operate these. The main feature of this system is that a slave computer can be dedicated to the experiment of interest. This slave computer will be able to acquire data from an experiment and exercise some control over that experiment if necessary. Later, the data can be transferred to the central computer for further data processing.

The hierarchical system enables the user to have the "best of both worlds"; that is, it is possible to have a computer which is dedicated to one or, at worst, only a few experiments, thus giving it rapid response compared with a single computer real-time system. However, the system offers all the advantages of a time-sharing system, such as mass storage, large computing power, and high-speed output devices. The hierarchical system, however, is the most expensive type of system because it incorporates a large amount of hardware. The advantage comes from the fact that these peripheral computers do not have to be powerful computers in their own right to accomplish very powerful operations. In fact, the peripheral computers can be stripped to the barest minimum needed for data acquisition and control functions; while the central computer can do all the complex computation.

The final decision as to the type of computer system which will be used for a given application is dictated by the nature of that application. If the application is to acquire and reduce data for a large number of slow devices in a well-defined environment, the best system would probably be a single computer time-sharing system. If it is necessary to control a few higher-speed experiments, it would probably be best to operate with dedicated computers. If a larger number of rapid devices are to be controlled and complex calculations are then to be made, a hierarchical system is probably best.

In all cases, however, the system should be tailored to the experiments to be placed on-line rather than one trying to force experiments to fit the ability of the computer system.

11
High-Level Programming Languages for Laboratory Experimentation[1]

In earlier chapters, we have described the relationships between various levels of programming languages. The most fundamental programming language is the binary-coded machine language. The type of programming language which most of the text has focused on is the assembly language, which has nearly a one-to-one correspondence with the machine language. Most scientists are also familiar with the high-level data processing languages such as FORTRAN, ALGOL, and others [1]. Because these high-level languages provide *macro* instructions allowing the user to represent highly complex data-handling and data processing functions in brief statements, they are the preferred languages for data processing. It was pointed out in earlier chapters, however, that high-level languages are not generally appropriate for on-line experimental applications. The most fundamental reason for this is that no standard data-acquisition, control, or timing instructions are incorporated. Even if such instructions could be provided, programs generated from high-level languages are generally

---

[1] Taken in part from S. P. Perone and J. F. Eagleston, *J. Chem. Educ.*, 48:317 (1971), by permission of the copyright owner.

inefficient in the utilization of computer memory space and computer time. Thus, these languages are inappropriate for real-time computer control applications.

Despite the obvious limitations of high-level languages for real-time laboratory applications, the programming convenience offered by these languages for non-real-time on-line applications or less demanding real-time experimentation suggests that their use in a laboratory environment should be explored [2]. This chapter will consider some of the technical and practical aspects of utilizing high-level languages for on-line experimentation.

11-1 TYPES OF HIGH-LEVEL LANGUAGES

There are basically two types of high-level languages: One is the Compiler-based language such as FORTRAN or ALGOL. The other is the Interpreter-based language such as BASIC or FOCAL.[1] A Compiler-based language requires the use of a translating program called a *Compiler*, which takes a complete program expressed in macro statements and translates it into a unique and self-contained machine-language program. The resultant compiled program can then be executed after it is loaded into the computer. The important point here is that the compiling and executing processes occur in two distinctly separate steps. In this sense, the compiler simply acts like a super assembler.

By contrast, the Interpreter-based languages have the characteristic that translation and execution are simultaneous processes. These languages are based on the characteristics of an *Interpreter* program, which remains core-resident during execution. This program has the capability of scanning a macro program and interpreting and executing program segments line by line. Thus, function statements in the programs are essentially subroutine calls which are linked together by the executive part of the Interpreter in a fashion demanded by the overall statement sequence in the program. Interpreter-based languages allow very convenient user functions. For example, these languages can be oriented toward the user's terminal and can readily allow on-line editing of programs through the Teletype printer. Moreover, because the Interpreter is always core-resident, the immediate execution of an entered program is feasible. Thus, rapid turn-around on executed programs is an inherent advantage of Interpreter-based languages. On the other hand, because each program statement must be translated, as well as executed, during the execution stage, execution time is relatively slow. In addition, a sizable amount of memory space must be allocated to the Interpreter itself at all times. This places restrictions on the size of the executable program and data banks that may be allowed.

Many of the convenient features of the Interpreter-based languages can also be provided with Compiler-based languages. Usually, however, this requires

[1] FOCAL is a registered trademark of Digital Equipment Corp., Maynard, Mass.

more than just a minimal on-line laboratory computer system. For example, with a magnetic disc and an appropriate operating system, it may be possible to implement a load-and-go FORTRAN Compiler, where an entered FORTRAN program is completely compiled and then loaded for immediate execution. This would allow the rapid turn-around feature of the Interpreter-based language while providing the more efficient use of computer space and time obtainable with the Compiler-based language.

11-2 COMBINING LANGUAGES

In earlier chapters, we have shown how assembly-language programming is essential to effective utilization of the computer's capability for on-line data acquisition and control. However, it should be equally obvious by now that assembly-language programming for sophisticated data processing or for formatting of typewritten report generation is extremely awkward and tedious. Thus, it would appear advantageous if the best features of high-level languages and assembly language could be combined. Thus, a seemingly ideal language for a laboratory application would be one where the data processing and standard I/O could be handled by high-level instructions; whereas the critical experimental control and data-handling functions could be handled by assembly-language program segments.

Combining high-level and low-level language program segments is a perfectly feasible approach. Obviously, since any level of language ultimately is reduced to the common denominator of machine language, all languages are inherently compatible. The only requirement is that some convenient mechanism be provided for switching from one language to the other within the framework of an overall programming objective. This is best accomplished by using the high-level language as the fundamental framework within which low-level program segments can be introduced. A specific discussion of how this might be accomplished is described in Sec. 11-3.

The primary requirement for linking a machine-language subroutine or subroutines to a program written in high-level language is that the Compiler or Interpreter be capable of recognizing a subroutine call within the high-level language and capable of generating the appropriate linkages to and from the subroutine. That is, it must be possible to input to the subroutine the linkages to variables generated in the high-level program as well as to provide a return address to the main program. With FORTRAN as an example of the high-level language, Fig. 11-1 shows the format which the Compiler might utilize for recognizing and translating subroutine calls. The CALL statement itself includes the label or name of the entry point into the subroutine as well as a specification of the variables to which the subroutine must have access. The Compiler then translates this subroutine CALL statement into a machine-language CALL sequence or segment which includes (1) the JSB instruction, (2) the return address to the

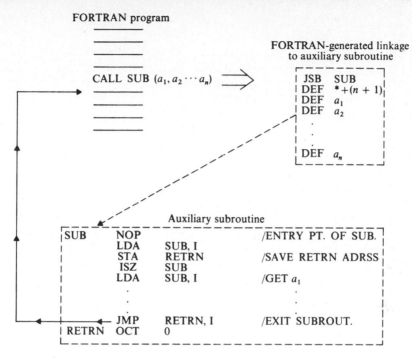

FORTRAN program

CALL SUB $(a_1, a_2 \cdots a_n)$ $\Longrightarrow$

FORTRAN-generated linkage
to auxiliary subroutine

```
JSB  SUB
DEF  *+(n + 1)
DEF  a₁
DEF  a₂
  .
  .
DEF  aₙ
```

Auxiliary subroutine

```
SUB    NOP                    /ENTRY PT. OF SUB.
       LDA    SUB, I
       STA    RETRN           /SAVE RETRN ADRSS
       ISZ    SUB
       LDA    SUB, I          /GET a₁
        .                      .
        .                      .
       JMP    RETRN, I        /EXIT SUBROUT.
RETRN  OCT    0
```

Fig. 11-1   Linkage of auxiliary subroutines to FORTRAN programs.

main program, and (3) a sequence of addresses corresponding to the addresses of each of the variables specified in the CALL statement.

An example of a typical application might be to provide a machine-language subroutine for data acquisition which will be combined with a FORTRAN program to process the acquired data. In this case, the subroutine might require the address of the program variable assigned to the input data and the number of data to be acquired. Both of these pieces of information will be transmitted by providing the appropriate addresses in the CALL sequence. Obviously, the subroutine can access information generated by the FORTRAN segment or transfer data for access by the FORTRAN program by using the linkages included in the CALL sequence. (See Chap. 3.)

The kind of linkage generated for an Interpreter-based programming language is similar to that shown for FORTRAN except that the details are somewhat more complex because of the dynamic nature of the interpretive process itself. The linkage format and the steps involved in the execution of an auxiliary machine-language subroutine using the Hewlett-Packard BASIC language are illustrated in Fig. 11-2. Each subroutine is designated by a unique integer number from 1 to 64_{10}. The CALL statement must first specify the subroutine number; subsequently, several variable or constant parameters may

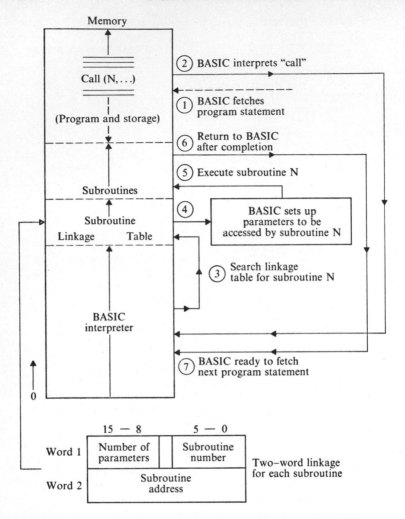

Fig. 11-2   Linkage and execution of auxiliary subroutines with BASIC.

be included in the CALL statement. For example, if subroutine 5 is a data-acquisition segment requiring certain initialization information—like the number of data points to take, the clock frequency, and the address of the first word of the data storage buffer—these parameters could be transferred through the CALL statement as shown below:

CALL   (5,T,F,A(1))

Before executing subroutine 5, BASIC will evaluate each parameter and set up an *address stack*. Upon entering the subroutine, the address of the last location in the stack will be in the A register. The stack itself will contain the addresses of each of the parameters in the CALL statement. Thus, the A register initially contains the address of an address. The last location in the stack contains the address of the first parameter, the next to the last contains the address of the second parameter, etc. Thus, the subroutine should be written to obtain parameters or addresses through the A-register linkage. The segment below illustrates what the initialization part of our data-acquisition subroutine might look like:

```
DATAC    NOP
         STA  B              /SAVE STACK POINTR IN "B" REG.
         LDA  B,I            /GET 1ST PARAMTR ADRSS
         STA  TEMP           /SAVE ADRSS
         LDA  A,I            /GET 1ST WORD OF FL. PT. VALUE
         STA  DCNT           /SAVE 1ST PRT.
         LDA  TEMP           /GET ADDRSS
         INA                 /INCRMNT TO GET 2ND WRD
         LDA  A,I            /GET 2ND WRD OF FL. PT. VALUE
         STA  DCNT+1         /SAVE 2ND PRT
         ADB  DCRMT          /DECRMNT STACK POINTR
         LDA  B,I            /GET 2ND PARAMTR ADRSS
         STA  TEMP           /SAVE
         LDA  A,I            /GET 1ST WRD OF FL. PT. VALUE
         STA  CLOK           /SAVE 1ST PRT
         LDA  TEMP           /GET ADRSS
         INA                 /INCRMNT TO GET 2ND WRD
         LDA  A,I            /GET 2ND WRD
         STA  CLOK+1         /SAVE 2ND PRT
         ADB  DCRMT          /DECREMENT STACK POINTER
         LDA  B,I            /GET 3RD PARAMTR ADRSS
         STA  PBFFR          /SAVE AS POINTR FOR LATER DATA
                             STORAGE
           .                   .
           .                   .
           .                   .
                             /DATA ACQUISITION AND STORAGE
         JMP  DATAC,I
DCRMT    OCT  -1               .
DCNT     BSS  2                .
CLOK     BSS  2                .
PBFFR    OCT  0
A        EQU  0              /SET A,B = 0,1; ADRSS'S OF A,B REGS.
B        EQU  1
```

Note: When numerical data are transferred, the data must be handled as two-word values. This is because all BASIC variables are represented in floating-point form. In order to use the numerical values within the subroutine, they may have to be "integerized." Also, data transferred back to BASIC will have to be "floated." Conversion from floating-point to integer values (or vice versa) can be accomplished by calling standard subroutines. (Often the desired service subroutine is contained within BASIC and can be utilized. For example, the IFIX subroutine to convert floating-point to integer values can be called directly from the auxiliary subroutine once the programmer knows where IFIX is located in his particular version of BASIC.)

Further details regarding the use of auxiliary subroutines with BASIC (or FORTRAN) should be obtained from the manuals provided by the computer manufacturer. Although details will vary from one computer system to another, the general approach described here is followed.

11-3   DEVELOPMENT OF A HIGH-LEVEL ON-LINE PROGRAMMING LANGUAGE

Because it had been desired to incorporate digital-computer technology into the undergraduate chemistry laboratory, a high-level programming language for on-line experimentation was developed at Purdue University. The objectives were to provide a laboratory computer system for use by undergraduate students which would require a minimum of course time for introduction. The programming language developed was built around the BASIC language [1, 3]. In addition, a general-purpose, convenient patchboard interface panel was part of the system. The system developed provides an excellent illustration of the principles discussed in this chapter, and for this reason a more detailed description follows. (See Ref. 4 for additional details.)

The software system developed at Purdue University will be referred to as *Purdue Real-time* BASIC (PRTB). The software was developed by modifying the BASIC Interpreter available from Hewlett-Packard (H.P. 20392). The modifications generated have involved the development of a series of machine-language subroutines which are directly callable from the BASIC software and which are designed to communicate in a variety of ways with experimental systems. The PRTB data-acquisition and control software are described below.

In order to operate the communication link between the computer and the experimental system, an electronic interface must be established. This interface accomplishes the functions of translation (analog-to-digital, digital-to-analog, decoding, logic conversions, etc.), timing, synchronization, and logical control. The programming language must take into account the nature of the electronic interfacing, and the interfacing must be designed with the fundamental characteristics and capabilities of the programming language in mind. The following section describes the subroutines developed for on-line experimentation with the PRTB system. A summary of these subroutines and characteristics is given in Table 11-1.

Data acquisition   Two different types of data-acquisition subroutines are available within PRTB. They both work in conjunction with an external clock-controlled analog-to-digital converter (ADC). One is a subroutine (SB3) which should be called by the user's program whenever the computer should be waiting for the next data point to become available. When the next datum is digitized, SB3 takes the data point from the data-acquisition device (ADC), converts the datum to floating-point format, and stores it in the appropriate

Table 11-1   Data-acquisition and control subroutines for PRTB

Subroutine	Function	Call format
SB1	Sets up data storage address and acquisition rate for other subroutines. (Encode and flag bits on all channels are unaltered.) X = data variable. F = data-acquisition frequency (0.001 to 100 kHz in 1, 2, 5 steps). Frequency may also be expressed as a constant.	CALL (1,X,F)
SB2	Is used to start the clock. (On ADC channel, encode bit is set, flag bit is cleared.) The output control bit (encode) must be connected to some external logic preceding the clock-enable input. This external logic may include a flip-flop and a NAND gate. (See Fig. 11-3.) The NAND gate can be used for synchronization with an external event such as the start of the experiment or have one input at logical 1 for synchronization by the computer's encode command. Additionally, the output of the flip-flop can be input to the NAND gate to simultaneously start the clock and the experiment. The clock will be disabled by calling SB10. (*Note*: SB2 *must* be preceded by SB1.)	CALL (2)
SB3	Waits for ADC flag bit, indicating conversion completed. (See Fig. 11-4.) One datum is then taken from the ADC, converted to millivolts in floating-point values, and saved as X. Data range depends on system being used. (*Note*: Flag bit is cleared; encode bit is unchanged.)	CALL (3)
SB4	Allows data to be taken from the switches and saved as a floating-point number S. Alternatively, if switch 15 is *set*, SB4 waits for switch 14 to be set and returns directly to the BASIC program. Since switches 14 and 15 are used for this timing routine, they are not input as S. *Note*: S must always be specified in the CALL even if SB4 is only used for timing (SW15 set).	CA`L (4,S)
SB5	Outputs one control bit (encode) and then waits for an event flag on the specified I/O channel C before continuing. Flag and encode bits are cleared before exit.	CALL (5,C)

Table 11-1   *Continued*

Subroutine	Function	Call format
SB6	Waits for flag on the ADC channel. When FLAG bit is set, the next statement in the BASIC program is executed. (Flag bit is cleared before exit; Encode bit is unchanged.) This subroutine can be used as a general-purpose timing routine for other external operations such as DAC output. It must be preceded by SB1 and SB2 if internal clock is used to generate flags. In this case, clock output should bypass ADC and be jumpered to flag.	CALL (6)
SB7	Takes in complete block of data before returning to BASIC program. T sets total number of data points taken (maximum of 250 points). F determines clock frequency (as in SB1), and X(I) specifies the array element assigned to first datum. Data are saved as millivolt values. Data range depends on system being used. The data-acquisition time base can be synchronized with the start of experiment through external gating of encode output as described for SB2. Flag and encode bits are cleared on exit. Up to a 20-kHz data-acquisition rate is possible.	CALL (7,X(I),F,T)
SB8	Causes bit Z to be set to 1 and output on specified I/O channel C. Z and C are specified in decimal. Encode and flag bits are unaltered.	CALL (8,Z,C)
SB9	Causes output of analog voltage through the DAC. The value is specified through CALL statement. D = millivolt output. Timing can be achieved for repetitive outputs by starting clock w/SB2 and waiting for flags w/SB6.	CALL (9,D)
SB10	Clears ADC channel. (Encode and flag bits are cleared.) External logic driven by the encode bit can be cleared also. This can be used to turn off the clock, stop the experiment, etc.	CALL (10)
SB11	Sets the encode bit and causes output of binary voltage pattern Z on specified output channel C and then waits for the flag bit. Flag and encode bits are cleared before exit. Binary pattern and channel are specified in decimal.	CALL (11,Z,C)

Table 11-1   *Continued*

Subroutine	Function	Call format
SB12	Sets encode bit and waits for the flag bits before inputting the status of the 16-bit register on channel C. Flag and encode bits are cleared before exit. The floating-point equivalent is saved in memory as Z.	CALL (12,Z,C)
SB13	Sets the encode bit and waits for the flag bit before inputting status of a specific bit Z on channel C. Value of S is made 1 or 0 accordingly. Flag and encode bits are cleared before exit.	CALL (13,Z,S,C)

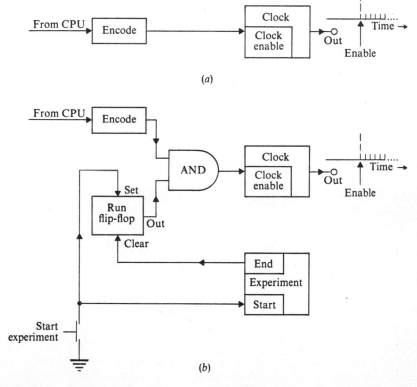

Fig. 11-3   Synchronization of clock output with computer and external events. (*From Perone and Eagleston* [4].)

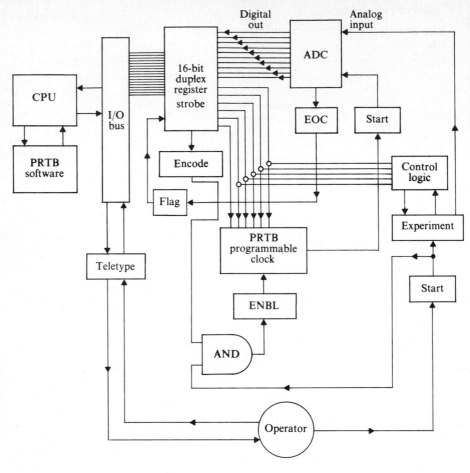

Fig. 11-4 Typical PRTB experimental data-acquisition setup. (*From Perone and Eagleston* [4]).

memory location for subsequent reference by the BASIC program. When SB3 is exited, the computer returns to the next statement in the BASIC program for execution. The new datum can be operated on, but before the next data point can be acquired, SB3 must be called again.

A second type of data-acquisition subroutine (SB7) is one which allows for the acquisition of a complete block of data before exit. The external clock is started and synchronized with data acquisition within the subroutine. The basic difference between SB7 and SB3 is that SB3 is called to acquire one data point at a time and therefore allows for program statements to be executed during the time between acquired data points. Thus, an experimenter could devise a program which can process experimental data in real time. However, because the computer program must involve the relatively inefficient execution of BASIC statements between data points, there is a more severe limit on the speed with which data can be acquired without the computer becoming out of synchronization with the experimental timing. In fact, SB3 is designed to detect when the computer has become out of synchronization; an error message will be typed, and the computer will halt. The user will then have to revise his program to require less real-time processing. With SB7, on the other hand, a complete block of data points is acquired before the subroutine is exited and control returned to BASIC. Therefore, the timing is limited by the efficiency of the machine-language programming developed within SB7. It is possible to acquire data at rates as great as 20 kHz with SB7, but no real-time data processing is possible.

The limiting data rate when using SB3 is about 500 Hz, with a minimum of real-time data handling. If several BASIC computations are to be executed between data points when using SB3, the limiting data-acquisition rate may be on the order of 1 to 10 Hz. However, this is generally more than adequate for most experiments in the chemistry laboratory.

Experimental control and logic   In addition to data acquisition, there are several other ways in which the computer can communicate with the experimental system. One of these is through a digital-to-analog converter (DAC), where the digital output of the computer is converted by the DAC to an equivalent analog voltage level. One of the PRTB subroutines (SB9) provides the capability for driving the DAC, the output of which can be connected to external experiments. It is possible to generate a continuous voltage waveform output from the DAC by mathematical generation within BASIC of the discrete points making up the waveform and to transmit these through SB9 to the DAC in a repetitive fashion synchronized with an external clock.

Subroutines SB8 and SB11 allow the programmer to utilize specific output bits on a selected I/O channel to control external devices. There are 16 output bits available for this function. With SB8, the user can select which bit he wants to set by specification of the bit number. The setting of one of these output bits causes a corresponding binary voltage level change at the specified output terminal, and this can be used to close or open switches, start or stop experimental events, light indicator lamps, etc. SB11 allows the programmer to output a 16-bit binary voltage *pattern* with any simultaneous combination of 1's and 0's he chooses. This binary pattern is selected by including in the subroutine CALL the decimal equivalent of the binary number to be generated. Thus, more than one event can be controlled simultaneously.

SB12 and SB13 provide digital *input* information for program sensing of external situations. Thus, 16 binary voltage input terminals are available to the user, the status of which is acquired by either subroutine. SB12 transmits to the BASIC program the numerical equivalent of the input 16-bit binary voltage pattern. Because the status of these bits can be set by external events, the computer could use this information to make appropriate changes in the data processing or control programming.

SB13 is similar to SB12, except that only the status of a single specified input bit is determined. This subroutine is useful because the operator need only specify a bit number in the CALL statement to check the status of a bit.

Timing and synchronization   The most fundamental operation which must be accomplished by the PRTB system is the generation of a time base for all experimental functions. This is accomplished by the incorporation of a fixed-frequency crystal clock (10 MHz) into the interface hardware. Also included is electronic countdown logic to scale the output clock pulses down to a usable frequency range for chemical experimentation (100 kHz to 0.01 Hz). The countdown logic can be modified under program control to select any frequency which can be generated by a decade and/or a 1,2,5 countdown sequence. The programmed clock output pulse train is then available to control the timing on the ADC, DAC, or any other external hardware. Also available simultaneously are synchronous clock pulses representing frequencies at the various stages of countdown.

The clock is controlled by the PRTB software through subroutines SB1, SB2, and SB10. SB1 is the *initialization* subroutine. Through it the programmer specifies the clock frequency, within the limits outlined above. (SB1 is also used to specify the symbol assigned to the variable which will take on the values of the digital data acquired in SB3. Note that initialization of the clock is also accomplished independently in SB7.)

SB2 is called when the computer program decides that it is time to *enable* the clock. That is, an encode bit is set to a 1 state on the data-acquisition channel. This bit can be connected externally with a patchcord to the enable terminal on the clock module. (See Fig. 11-3*a*.) Alternatively, the encode bit may be brought to some external logic, the output of which will set the enable clock input when other external events have occurred—like the start of the experiment. One possible external logic configuration is shown in Fig. 11-3*b*, where the encode bit conditions one input of an AND gate. The other gate input is conditioned *true* when the experiment has been initiated. The AND gate output will then enable the clock. At that point, the countdown logic will be enabled and programmed clock pulses will begin to appear at the available outputs.

SB10 is called to disable the clock when the time base is no longer required. It simply clears the encode bit on the data-acquisition channel which has been used externally to enable the clock.

Another type of communication required between the experiment and computer is associated with *synchronization* between computer operations and external events. For example, if the computer has completed all preliminary program execution required before being able to accept experimental data, the user may choose to receive an output from the computer which may tell not only the experiment to start but may also initiate external events associated with the experimental system. One subroutine available for this kind of operation is SB5. SB5 causes the output of an encode bit on a selected I/O channel to change state (go to the true level) when SB5 is called. Then the computer waits within SB5 for an external event to occur which will cause a flag bit on the same I/O channel to change to a true state. The subroutine detects this event and then allows the next sequential program statement to be executed. This next statement might conceivably be a CALL to SB2 which initiates data acquisition.

A simpler, but less precise, means of communication with the computer for the purpose of synchronization is provided by SB4. When this subroutine is entered, the computer simply waits until the operator flips the appropriate toggle switch on the computer console switch register (SR). The computer then exits SB4 and executes the next program statement—which might be to call SB2 and start data acquisition.

Finally, a subroutine (SB6) is available which allows general-purpose timing functions. This routine simply waits for the flag bit to be set on the data-acquisition channel, clears the flag bit, and exits the subroutine; the next sequential program statement is then executed. The purpose of this subroutine is synchronization of program segments with the external time base. (Note that SB5 can be used, also, for general-purpose timing functions if the data-acquisition channel is needed for other purposes. The user must simply connect the clock output to the flag input of the alternative I/O channel.)

Computer instrumentation   The digital-computer system in this work was an H.P. 2115A, equipped with 8K of core memory. The computer has an interfaced data-acquisition system and a general-purpose experimental interface capability.

A schematic diagram of the data-acquisition system operated in conjunction with PRTB is shown in Fig. 11-4. In addition to the ADC and programmable clock, an encode output bit and a flag input terminal were available on the data-acquisition panel. The end-of-conversion (EOC) flip-flop of the ADC was normally connected to the flag input; however, it was possible to connect any appropriate externally generated signal to the flag input. Six bits of digital information could be transferred to or from the computer through the data-acquisition panel using patchcord connections on the panel. Other generally useful functions available on the data-acquisition panel included patchcord connection to various logical devices such as sixteen AND, OR, and NOT gates, four flip-flops, two one-shots, four analog switches, four relay drivers, a Schmitt trigger, a T/H amplifier, six indicator lamps, and two push-button switches with Schmitt trigger outputs.

An interfaced digital I/O module was also available for general-purpose digital communication between computer and experiment. This digital I/O module provided 16 bits of input and 16 bits of output and was connected to a different I/O channel than the data-acquisition panel. Thus, a separate independent encode output and flag input were available through this module.

11-4   Examples

Example 11-1   Take in 100 digitized data points at a rate of 1 point per second. Sum those which are greater than +10 mV. Synchronize the start of data acquisition with SB4.

Basic program			Comments
10	LET	X = 0	/DEFINE DATA VARIABLE
20	LET	F = 1	/SET FREQ. = 1 Hz
30	LET	S = N = 0	
50	CALL	(1, X, F)	/INITIALIZE
60	CALL	(4, S)	/WAIT FOR SWITCH CLOSURE. SW15 MUST BE SET BEFORE RUN.
70	CALL	(2)	/START CLOCK.
80	CALL	(3)	/WAIT FOR AND TAKE ADC DATUM
90	IF	X>10 THEN 150	/CHECK NEW X
100	LET	N = N + 1	
110	IF	N = 100 THEN 200	
120	GO TO 80		
150	LET	S = S + X	
160	GO TO 100		
200	PRINT S		
210	END		

Several aspects of the program of Example 11-1 should be discussed further. First of all, the execution of SB2 [CALL (2)] takes place within a short, but finite, time after the switch closure recognized by SB4 [CALL (4)]. This time delay is imposed because the next BASIC statement must be accessed before moving on. Perhaps 0.5 msec transpires before the clock is started by SB2. Thus, the synchronization of the clock with the start of the experiment can be in error by at least this amount. More importantly, the manual switching operation can probably be no more precisely synchronized with an external event than to ±0.2 sec. For data-acquisition rates up to about 1.0 Hz, this error may be insignificant, but it becomes serious for higher data rates. Thus, for higher data-acquisition rates, a more exact synchronization routine (SB5) should be used which limits the error to about 0.5 msec. For still more precise synchronization, SB2 could be used alone with the external logic of Fig. 11-3*b*. Alternatively, one could use this external logic and acquire data with SB7, which is designed for high-speed data acquisition.

A second aspect of Example 11-1 to be pointed out is that in SB3 [CALL (3)], the computer is waiting for the right time to take the next data point.

Thus, it is called immediately after starting the clock, and the computer waits for about 1 sec for the first data point. After the first data point is taken in, however, it can be processed as X until the next data point becomes available. At this later time, the computer must have completed all processing and simply be waiting for the next data point. In other words, program control should have been returned to statement 80. Thus, for this example, statements 90, 150, 160, 100, 110, and 120 must be executed within less than 1 sec. This is reasonable to expect since each BASIC statement is accessed and executed in the order of 0.5 to 10 msec. (Longer times can be required for complex arithmetic statements.)

If, however, the data rate is set higher or more processing is attempted between data points, program control may not return to SB3 before the next data point is available. If this occurs, an error routine will cause an error statement to be printed, and the computer will halt.

Example 11-2   Take a block of 200 digitized data points from an experiment at a rate of 1 kHz. Process the data as in Example 11-1 after acquisition is complete.

BASIC program

```
20    DIM X(200)
30    LET S = 0
40    LET F = 1000
50    LET T = 200
70    CALL (7, X(1), F, T)
80    LET I = 1
90    IF X(I)>10 THEN 150
100   LET I = I + 1
130   IF I>T THEN 200
140   GO TO 90
150   LET S = S + X(I)
160   GO TO 100
200   PRINT S
210   END
```

Note: SB7 was called without any other initialization steps (such as SB1 and/or SB2). This is possible because SB7 incorporates the functions of both SB1 and SB2; that is, SB7 initializes the data array, interprets the frequency request, starts the clock, takes in the data, and converts the data to millivolts (floating-point) before exiting.

The program of Example 11-3 and the one following it are useful for taking data from an experiment where the data are changing rapidly at first but where the rate of change slows down as the initial stages of the experiment are passed. Neither is appropriate for very high data rates because of timing errors introduced when changing clock frequencies. When using SB3, however, no timing error is introduced when changing frequencies in real time *unless* the time required to change the frequency exceeds one time period of the prior clock frequency. If the time required *is* excessive, SB3 will print an error statement and the program will halt.

Example 11-3   Take 100 data points at variable rates: 10 points at 10 Hz and the next 90 points at 1 Hz. Synchronize the start of data acquisition with an external event.

BASIC program

```
20    DIM D(100)
30    LET X = 0
40    LET I = 1
50    LET F = 10
60    CALL (1, X, F)
70    CALL (5, C)
80    CALL (2)
90    CALL (3)
100   LET X = D(I)
110   LET I = I + 1
130   IF I = 10 THEN 200
140   IF I>100 THEN 250
150   GO TO 90
200   LET F = 1
210   CALL (1, X, F)
220   GO TO 90
250   (process data)
          .
          .
          .
999   END
```

Note: The clock frequency does not change in Example 11-3 until *after* the tenth datum has been acquired, even though SB1 was called between the ninth and tenth data samplings. This is an important feature of the PRTB clock. The frequency cannot be changed during a clock period.

Alternatively, the program for Example 11-3 could be written using SB7 as long as no data processing is required between data points:

```
20    DIM X(100)
30    LET T = 10
40    LET F = 10
60    CALL (7, X(1), F, T)
90    CALL (7, X(11), 1, 90)
110   (process data)
          .
          .
          .
999   END
```

Note: Statement 90 refers to the eleventh location in the data block assigned to the parameter X. This is because the first 10 locations have been filled with the first 10 data points at this point in the program.

Example 11-4   Use the DAC to output a ramp voltage varying from 0 to +5.00 V at a rate of 100 mV/sec in 10-mV steps.

```
      BASIC program
10    LET X = 0
20    LET D = 0
30    LET F = 10
40    CALL (1, X, F)

50    CALL (2)
60    CALL (6)
70    CALL (9, D)
80    IF D = 5000 THEN 150
90    LET D = D + 10
100   GO TO 60
150   END
```

SB11 could be used to control an external multiplexer that runs on a simple binary count. To sample four channels, for example, one would program

```
1     LET C = N          /SELECT I/O CHANNEL NO. N
5     LET X = 0
10    LET F = 10
20    DIM D(4)
30    CALL (1, X, F)
40    CALL (2)
50    FOR I = 1 TO 4
60    CALL (11, I, C)
70    CALL (3)
75    LET X = D(I)
80    NEXT I
90    CALL (10)
         .
         .
         .
200   END
```

SB12 could be used to bring in and print an external counter value and would be programmed as

```
50    LET C = N
60    CALL (12, Z, C)
70    PRINT Z
         .
```

SB13 could be used in program branching as here, where the status of bit 1 is checked:

```
10    LET Z = 1
20    LET C = N
30    CALL (13, Z, S, C)
40    IF S = 1 THEN 60
50    GO TO 10
60    .
      .
      .
```

11-5 COMMENTS

It should be reiterated here that the PRTB system and language described in the previous sections are presented only as an example of one of the many possible implementations of a high-level laboratory computer system. We have chosen this implementation because students can learn to use it quickly and we are willing to live with the limitations. These limitations are primarily related to the inherent features of BASIC as a computer language. It is not as concise as others (such as FORTRAN or ALGOL); moreover, its interpretive nature seriously limits real-time on-line data processing. It should be obvious, however, that an alternative high-level language could be modified as described above.

The list of subroutines described in Table 11-1 comprises what appear to be the minimum required *functional* on-line capabilities of a high-level language modified for general-purpose laboratory applications. This particular set was generated after a great deal of trial and error. Thus, it should provide a good starting point for anyone planning to implement a similar high-level system.

It should be pointed out also that the PRTB system is being used currently in junior- and senior-level chemistry-laboratory course work at Purdue University and that formal experiments utilizing the system for on-line applications have been developed and published [4-6]. In addition, the system is being used in the graduate research program for relatively straightforward automation and digital data-acquisition applications. Because a user need not be an expert in computer technology to use the PRTB system, it becomes a valuable routine laboratory tool. For the more expert user, many extensions and modifications have been generated to provide more specific applicability to particular research problems. In addition, a modified FORTRAN language has been developed for more demanding real-time research applications, incorporating essentially the same functional capabilities as PRTB.

Finally, a significant observation is that, at this writing, computer systems similar to the PRTB system in capability are already commercially available. More developments along these lines are sure to follow. For many scientists, this fact may appear to relieve them from the need to master those aspects of on-line computer technology presented in this text. However, it has been our observation that, for effective utilization of a high-level system in a research environment, a thorough understanding of the inner workings of such a system is required. Also required still is an appreciation for digital logic, timing, synchronization, interfacing, and other concepts essential for proper on-line applications of the digital computer.

BIBLIOGRAPHY

1. Sanderson, P. C.: "Programming Languages," Philosophical Library, New York, 1970.
2. Anderson, R. E.: *J. Chromatogr. Sci.,* 7:725 (1969).
3. Kemeny, J. G., and T. E. Kurtz: "BASIC Programming," Wiley, New York, 1967.
4. Perone, S. P., and J. F. Eagleston: *J. Chem. Educ.,* 48:317 (1971).
5. Perone, S. P., and J. F. Eagleston: *J. Chem. Educ.,* 48:438 (1971).
6. Jones, D. O., M. D. Scamuffa, L. S. Portnoff, and S. P. Perone: *J. Chem. Educ.,* 49:717 (1972).

12

A Survey of Computer Applications in the Chemistry Laboratory[1]

This chapter attempts to survey laboratory applications of digital computers in chemistry. Section 12-1 considers the historical development of laboratory computer instrumentation in chemistry. The subsequent discussions will cover the following areas: (1) categorization and definition of various ways to use a dedicated on-line digital computer in the chemistry laboratory, (2) a review of published applications within the various categories, (3) a description in some detail of selected applications in each category, (4) the extension of dedicated computer applications to nondedicated systems, and (5) possible future applications of on-line computers in the laboratory.

12-1 HISTORICAL DEVELOPMENT

It appears that it was in the middle 1960s when scientists first realized that computer handling of experimental data would soon be an absolute necessity

[1] Taken in part from a review article by S. P. Perone, *Anal. Chem.*, **43**:1288 (1971), with permission of the copyright owner.

because of the large volumes of raw data being generated by modern instrumental techniques. This was particularly evident to researchers in the area of mass spectrometry, where large numbers of complex spectra could be generated in a short period of time. In addition, the combined techniques of gas chromatography and mass spectrometry dictated that a large number of different sequentially acquired mass spectra would be required for proper interpretation of analytical information. The earliest solutions to these data-handling problems involved the construction of off-line digital data-acquisition and data-logging hardware. One approach involved the recording of sequential mass spectra on photographic plates followed by automated densitometer scanning of the plates with digital readout stored on magnetic tape [1]. An alternative approach [2,3] involved recording on analog magnetic tape the output of an electron multiplier on the mass spectrometer system. The data could be digitized later for computer processing. Direct digitization of the mass spectrometer output and recording on digital magnetic tape were also reported [4].

The automatic digitizing of electroanalytical data was also shown to provide distinct advantages over analog data-acquisition techniques. For example, Booman [5] demonstrated that precise electrochemical data could be acquired with rapid time resolution (50 μsec) and a large dynamic range (five decades) within the same experiment. Brown, Smith, and DeFord [6] showed that digitizing and multiplexing could provide data from complex experiments in a format readily compatible with the large processing computer. In the area of gas chromatography, Oberholtzer and Rogers [7-9] described the advantages of digital data-acquisition techniques with digitized data stored on some buffer medium compatible with introduction to the central computer system.

As a result of these early efforts at digital data acquisition, many researchers observed that the weakest point in the whole system was the communication between the laboratory and the central computer facility. Data buffers included punched cards [6], paper tape [7], magnetic tape [1,4], and multichannel analyzer core memory [10]. The inherent delay in transmission of the data to the computer center and the relatively slow batch turn-around time became a frustrating fact of life for the researcher who could turn out mountains of data in minutes but then had to wait as long as a few days in order to observe the results of those experiments. At about the same time as it became obvious that a more direct line of communication between the experiment and the processing computer would be necessary, the availability of reasonably priced laboratory computer systems became a reality. Several early publications recognized the potential for the marriage of computers and instruments in the scientific laboratory [11-14].

The earliest extensive use of on-line computers was in the field of nuclear chemistry. Reviews of such applications appeared as early as 1966 [15,16]. Allen and Bird [17] provided an introduction to on-line methodology in nuclear and reactor physics experiments, and Mollenauer [18] described on-line computer systems at the Brookhaven cyclotron and Rutgers-Bell accelerator. Several articles describing applications in this field have appeared since, and these will be described below in the appropriate categories.

Another area of chemistry where on-line computers were applied early and extensively is in process control. Williams has reviewed this field annually for *Ind. Eng. Chem.* since 1956 [19,20]. No attempt to include discussions of process control applications will be made here. However, Williams' reviews since 1964 generally include references to papers describing laboratory computer applications outside of the process control area.

12-2 CATEGORIES OF DEDICATED ON-LINE LABORATORY COMPUTER APPLICATIONS

Table 12-1 summarizes the various categories of laboratory computer applications considered here. We will attempt to define the various types of applications listed. It appears that all the applications which have appeared to date fit within these categories.

The two basic distinguishing characteristics of laboratory computer applications are described by the terms *passive* and *active*. A *passive* application is one where there is no significant computer control of the experiment involved. An *active* application, on the other hand, is one where the computer is invoived to some significant extent in the control of the experiment.

There are several subcategories within the major classification of a passive

Table 12-1   Categories of on-line digital-computer applications

---

Passive

---

1. Simple data handling and processing
2. Complex data processing (deconvolution, curve fitting, etc.)
3. File search or pattern recognition
4. Display

Active

---

1. Automation
    a. Routine
    b. Non routine
2. Real-time computer interaction
3. Iterative optimization of experiments
4. Computer-user interactive experimentation
5. Instrumentation design

---

application. In general, these applications reflect situations where the digital computer is used primarily as a data-logging device. Quite often these applications simply take advantage of the high-speed data acquisition, large data storage capacity, and rapid data-handling or data processing capabilities of an on-line computer. In addition to the on-line data-logging characteristics, these applications fall into the various subcategories including (1) simple data handling and processing, (2) complex data processing, (3) file search or pattern recognition, and (4) display of raw or correlated data. Some applications, of course, will involve combinations of features inherent in more than one of the above subcategories.

Those applications which fall in the active category generally include those where the digital computer exercises at least a minimum amount of control of the experiment and/or of experimental parameters.

This control may be preprogrammed (simple automation), or it may be exercised with computer decision-making and real-time modifications. Various sub-categories include (1) automation (routine and non-routine), (2) real-time computer interaction, (3) iterative optimization of experiments, (4) computer-user interactive experimentation, and (5) instrumentation design.

The earliest types of laboratory computer applications fell into the passive category. However, the preponderance of early work where the computer has been used as a passive element in laboratory instrumentation reflects the fact that the primary objective of the early workers was to handle voluminous amounts of data and simply provide a convenient communication link to a processing computer.

12-3   DISCUSSION OF PASSIVE LABORATORY COMPUTER APPLICATIONS

Simple data handling or data processing   One early publication describing the incorporation of a small digital computer into laboratory instrumentation involved application to an electrochemical system [13]. These workers used the computer system for on-line data acquisition and analysis of chronocoulometric experiments. This type experiment involves the application of a controlled-potential step to a microelectrode in solution and the measurement of the integrated current (coulombs) which flows as a function of time. All the potential control, current measurement, and integration operations were carried out by analog instrumentation. The computer caused the potentiostat to impose the desired potential function on the electrode, activated the ADC to digitize and transmit data, stored the digitized data in memory, analyzed the stored data to fit theoretical equations, and output the results of the analyzed data on an electric typewriter. Analyzed data were available within 10 to 100 sec after completion of each experiment. A typical experiment might have lasted 40 msec during which 200 data points might have been taken at 200-μsec intervals. The authors estimated that the computerized instrumentation reduced the time involved in executing experiments and analyzing results by a factor of at least 20. The same system was used for later work involving the study of anion-induced adsorption of cadmium on mercury from iodide and bromide media [21].

Another early publication describing an on-line laboratory computer application involved high-resolution mass spectrometry instrumentation [22]. The computer system was linked to a Consolidated Electrodynamics Corporation 21-110B mass spectrometer. Scanning rates from 15 sec to several minutes for m/e equal to from 800 to 20 could be chosen. The output of the electron multiplier was fed to an amplifier system, and the amplified signal was provided to a high-speed ADC (25-kHz conversion rate, 15 bits). Two other important parts of the overall system were a digital magnetic-tape data storage buffer and an interfaced oscilloscopic display device. To initialize the program, the operator specified a threshold value and memory storage area. Data acquisition was synchronized with the start of a spectrum scan by means of a clock interrupt. The data-acquisition process was terminated by an interrupt at the end of the spectrum scan. At the conclusion of a scan, a block of data was available in which up to 500 individual peak profiles could be stored along with the times associated with the beginning of each peak. The operator could then have had the data transmitted to magnetic tape, or he may have chosen to display the data. The operator could have chosen to display the whole spectrum at once on the cathode-ray tube (CRT), or the time scale could have been expanded to display one peak envelope at a time. After the storage or display operations had been completed, the computer was ready to accept the next scan. The investigators indicated that the CRT display module was one of the most important parts of the system. It provided a "window" which allowed the operator to be aware immediately of the operational acceptability of the overall system and allowed rapid adjustment of instrumental parameters to obtain optimum resolution, sensitivity, and proper beam of calibration compounds. Data acquired by the on-line computer system could be transmitted via magnetic tape buffers to a CDC 6600 computer system for off-line data processing.

Other articles have appeared describing mass spectrometry-computer marriages [23,24]. In addition, the applications of on-line computer systems for combined gas chromatographic-mass spectrometry (GC/MS) instrumentation systems have been reported [25,26]. Prior to this work, the GC had been recognized as a sophisticated sample introduction system for a mass spectrometer. However, the utility had been limited by the large volume of data produced. For example, it was a common practice to record a mass spectrum during each observed chromatographic peak of interest, as well as some spectra between fractions to obtain a representative background. Such a procedure can result in the order of 20 to 100 spectra representative of a single chromatographic sample. Conversion of these raw data into interpretable form is extremely time consuming in comparison with the time required to record the data. Hites and Biemann [25] pointed out in their work that even when spectra are recorded selectively, the utilization of spectra cannot keep pace with their accumulation. Moreover, they point out, the selective process places the burden of constant decision on the operator and can result in the loss of important information from those parts of the chromatogram allowed to pass without recording the mass spectrum.

To minimize the problems of selection and possible loss of valuable information, Hites and Biemann [25] proposed and developed a computerized system which provided continuous recording of mass spectra during an entire chromatogram. The GC/MS system used in their work included a Varian Aerograph model 600 GC and a Hitachi Perkin-Elmer RMU6-D magnetic scanning mass spectrometer. The spectrometer's magnetic field was swept from mass 20 to 600 in 3 sec and allowed to collapse for 1 sec, with this cycle repeated continuously. The interface to the mass spectrometer was simply a pulse generator which generated data-acquisition service interrupts at 350-μsec intervals whenever the mass spectrometer scan was on. Thus, data were digitized at about a rate of 3,000 per second when the mass spectrometer was operating. Resolution of the ADC was 15 bits. One complete mass spectrum was run every 4 sec, regardless of the emergence of chromatographic peaks. Thus, for a 30-min chromatogram, for example, a total of about 450 spectra were recorded. Mass spectrometric peak centers were located during data acquisition.

In this earlier publication [25], Hites and Biemann did not go much beyond providing the data-acquisition and data-retrieval system. They recognized that if the raw data points for each spectrum are summed, the result will be directly proportional to the unresolved ion beam current as measured by the beam monitor of the mass spectrometer. With each spectrum indexed serially, a plot of the summed intensity versus index closely resembles the gas chromatogram itself. Thus, they provided for the computer to generate this kind of display on an incremental plotter. The operator then could evaluate the overall chromatographic output and select from this plot the indices representing the GC fractions of most interest. By entering these indices using the computer keyboard, the operator could have the mass spectra corresponding to these individual indices displayed or provided in tabular form. Computer-aided correlations between mass spectra and GC sample characteristics were not provided in this work. However, the instrumentation developed in this earlier work provided the basis for later work [27-29] where computerized correlations were made.

The article by Sweeley et al. [26] describes an alternative approach to providing on-line computer data acquisition and processing for high-speed mass spectrometry with GC inlet. These workers used an LKB 9000 GC/MS instrument. The output of this system was obtained from a 14-stage electron multiplier. A Hall effect probe provided effective magnetic field strength measurements. Data from the electron multiplier and the Hall effect probe could be fed to the computer through a multiplexer input. The maximum data-acquisition rate available was 25 kHz. Data acquired during a spectrum scan were read directly onto a 64K-word magnetic disc, which could accommodate from 31 to 62 scans with a maximum of 1,000 peaks per scan. The computer achieved synchronization of data acquisition with the mass spectrometer scan by initiating and terminating the mass spectrometer scan as well as data acquisition. The complete system had the capability to monitor mass spectrum scans of 1-sec duration with m/e equal to from 0 to 500. Peak intensities and magnetic field strength at the summits of peaks were determined in real time. Data stored on disc could be processed, subsequently, by the on-line computer system. Processing included calibration, conversion of magnetic field strength to exact mass, conversion of exact mass to nominal mass, normalization, output to magnetic tape, and final output on Teletype or plotter. More sophisticated processing could have been carried out at a large computer system by transporting the magnetic-tape output.

An interesting application of an on-line computer was reported by Frazer et al. [30]. They described a method for determining microgram amounts of nitrogen using distillation and flow spectrophotometry with data acquisition by a small digital computer. Distillation was used as a method for prior separation of the ammonia from the sample solution, which interfered with the direct colorimetric measurement. The distillate was mixed in a Technicon Auto-analyzer with color-developing reagents (hypochlorous acid and sodium phenate). The distillate and reagents were mixed and passed through a 210-sec coil in the Autoanalyzer system to allow complete color development. The solution then passed through a 10-mm absorption cell, and the absorption was monitored continuously at 620 nm. The photometric signal was converted to absorbance by a logarithmic amplifier circuit and introduced to the computer data-acquisition interface. This system was programmed for the computer to start data acquisition 100 sec after the start of sampling from the collection flask. The program idled for 100 sec, took a base line for 50 sec, integrated for 500 sec, then took another base line for 50 sec. The integrated data were corrected for base line. The authors reported that this approach using flow spectrophotometry with data acquisition and integration by a digital computer improved the sensitivity of the method for nitrogen by about a factor of 10. In addition, the standard deviation of the flow method was about 0.1 μg of nitrogen compared to about 1 or 2 μg of nitrogen in the single-measurement method.

The use of an on-line digital computer for data acquisition and data handling in fast-sweep derivative polarography was described by Perone et al. [31]. The fast-sweep derivative polarograph used in the work could impose a rapid linear voltage sweep on an indicator microelectrode in solution and measure the direct electrolysis current as well as the first and second derivatives of that current. A dropping-mercury electrode was used as the indicator electrode, and the cell was equipped with a mechanical drop dislodger controlled by the timing circuits of the polarograph. The potential sweep was actually applied during the last second of drop life, just before dislodgment, when the electrode area was not changing very rapidly. Thus, a new fast-sweep experiment could be run every few seconds on each new droplet from the dropping-mercury electrode. The initiation of each sweep was accompanied by a pulse output from the polarograph which could be used for synchronization with the digital data-acquisition system. Data were taken at a 500-Hz rate during each 1-sec sweep to provide a total of 500 data points per experiment. After each experiment, the stored data could be processed by a program which provided for the location of reduction peaks and the measurement of peak heights. In addition, the operator could instruct the computer to ensemble average data for up to 2,048 repetitive sweeps on the same sample. The investigators found that the ensemble-averaging technique provided an enhanced signal-to-noise ratio and about an order of magnitude increase in sensitivity. Moreover, because each experiment could be repeated on the order of every three to six sec, a large number of runs could be averaged in a relatively short period of time.

Although the laboratory computer has been applied more extensively to gas chromatography than any other single instrumental technique, *dedicated* applications for simple data acquisition and data processing are relatively rare. Some fundamental data-handling and data processing considerations have been presented by McCullough [12], Baudisch [32], and Perone and Eagleston [33]. An early application of a simple dedicated computer data-acquisition system that provided base-line correction and peak-area calculations was discussed by Simon et al. [34]. Computer monitoring of multiple chromatographs has been described [35-37].

Several publications have appeared which describe the features of commercially available dedicated computer-GC systems [38-40]. Baumann, Brown, and Mitchell [38] provide a detailed evaluation of a dedicated computer system for chromatography in the research and development laboratory. These workers used a Varian Aerograph Chromatography Data System 200 and a variety of chromatographs in their studies. Two versions were used in their studies: the Varian 210-8, which handles 10 simultaneous input channels with 8K of core memory, providing for 240 peaks and 100 peak factors and the Varian 210-12 with 12K of core memory providing for 960 peaks and 480 peak factors. Both systems were equipped with a control console designed for easy and fast interaction with the system by the chromatographer. These investigators focussed their attention on several factors important to chromatographic analysis. These factors include precision, accuracy, resolution of fused peaks, signal transmission from remote location, and interaction with the computer system.

The computer-system precision was evaluated by a novel approach involving multichannel analyses. Essentially, their approach involved the simultaneous acquisition of data from the same experiment through more than one data-acquisition channel. The data were processed independently, as though obtained from separate chromatographs, and compared. Any significant deviation of results from channel to channel would have been indicative of problems with the data-acquisition, data-handling, or data processing operations of the computer system. The investigators pointed out that the multichannel technique could be used to determine the relative precisions of the GC and the computer system. Total system precision is measured by the standard deviation from multiple analyses (multiple injections), and the computer precision is indicated by the standard deviation of multichannel comparisons for a single injection. The accuracy of the computerized system was evaluated by using samples of carefully weighed normal paraffins. The computer results were compared with simultaneous digital-integrator results. Computer-processed data for percent composition agreed within about ±0.15 percent with the standards. The accuracy was somewhat better than obtainable with the digital integrator, and the investigators projected that the ultimate accuracy will probably be better than ±0.1 percent.

Their work involving resolution of fused peaks utilized a tangent method intended for cases where small peaks appear on the back side of larger peaks. A detailed description of the logical decisions and results obtainable in the tangent method was presented. The handling of signal transmission problems was illustrated by describing an application where the output of a thermal conductivity bridge was transmitted over 350 ft from instrument to computer. These microvolt-level signals were transmitted over twisted-pair cable with aluminized Mylar material shield. Both the signal low and the cable shield were grounded to the same point at the chromatograph chassis. Hardware filtering included a low-pass filter with roll-off at 5 Hz and an attenuation of over 30 db at 60 Hz. They reported that satisfactory precision was obtained in all cases despite the fact that high-amplitude noise spikes did appear on the transmission line. Finally the authors evaluated the user-oriented control console for convenient interaction with the computer system for the establishment of appropriate data-acquisition and data processing parameters in the research and development environment.

The use of an on-line digital computer for data acquisition and processing in stopped-flow spectrophotometry has been demonstrated [41,42]. Because stopped-flow data can involve millisecond time scales, oscilloscopic recording followed by visual data-reduction techniques have been employed conventionally. The on-line computer system eliminates the tedious conventional data-handling problems by providing direct digitization and storage of data. Moreover, Willis et al. [42] demonstrated that ensemble averaging could be applied to advantage in stopped-flow spectrophotometry. These authors investigated particularly the application to quantitative kinetic analyses. Their system included an oscilloscopic display to project raw or processed kinetic data. On-line processing included real-time least-squares smoothing [43], calibration with a standard sample, and evaluation of rate constants and/or concentrations from first-order decay curves.

Toren et al. [44] have also described a computer-spectrophotometer interface for reaction rate measurements. Klauminzer [45] presented an inexpensive on-line computer system for spectrophotometry. Dedicated computer data-acquisition systems for nmr [46], epr [47], and Mössbauer spectroscopy [48,49] have been described. Goodman [49] also reported the extension of his system to the time sharing of three Mössbauer spectrometers.

An on-line computer has been used also in phosphorescent decay studies [50]. In this work, the computer controlled data acquisition with a precise time base over several decades of time. Real-time data smoothing, as well as ensemble averaging, could be provided. Computer monitoring of luminescent output was synchronized with rotation of a dual shutter which controlled the excitation and luminescent light paths. Chemical mixtures with half-lives ranging from milliseconds to seconds could be studied in a single experiment.

Many other on-line computer applications can be found which fall into this first category. In the nuclear chemistry field, for example, computer monitoring of bubble-chamber experiments has been reported [51]. These studies demonstrated how efficiency can be improved by computer elimination of redundant measurements. In addition, on-line computers have been used for acquisition and processing in pulse-height analysis [52,53] and for interpretation of gamma-ray spectra [54]. Other specially designed systems for nuclear instrumentation methods incorporating on-line computers have been reported [15,16,55-64]. Also, Fujiwara et al. [65] described an on-line computer system for conventional polarography. They demonstrated the value of digital data acquisition over a wide time domain in allowing analysis of abnormal current-time curves at the dropping mercury electrode.

Complex data processing   Numerical deconvolution techniques refer to mathematical methods applied to allow detection and extraction of quantitative data from severely overlapping waveforms. The implication is that simple techniques such as simultaneous equations are not applicable in such cases, either because overlapping is so severe as to preclude recognition of individual contributions to the resultant waveform or because the experimental phenomena which caused the overlap effect distortions in the contributing waveforms which do not appear in the isolated case.

Deconvolution approaches have been applied in *on-line* computer systems. Scott, Chilcote, and Pitt [66] have described a small on-line computer system for peak detection and deconvolution of GC peaks. Other articles [67,68] have appeared describing deconvolution techniques for computer systems designed for the time sharing of several concurrently operating on-line GCs. This work will be discussed in a later section.

Gutknecht and Perone [69] have applied numerical deconvolution techniques to overlapping stationary-electrode polarographic curves using a dedicated on-line laboratory computer. The computer system used had a 16-bit word size and an 8K-word core memory. Data acquisition was provided through a 10-bit 33-μsec conversion time ADC at a 1-kHz experimental rate. An interfaced oscilloscope display device was also included in the system. The computer was interfaced to a fast-sweep differential polarograph which featured potentiostatic control of matched sample and reference cells. The differential electrolysis current measured during a linear voltage-sweep experiment was amplified and output to the ADC. In their work, the investigators developed an empirical equation designed to fit a large variety of polarographic waveforms by adjustment of appropriate constants. This function was first fit to a number of different standard polarograms, and the constants of the function for each species were stored in memory. Upon analysis of unknown mixtures, these constants were used to regenerate the standard curves, a composite of which was then fit to the unknown signal. The technique used for curve-fitting to standards and unknown mixtures was the nonlinear residual least-squares method [70,71]. The investigators demonstrated that this technique had the capability for qualitative identification and quantitative analysis of mixtures where reduction peaks were separated by as little as 40 mV with concentration ratios ranging from 1:5 to 5:1. Concentration values were obtained with relative errors on the order of 0.1 to 2 percent, as compared to relative errors of 10 to 40 percent using simple simultaneous equations.

Another illustration of complex data processing handled by a small on-line digital computer was provided by Margerum, Pardue, and coworkers [72]. They described multicomponent kinetic analyses using a linear least-squares method for fitting the entire response curve for up to three simultaneous first-order reactions with a common reagent. Turn-around of the order of three to five min was attained, which included data acquisition, data display, data processing, and printout of results. The authors estimated that their system was applicable for mixtures with rate-constant ratios in the range of 1.5:1 to 120:1.

Another application involving complex data processing is in the area of Fourier transform spectroscopy. Low [73] has discussed the application to infrared (IR) spectrometry. The output of the optical system of an IR Fourier transform spectrometer is essentially an interferogram recorded in the time domain. Fourier transformation of these data results in a conventional frequency spectrum with the possibility of considerably enhanced sensitivity over conventional dispersive spectrometers. The use of an on-line computer for data acquisition and data processing makes this a feasible laboratory technique, particularly because a "fast" Fourier transform algorithm can be used to provide rapid processing with a small computer [74,75]. Horlick and Malmstadt [76] have discussed some very practical considerations for sampling and digitizing interferograms for Fourier analysis. The dramatic effects of inadequate sampling rate and missed data points were illustrated. Ridyard [77] has described a far IR

Fourier spectrometer system incorporating a dedicated special-purpose computer.

Farrar [78] has discussed pulsed or Fourier transform nmr. The *pulsed* nmr technique involves the application of a short, intense pulse of radio-frequency energy followed by the time-domain measurement of the free-induction decay signal from the nuclear spins. A type of interferogram is obtained. Fourier transformation results in a typical high-resolution nmr spectrum. The approach is advantageous in that the decay signal is obtained rapidly; the experiment can be run repetitively to allow signal averaging. Here, as with IR Fourier transform spectroscopy, the use of an on-line computer results in a feasible laboratory technique. Several manufacturers have developed Fourier transform spectrometers utilizing built-in computer systems. (See the discussion under "Automation," Ref. 101.)

Many other studies have been reported describing *off-line* complex processing of experimental data which might be extended readily to on-line laboratory computer implementation. Some of these include methods for resolving overlapping peaks in gas chromatography [79,80], mass spectroscopy [81], nmr spectroscopy [82], EPR spectroscopy [83], and IR spectroscopy [84-86]. Aczel et al. [87] described a data processing approach for handling quantitative, high-resolution mass spectral analyses of complex hydrocarbon mixtures. They used an on-line laboratory computer for data acquisition and storage of digital data on magnetic tape for later processing with a large central computer. Morrey [88] and Margoshes and Rasberry [89] have discussed, respectively, computer resolution of overlapping spectral peaks and the fitting of analytic functions in spectrochemical studies. We should also mention again here the important work of Savitzky and Golay [43] on data-smoothing techniques.

File-search and pattern-recognition applications   A file-search application of the on-line digital computer involves the use of high-speed access to large volumes of digitized, standard qualitative information to match up with experimental data. This will normally require magnetic-disc or magnetic-tape file storage [90]. Although no articles have appeared to date describing an *on-line* application of a dedicated laboratory computer system, recent articles indicate that such an application is feasible [91-93]. For example, Lytle [91] described the use of a *small* computer for rapid file searching, using a punched paper-tape inverted file. Thirty thousand spectra could be searched in less than 60 sec. Lytle and Brazie [92] described a data compression technique which allowed searching of IR data at 18,000 spectra per second, using a 32-bit computer with magnetic-disc input. Jurs [93] described a hash coding technique which allows searching of 16-bit spectra for a single unknown with an average search time of 40 msec for 20,000 spectra.

Several articles have described computerized pattern-recognition methods applied to chemical analysis. These approaches have involved *off-line* computer operations using programs generally referred to as *learning machines*. Binary pattern classifiers were used for molecular-formula determination from low-resolution mass spectra [94,95], for classification of IR spectral data according to chemical class [96], and for the interpretation of combined data from mass spectra, IR spectra, and melting and boiling points [97]. Multicategory pattern classification by the least-squares method has been applied to molecular-formula and molecular-weight determination from mass spectra [98]. The computerized learning-machine approach has also been applied to semiquantitative analysis of mixed gamma-ray spectra [99] and to qualitative analysis of complex mixtures by stationary-electrode polarography [100]. This latter work involved the use of a small laboratory-scale digital-computer system. Obviously, the learning-machine approach can be extended to *on-line* systems for routine laboratory analyses.

Data-display applications   This type of passive computer application utilizes the high-speed data-fetching and correlation capabilities of the digital computer to provide visual displays of selected or correlated data with rapid turn-around. This computer function often allows operator interaction with the experiment which could not be possible ordinarily.

We have already mentioned a couple of applications which utilize oscilloscope or plotter displays of data [22,25]. The work of Gutknecht and Perone [69] utilized an oscilloscope display system to provide visual observation of the progress of curve-fitting routines on-line. Hites and Biemann [29] effectively illustrated the utility of the interfaced oscilloscopic display system for correlation and interpretation of GC-mass spectral data. One of the most useful ideas developed in that work was the provision of a display of the change in abundance of certain ions during the gas chromatogram. This display was termed a *mass chromatogram*. The computer and analytical instrumentation used in this work were the same as reported in earlier work [25]. That is, the computer-mass spectrometer system records spectra continuously, regardless of the emergence of GC fractions, and stores all spectra on a magnetic disc. A plot of computer-calculated total ionization versus spectrum index number provides a display similar to the original gas chromatogram. In addition, the operator can select the mass spectrum corresponding to any particular index number and have it displayed. The major extension of this later work was to allow the operator to request a display of the change in abundance of ions of a particular m/e ratio during the chromatogram. The authors showed that this technique permitted detection of the presence or absence of a homologous series of compounds as well as specific substances of known or predictable mass spectra. One particularly striking example of the added interpretive and correlative power of this system was described by the authors: In a study of a chloroform extract of human blood, it was noted that the list of significant peaks showed m/e ratios of 178, 202, and 228 to be the most intense in several spectra. Mass chromatograms revealed that the exact locations of these intense mass spectral peaks with regard to the chromatogram were as unresolved shoulders on small chromatographic peaks. The investigators pointed out that these particular components would not likely have been detected had the mass spectra not been collected continuously during the chromatographic process. Moreover, the mass chromatogram display technique provided an invaluable perspective on the voluminous experimental data.

12-4   ACTIVE ON-LINE APPLICATIONS OF THE DIGITAL COMPUTER

As defined above, active applications of the on-line digital computer include those where the computer exercises at least some minimum amount of control of the experiment and/or of the experimental parameters. We will describe several typical applications in this category below.

Automation   Within this category, one can consider two distinct kinds of automation applications—*routine* and *nonroutine*. Routine applications include those where the primary objective is to handle automatically large volumes of analytical data and/or to minimize manual operations for more or less conventional experimentation. That is, the normal features of conventional instrumentation and methodology are used with computer monitoring and control of the experimentation. Typical examples of this kind of automation application of computers are found in the commercially available package systems for spectrophotometry, gas chromatography, mass spectrometry, x-ray crystallography, nmr and epr spectroscopy, and other instrumental methods. The features and capabilities of these various systems are listed and described in detail in the various manufacturers' literature. The reader is referred to those sources for more extensive discussions [101]. In addition, discussions of systems for routine automation in the clinical and biochemical laboratories have been presented [102-104]. Gambino [105] has discussed the problems and needs in automated clinical laboratories.

The nonroutine computer automation applications can be defined similarly to the routine applications except that the types of instrumentation or experimental problems to which the computer is applied are categorized as unique, innovative, or unconventional research systems. In addition, such applications might include exploitation of measurement methods uniquely compatible to the digital computer.

Two articles describing general-purpose laboratory computer systems for data acquisition and control are those by Ramaley and Wilson [106] and Lauer and Osteryoung [107]. The systems described in these articles are particularly appropriate for variable nonroutine laboratory automation applications. The systems described include general-purpose data-acquisition, timing, interfacing, and digital control features with a great deal of flexibility for application to a variety of instrumental situations. Lauer and Osteryoung [107] mentioned several specific applications of their laboratory computer system, including the various electroanalytical methods of chronocoulometry, pulse polarography, ac polarography, chronopotentiometry, and cyclic voltammetry. The approach described by these authors involved allowing the computer to generate all the control functions in the various electroanalytical techniques, varying the specific technique employed primarily by software changes. Thus, extremely versatile electroanalytical instrumentation in which the digital computer became an integral part was developed.

An interesting nonroutine automation application was reported by James and Pardue [108] in their studies of reaction-rate methods of analysis. Those authors pointed out several significant limitations to conventional analog instrumentation approaches to automation of kinetic methods for quantitative analysis. For example, in most cases final results are based on a two-point measurement scheme, and no provision is generally made for providing multiple measurements on a single sample. In addition, they point out that most approaches are limited to a relatively narrow range of reaction rates which can be tolerated without operator intervention to modify instrumental or experimental parameters. In their work, they utilized the rapid response and data storage capability of an on-line digital computer to minimize many of the limitations specified above. Preliminary rate measurements were utilized to adjust data-acquisition rates and other operating parameters to optimum values for each experiment, regardless of the reaction rate. Consequently, a wide dynamic range of analyses could be performed without any hardware or experimental changes. In addition, multiple rate measurements were performed during each experiment, resulting in considerably enhanced precision for rate determinations. They reported results for the determination of alkaline phosphatase activity in reconstituted serum and saline and for determination of osmium at concentrations between 10^{-9} and 10^{-11} molar. For normal conditions, relative standard deviations and deviations from linearity were below 1 percent.

Hicks, Eggert, and Toren [109] have reported their efforts to provide on-line computer automation of analytical experiments in a clinical laboratory environment. The computer system was designed to control a variety of analytical instruments and to provide on-line data analysis. In addition, the operator could cause raw data or correlated information to be displayed during or after a given experiment had been executed. Because of this capability, the experimenter could observe subtle effects, such as slight losses of enzyme activity, which would not have been obvious by looking only at numerical data or crude manual plots. Most important, the laboratory system described incorporated the computer as an active element in the experimental procedure, performing certain operations like range selection and calibration automatically while responding to various instructions provided by the operator.

The authors [109] describe application of the system to kinetic methods of quantitative analysis. They report several significant observations of the effectiveness of their automated laboratory system. For example, they point out that in a dedicated configuration only about five to 10 percent of the running time of the computer was being used. Thus, they predicted that the system could be extended quite readily to provide data acquisition and data analysis in a time-sharing mode. In addition, they illustrated the considerable advantage of on-line data analysis and display by discussing an experiment where temperature sensitivity of reagents was detected and verified because on-line analysis showed a problem existed while the solution was still available. Finally, the authors point out that because the computer freed the operator from performing the tedious manual instrumental operations and data processing functions and, moreover, provided sophisticated on-line data handling and display, it became feasible for the operator to examine more rigorously and carefully experimental data. They report, as a result of this, the development of a new graphic approach to routinely analyze data from enzyme substrate studies, which would not be feasible without the automated system.

A novel approach to computer automation in mass spectrometry was reported by Reynolds *et al.* [110]. These workers developed an MS-computer system in which the computer controls the operation of a quadrupole mass spectrometer. The unique advantage of this system is that the quadrupole mass spectrometer does not require a time-base scan of the spectrum. Rather, any particular *m/e* ratio can be selected for detection by the appropriate application of a low-level control voltage. The computer system is programmed to supply this control voltage through a DAC interface. One advantage of this approach over continuous scanning techniques is that the region between the peaks need not be monitored, resulting in a considerable saving of computer time. Moreover, a larger duty cycle of analyzer *on-peak* time is obtainable, resulting in the detection of more ions for a given mass position than possible in conventional scanning instruments. The system is pretty much independent of the particular computer used. The workers report that they have used a LINC computer [111] with 2K words of core memory and a 12-bit word size as well as a time-sharing IBM 360/50 buffered with an IBM 1800 computer. In all cases, their system was operated in a remote location with respect to the computer system. They indicate that any small general-purpose computer should be capable of providing the described functions. In addition, they suggest that it would be feasible to build a limited-purpose computer into a quadrupole mass spectrometer, forming a compact instrument incorporating the features described in their research publication.

Housemann and Hafner [112] have also discussed the design and development of a computer-controlled, quadrupole mass spectrometer system. Jones et al. [113] described a portable computerized quadrupole mass filter. Bonelli [114] has described some useful applications of a commercially available quadrupole MS-GC system. Lagergren and Stoffels [115] reported the development of a computer-controlled three-stage mass spectrometer.

Several other examples of nonroutine computer automation systems can be discussed. For example, Stephens et al. [116] have used an on-line digital computer to monitor and control coulometric analyses. A unique feature of their application is that the computer was used to analyze coulometric data in real-time and continually project the final quantitative result. When the repetitive projections agreed within a specified degree, the experiment could be terminated before reaching conclusion. Thus, they were able to obtain acceptable quantitative analyses without carrying coulometric experiments to their normal stoichiometric conclusion achieved by exhaustive electrolysis of the sample. Analysis times were cut at least in half by this procedure with no sacrifice in accuracy or precision.

Ramaley [117] has described a computer-controlled sweep generator for electrochemical investigations; while Keller and Osteryoung [118] have extended computer control in electrochemistry to the generation of pulse polarographic experiments at a hanging mercury-drop electrode. Precise and variable control of pulse width and frequency allowed convenient exploration of stationary-electrode effects. Although some nonlinearity in the concentration dependence was observed, the potential advantage of considerably more rapid analyses should offset this problem.

Another example of nonroutine automation is provided by the work of Anderson [119], where a Cary model 14 spectrophotometer was modified for computer control of an ensemble-averaging experiment. The objective of the experimental modification was to allow analysis of dilute samples of berkelium. Johnson et al. [120] have discussed and demonstrated the common computer-control and data-acquisition considerations for various forms of spectroscopy. Grant [121] has described computer control of sample and detector positioning as well as wavelength and data acquisition for reflectance measurements. Other articles have described various aspects of computer control in visible [122] and vacuum ultraviolet [123] spectroscopy as well as in nmr [124,125] and x-ray fluorescence studies [126].

Boyle and Sunderland [127] demonstrated striking improvement in the precision of atomic absorption experiments when computer control was used to randomly sample standard and unknown solutions from a sampling turntable. The computer also controlled data acquisition and determined concentrations of unknowns from a calibration curve computed on-line. Repetitive sampling with base-line corrections and ensemble averaging minimized the problems of instrumental drift common to atomic absorption measurements.

Perone and Kirschner [128] have developed computer-automated flash photolysis instrumentation. The approach taken was to allow the computer to repetitively execute the flash experiments with a programmed sequencing of experimental parameters from one experiment to the next. For example, with electrochemical monitoring, the potential at which electrolysis currents are measured after the flash is incremented or adjusted between experiments. All the experimental adjustments are automatically initiated or monitored by the computer except for the solution-handling operations. One important feature of the automated system is that all data can be stored either on magnetic-tape or punched paper-tape buffers for later correlation, data processing, or graphic display. Another feature is the inclusion of an interfaced oscilloscopic display device to provide instantaneous retrieval of transient data or display of correlations with theoretical equations. This approach allows new perspectives on the fundamental data which were not previously feasible. Moreover, the ability to rapidly complete a series of experiments and obtain qualitative and quantitative correlated data regarding a photochemical system is extremely advantageous. For example, considerably less fluctuation in experimental conditions over the entire data-acquisition period is encountered because the total time required to complete an experimental study is diminished. In addition, because computer assimilation and processing of data can be carried out rapidly, the experimenter can conveniently interact with the experimental system to assure that satisfactory experimental conditions exist at all times.

Real-time computer interaction The types of on-line computer applications included in this category are those where the digital computer can monitor experimental outputs, evaluate data in real time, and exercise real-time control of experimental parameters to optimize or enhance experimental measurements. A detailed discussion of this type of computer-controlled experiment and the limitations on computer feedback in a *closed-loop* system has been presented in Chap. 9 and Ref. 129.

Some of the earliest work involving real-time computer interaction with experiments is found in the descriptions of computer-controlled x-ray crystallographic measurements [11, 130-135]. Computer-controlled x-ray diffractometer systems have been constructed for single-crystal [11, 130-132, 134, 135] and polycrystalline [133] samples. In these systems, the sample is mounted on a conventional goniometer head, and the computer controls the various motors determining the sample orientation with respect to the x-ray beam. In addition, the computer is programmed to search for specific reflections, determine intensities and angles, compute various integrals in real time, and optimize data collection by collecting intensity data only when reflections are found and by counting to preset values or times. A review of off- and on-line computer applications in crystal structure determinations was published by Jitaka [136].

Burke and Thurman [137] have described a computer-controlled GC system, which they designed for real-time computer-optimized GC experiments. These investigators used a GC system built up from selected high-quality components. These components included flow controllers, where flow rates could be controlled to better than 0.4 percent over a range of 20 to 100 cc/min.

The sample introduction system was a Carle micro-gas-sampling valve, model 2014 with automatic actuator. The valve was operated by a solenoid-actuated pneumatic system. The solenoids could be controlled by the digital computer. The maximum data-acquisition was 60 points per second. A versatile digital interface was provided for computer control of external chromatograph hardware. The investigators demonstrated the utility of the system for providing excellent control and a precise time base for multiple-injection studies. In addition, they demonstrated the usefulness of the real-time decision-making capability of the computer system for controlling multiple-sample experiments. Data processing features of the programming were minimal as the investigators were attempting to demonstrate only the degree of precision with which computer operations could be controlled and monitored. Later work [138] described a system for complete real-time control of a GC. This included computer control of a differential flow controller and the temperature of the column oven. Real-time interaction allowed optimization of carrier-gas flow rate and column-oven temperature. They provided an example of an automatic *van Deemter* experiment as a function of temperature.

Perone, Jones, and Gutknecht [139] reported an example of real-time computer optimization of electroanalytical measurements. The investigators pointed out that in the conventional implementation of stationary-electrode polarography, electrolysis currents from easily reducible species continue to flow and distort or mask currents measured for more difficultly reducible species. To minimize this serious interference and limitation of the method, they constructed a computer-controlled fast-sweep polarograph and programmed the computer to impose a *discontinuous* sweep to the electrochemical cell. That is, the computer was programmed to monitor the polarographic output continuously and interrupt the voltage sweep whenever a reduction peak was observed.

Based on real-time computer evaluation of the characteristics of an observed peak, the specific potential and the length of time of the interrupt are computed for maximum effect. During the interrupt period, the electrolysis of the material causing the preceding peak continues at the microelectrode and depletes that material in the vicinity of the electrode. Thus, after an appropriate delay time, the sweep can be continued, and reduction steps for more difficultly reducible species can be observed relatively free of contributing current due to the preceding reduction steps. The investigators reported about a hundredfold increase in quantitative resolution capabilities for the interrupted-sweep technique as opposed to the conventional polarographic method. The computer-optimized method was limited, however, to resolution of multicomponent mixtures where peak potentials were separated by more than 150 mV. They provided data showing quantitative resolution of 1,000:1 mixtures of thallium to lead where the peak potentials were separated by 280 mV.

Iterative optimization of experiments   In the preceding examples of real-time computer interaction for experimental optimization, it was required that the computer process data in real time to adjust experimental parameters before the completion of a given experiment. An alternative approach might be to allow the computer to take and store data from a given experiment, process the data, and then make decisions on how a new experiment should be run to obtain better data. Thus, the computer could modify experimental parameters for the next experiment, initiate the new experiment, and evaluate the experimental data again to see if improved results had been obtained. This procedure could be repeated until the computer decides that the optimum experimental conditions have been achieved and the best possible analytical information obtained. The advantage of the iterative, computerized, experimental design approach over the real-time interactive approach is that often it is not possible to make the best decisions regarding experimental parameters until all the data are available and analyzed.

An example of the iterative approach is described in the work of Perone and Zipper [140]. These workers recognized that the interrupted-sweep polarographic technique described above [139] could never involve the truly optimum experimental conditions because the computer could not look ahead to succeeding reduction steps and use that information in selecting the proper interrupt potential. Therefore, they developed an iterative computerized approach which involved, first, running an interrupted-sweep experiment with conditions adjusted so that the maximum amount of qualitative information could be obtained from a single sweep. That is, at least the locations and relative magnitudes of the various reduction peaks were detected in a single interrupted-sweep experiment. Then the computer was programmed to take that information and decide on the exact values of various potentials at which electrolysis current should be monitored in a succeeding experiment to provide the optimum quantitative resolution of data for each of the indicated components in the mixture. Thus, the second experiment applied was not an interrupted-sweep experiment but rather a succession of discrete steps to selected potentials at which electrolysis currents were monitored for computed lengths of time to obtain optimum resolution. After the second experiment, the computer analyzed the data, again made adjustments in the selected electrolysis potentials—if adjustments were needed—and reinitiated the experiment, comparing the quantitative results of the third experiment with those obtained in the second. This approach could be repeated until consistent optimum results were obtained.

Computer-user interactive experimentation   The types of active on-line computer applications discussed above have all involved preprogrammed computerized control of experimentation. Although the computer might be required to make decisions in real time over which the operator or programmer has no control, the basis on which the decisions are made is provided by the fundamental real-time programming algorithm. An alternative approach is to allow the experimenter to operate in parallel with the computer in the feedback loop of the overall experimental system. For many experiments, the experimental lifetime is so short that the operator could not effectively interact with the control or processing algorithms. However, for many experiments the experimental lifetime is long enough to allow such interaction. Moreover, because the computer can rapidly display not only raw data but correlations between various experimental parameters and raw data in real time, it becomes feasible for the experimenter to exercise some judgment in modifying experimental parameters on the basis of observed results.

The previously mentioned computer-controlled x-ray diffractometer systems provide excellent examples of real-time operator interaction with experimental control or data processing. Because of the long time scale of these experiments, such interaction is both feasible and effective. The computer is programmed to perform the tedious data-handling, processing, and sample-orientation tasks; while the crystallographer maintains overall supervision of the operation, provides initialization values, and can interrupt and redirect the computer-controlled functions in real time.

Deming and Pardue [141] have described computer-automated instrumentation which allows non-real-time operator interaction in fundamental studies of chemical reactions. They have applied the system to studies of kinetic processes for chemical analysis.

Two other publications describe a generally applicable system for operator interaction with computerized data processing programs [142,143]. Direct operator communications through a scope display with stylus, push-button control console, and Teletype printer achieve rapid and effective data acquisition and analysis. The articles by Frazer et al. [142] and Perone et al. [143] describe interactive data processing for spark-source mass spectrographic data, GC data, and electrochemical data. The experiments are on-line to the interactive computer system, and it appears that the principles established in this work can be extended readily to real-time situations.

Instrumentation design   The general-purpose on-line digital computer, such as described by Lauer and Osteryoung [107] and Ramaley and Wilson [106], can be used advantageously for the development of new experimental methods and the design of fundamentally new instrumentation. The flexible on-line computer system allows the convenient exploration of experimental control parameters and measurement methods not possible with conventional instrumentation. However, once the optimum control functions and measurement approaches have been defined, the experimental features can be implemented in special-purpose hardware instrumentation.

This approach to computerized instrumentation design has been demonstrated by Jones and Perone [144]. They constructed a hybrid analog-digital hardware device for fast-sweep derivative polarographic analysis based on the optimum experimental parameters defined in earlier work involving computerized investigations [139]. The earlier work explored the feasibility and capabilities of an interrupted-sweep polarographic method (as described above). The hardware-implemented instrument could not have been designed readily without the prior computerized studies. Moreover, the hardware device was constructed for about 5 percent of the cost of the computerized system yet provided essentially the same performance.

12-5   IMPLEMENTATION OF ON-LINE COMPUTER APPLICATIONS FOR TIME-SHARING SYSTEMS

Many of the applications described above, particularly in the area of routine automation, have been or can be implemented in a time-sharing computer environment. Many of these time-sharing extensions have been reviewed by Frazer [145,146]. The area of gas chromatography has probably received the most attention. Westerberg [147] has described a real-time sampling algorithm for a time-sharing on-line computer system for gas chromatography. The algorithm has been implemented in software for a CDC 1700 computer system to handle several concurrently operating GCs. The algorithm described guarantees each chromatograph a fixed frequency access to the digital computer up to the saturation capacity of the data-acquisition system. Each individual chromatograph can have a different data-acquisition frequency, where frequencies can be multiples of 2 of each other. In addition, the algorithm permits the data-acquisition rate for any chromatograph to be halved any number of times during a given run. One of the unique aspects of the algorithm described is that data-acquisition time slots can be dynamically reallocated as chromatographs come on- and off-line and as data-acquisition rates for a given chromatograph change with time. Conditions are defined for situations where two or more low-frequency sampling time slots can be combined to provide a single higher-frequency data-acquisition rate.

Another article by Westerberg [148] presents a discussion of the detection and resolution of overlapped chromatographic peaks with an on-line digital computer. This discussion is based on the software provided for a time-sharing GC computer system. In this paper, the author evaluates errors in common methods for quantitative resolution of overlapped GC peaks, such as triangulation and perpendicular-drop methods. The errors are correlated with the factors of peak overlap and height ratio. The author also examines the limitations of peak detection methods involving the recognition of inflection-point pairs. He considers the effectiveness of first filtering chromatographic data and then counting true inflection points. With regard to quantitative extraction of information for overlapped peaks, Westerberg describes numerical methods based on gaussian, modified-gaussian, and general tabular data models. The technique involves model curve fitting. Included in the paper are sample data analyzed by the computer system described.

Hancock, Dahm, and Muldoon [149] describe an alternative computer-programming algorithm for resolution of fused chromatographic peaks used in a commercial time-sharing computer system for gas chromatography. The algorithm is based on an abbreviated simultaneous-equation solution assuming symmetrical or asymmetrical gaussian waveforms. The approach is capable of real-time execution and applicable to a wide variety of fused waveforms. The authors compared their mathematical approach with common techniques for peak resolution like dropping a perpendicular or skimming peaks. Data for standard mixtures of methanol, propanol, and butanol were analyzed using the various approaches. Using their gaussian model for peak resolution, the authors showed that essentially equivalent accuracy and reproducibility could be obtained whether the components were base-line—resolved or fused together.

Other recent articles have described the experiences of groups who have implemented time-sharing computer systems in the nuclear laboratory [150, 151] and for general laboratory automation [152-157]. In addition, some very practical aspects of time-sharing chromatographic computer installations can be obtained from several articles [158-162]; while the detailed considerations of interfacing GCs have also been treated [163].

12-6 FUTURE WORK IN LABORATORY COMPUTER INSTRUMENTATION

In a field that is changing as rapidly as the area of digital instrumentation and laboratory computers, it is extremely difficult to project into the future and anticipate the kinds of applications that may be on the horizon. However, some trends do appear fairly clear at this time. One of these is that digital devices and digital computers are becoming considerably less expensive, faster, and generally more capable and reliable. On the other hand, the cost of generating software is not decreasing. Therefore, we will probably see more experimental control and simple data-handling and data processing functions being handled by sophisticated hardware devices rather than by programmable general-purpose computers. (See Sec. 13-5.) Large instruments with hard-wired computer-control and data-handling systems built in will probably be more common. These systems will probably be connected to large time-sharing central computer facilities for the handling of large volumes of data, data plotting, and sophisticated data processing. A recent paper by Gill [164] considers in more detail the directions in which computerized instrumentation may be going, particularly in the area of gas chromatography. In addition, Frazer [146], Klopfenstein [165], and Sederholm [166] have discussed the trend toward hierarchical computer systems, where several laboratory minicomputers can communicate with one or more central, more powerful computers. Thus, the laboratory computer can be dedicated to the experimentation but can take advantage of big-computer capabilities like large bulk storage, efficient Compilers, line printers, etc.

With regard to research applications, the small laboratory computer will probably be used more extensively by far than it has been in the past. However, more sophisticated on-line programming languages will be available, either through high-level interpreters or hardware-implemented macro instructions. Recent papers by Anderson [167] and Perone and Eagleston [168] consider the use of high-level on-line programming languages for laboratory applications. In addition, future work will probably see a shift in emphasis from *passive* applications to *active* applications of the on-line digital computer.

12-7 COMMENTS

The discussion in this survey of laboratory computer applications should not be considered comprehensive. However, the field has developed sufficiently at this point, so a fairly adequate coverage of all the *types* of applications categorized was possible here. The specific articles discussed extensively here were selected because they best illustrated each of the categories. Thus, the preceding discussions should provide a reasonable starting point for anyone entering the area of laboratory computer applications. Additional review articles are provided in the Bibliography [169-172]. In addition, introductory and survey articles are available in the Bibliography [107, 145, 170, 173-184]. Articles describing educational programs involving computer technology at the undergraduate, graduate, and postgraduate levels have also appeared [168,185]. Finally, several articles and books have appeared which, although not discussing specific on-line computer applications, provide useful background or otherwise pertinent information to the scientist using laboratory computers. Some of these are listed in the Bibliography [163, 186-194].

ADDENDUM

Between the preparation of a manuscript and its eventual publication, a considerable period of time may pass. The purpose of adding this Addendum to this chapter is to provide the reader with a minimal gap in the literature survey.

Although the objective of the preceding sections has been to give the reader a broad picture of types of laboratory computer applications, and not necessarily to present a comprehensive literature survey, the reader understandably should desire the literature references to be as up-to-date as possible. In this Addendum, we will attempt to provide references to articles which appeared between the time of completion of the main body of this chapter and the time of final production of the text.

Because the objective of the Addendum is only to maintain current coverage of each of the categories of computer applications defined in Table 12-1 and not to reillustrate each category, no detailed discussions will be presented. The coverage will be divided simply into four general areas: passive applications, active applications, off-line applications, and general discussions.

12-A-1   PASSIVE APPLICATIONS

A considerable amount of work has been done recently in the area of clinical chemistry laboratory automation [195-202]. A particularly noteworthy advancement has been in the development and exploitation of the GEMSAEC approach to clinical analyses developed by Norman Anderson and associates at Oak Ridge National Laboratory [195,196]. The approach involves using continuous rotors and allowing centrifugal force to move and mix several samples and reagents concurrently, allowing simultaneous multiple analyses. The automation and computerization of this approach have been made available commercially [105], and articles exploring the utility of the GEMSAEC system have appeared [197-199].

In mass spectrometry a display-oriented computerized system was reported [203]. It was built around an LKB-9000 GC/MS instrument using a DEC PDP-12 computer for data-acquisition and display functions. The application of another computerized GC/MS system to the detection and identification of abnormal metabolic products in physiological fluids was described [204]. Other computerized data systems for mass spectrometry have been reported [205, 206].

Dedicated laboratory computer systems for gas chromatography [207, 208] and applications of other systems [209,210] have been reported. One of these applications [210] involved on-line calculation of adsorption isotherms for a wide range of adsorbates and adsorbents using a frontal-analysis chromatographic apparatus. Computer data systems for gel permeation chromatography [211] and high-resolution liquid chromatography [212] have also been reported. A computer system for a laser-Raman spectrometer instrument has been described [213].

12-A-2   ACTIVE APPLICATIONS

Articles describing recent work with computer-controlled x-ray diffractometers have appeared [214,215]. A computer-controlled beam foil spectroscopy system [216] and other controlled instrumentation for nuclear chemical experiments [217,218] have been described. In addition, a hard-wire—programmed computer for particle identification was reported [219].

A general-purpose computer system for high- and low-resolution mass spectrometry and GC/MS instrumentation has been described [220]. This system—called LOGOS—is also capable of feedback control operations. A powerful set of operational programs was developed; scope display and hard-copy plotting were provided; digital magnetic-tape and disc units added bulk storage and, during operation, repetitive recycling of the magnetic field could be carried out with computer feedback to enhance the reproducibility of the magnetic scan.

A computer-controlled operating system for a quadrupole mass spectrometer was reported [221]. McLafferty and coworkers [222] noted that the sensitivity of conventional magnetic scanning mass spectrometers could be increased by using computer control to repetitively rescan each peak after it has been detected. While the magnetic scan continues between peaks, the potential on the electrostatic analyzer is adjusted incrementally to bring the preceding peak back in focus. At a resolving power of 16,000 about 100 replicate scans of each peak at any part of the spectrum are possible.

Swingle and Rogers [223] reported a GC system where sample injection, column temperature, and flow rate were controlled by an on-line minicomputer. Real-time readout of processed data as well as inlet pressure and mass flow rate was provided. Their results showed a standard deviation of less than 0.05 percent in retention-time computations. The investigators reported that the system was being used to provide automatic determination of optimum chromatographic operating conditions.

Several additional active computer applications have appeared. Cushley et al. [224] described a computer-controlled, pulsed system where the computer was used to generate a variety of perturbing pulses. Lawrence and Mohilner [225] described a computer-controlled capillary electrometer which utilizes closed-loop computer functions to accomplish surface-tension measurements on mercury electrodes. Mueller and Burke [226] discussed real-time control of a reagent addition system. An image orthicon spectrograph operated under computer control was also reported [227]. Computer-controlled experiments on muscle [228] and in microcalorimetry [229] have been described.

12-A-3 OFF-LINE APPLICATIONS OF INTEREST

Chesler and Cram [230] have investigated the effects of locating the limits of integration and of noise on the precision and accuracy of statistical moment analyses from digitized chromatographic data. Their results are meaningful for optimized data-acquisition algorithms in chromatography.

Computer identification of mass spectrometric data has been discussed [231]; a numerical deconvolution algorithm for overlapping spectral lines has been presented [232], and a course on computer applications to numerical analysis in chemistry has been described [233]. Much progress has been made in the application of computerized pattern-recognition methods in chemistry [234-238]. Also, many developments in file searching and information retrieval have been reported [239-242], and investigations of small computer systems for rapid searching continue [243].

12-A-4 INTRODUCTORY, SURVEY, AND GENERAL ARTICLES

Chilcote and Mrochek [212], in their article on a computer system for high-resolution liquid chromatography, present an excellent discussion of algorithms for peak identification, stripping, integration, etc., for chromato-graphic-type data. General discussions of computerized mass spectrometry systems have been presented [244,245]. Computer systems approaches for interferometry [246] and for x-ray spectroscopy [247] have been discussed. In addition, many introductory and survey articles have appeared [248-254].

BIBLIOGRAPHY

1. Venkataraghavan, R., F. W. McLafferty, and J. W. Amy: *Anal. Chem.,* **39**:178 (1967).
2. Merritt, C., Jr., P. Issenberg, M. L. Bazinet, B. N. Green, T. O. Merron, and J. G. Murray: *Anal. Chem.,* **37**:1037 (1965).
3. McMurray, W. J., B. N. Greene, and S. R. Lipsky: *Anal. Chem.,* **38**:1194 (1966).
4. Hites, R. A., and K. Biemann: *Anal. Chem.,* **39**:965 (1967).
5. Booman, G. L.: *Anal. Chem.,* **38**:1141 (1966).
6. Brown, E. R., D. E. Smith, and D. DeFord: *Anal. Chem.,* **38**:1130 (1966).
7. Oberholtzer, J. E.: *Anal Chem.,* **39**:959 (1967).
8. Oberholtzer, J. E., and L. B. Rogers: *Anal. Chem.,* **41**:1234 (1969).
9. Oberholtzer, J. E.: *J. Chromatogr. Sci.,* **7**:720 (1969).
10. Lauer, G., and R. A. Osteryoung: *Anal. Chem.,* **38**:1137 (1966).
11. Cole, H., Y. Okaya, and F. Chambers: *Rev. Sci. Inst.,* **34**:872 (1963).
12. McCullough, R. D.: *J. Gas Chromatogr.,* **5**:595 (1967).
13. Lauer, G., R. Abel, and F. C. Anson: *Anal. Chem.,* **39**:765 (1967).
14. Biemann, K.: *Adv. Mass Spectrom.,* **4**:139 (1968).
15. Lindenbaum, S. J.: *Ann. Rev. Nucl. Sci.,* **16**:619 (1966).
16. Jones, J. A.: *Trans. Nucl. Sci.,* **14**:576 (1967).
17. Allen, B. J., and J. R. Bird: *At. Energy (Aust.),* **9**:6 (1966).
18. Mollenauer, J. F.: *Trans. Nucl. Sci.,* **14**:611 (1967).
19. Williams, T. J.: *Ind. Eng. Chem.,* **61**:76 (1968).
20. Williams, T. J.: *Ind. Eng. Chem.,* **62**(12):28,94 (1970).
21. Anson, F. C., and D. J. Barclay: *Anal. Chem.,* **40**:1791 (1968).
22. Burlingame, A. L., D. H. Smith, and R. W. Olsen: *Anal. Chem.,* **40**:13 (1968).
23. Knutti, R., and R. E. Buhler: *Chimia,* **24**:437 (1970).
24. Johnstone, R. A. W., F. A. Mellon, and S. D. Ward: *Int. J. Mass Spectrom. Ion Phys.,* **5**:241 (1970).
25. Hites, R. A., and K. Biemann: *Anal. Chem.,* **40**:1217 (1968).
26. Sweeley, C. C., B. D. Ray, W. I. Wood, J. F. Holland, and M. K. Krichevsky: *Anal. Chem.,* **42**:1505 (1970).
27. Murphy, R. C., M. V. Djuricic, S. P. Markley, and K. Biemann: *Science,* **165**:695 (1969).
28. Althaus, I. R., K. Biemann, J. Biller, P. F. Donaghue, D. A. Evans, H. J. Foerster, H. S. Hertz, C. E. Hignite, R. C. Murphy, G. Preti, and V. Reinhold: *Experientia,* **26**:741B (1970).
29. Hites, R. A., and K. Biemann: *Anal. Chem.,* **42**:855 (1970).
30. Frazer, J. W., G. D. Jones, R. Lim, M. C. Waggoner, and L. B. Rogers: *Anal. Chem.,* **41**:1485 (1969).
31. Perone, S. P., J. E. Harrar, F. B. Stephens, and R. E. Anderson: *Anal. Chem.,* **40**:899 (1968).
32. Baudisch, J.: *Chromatographia,* **11-12**:443 (1968).
33. Perone, S. P., and J. F. Eagleston: *J. Chem. Educ.,* **48**:438 (1971).

34. Simon, W., W. P. Castelli, and D. D. Rutstein: *J. Gas Chromatogr.*, 5:578 (1967).
35. Fraade, D. J.: *Instrum. Chem. Pet. Ind.*, 4:53 (1967).
36. Mears, F. C.: *Hydrocarbon Process*, 46:105 (1967).
37. Vestergaard, P., L. Hemmingsen, and P. W. Hansen: *J. Chromatogr.*, 40:16 (1969).
38. Baumann, F., A. C. Brown, and M. B. Mitchell: *J. Chromatogr. Sci.*, 8:20 (1970).
39. Peterson, G. V., E. Zerenner, M. Eccles and W. Kapuskar; J. Poole; L. Mikkelsen; and L. Mikkelsen: *Adv.*, 3(1):1-28 (1970), Hewlett-Packard, Avondale, Pa.
40. Paul, G. T., and W. J. Slaughter; H. A. Gill, R. E. Lee, and R. D. Condon; D. J. Noonan and R. D. Condon; E. W. March; E. C. Pieper: papers presented at Pittsburgh Conf. Anal. Chem. Appl. Spectrosc., Cleveland, Ohio, 1971.
41. Desa, R. J., and Q. H. Gibson: *Comput. Biomed. Res.*, 2:494 (1969).
42. Willis, B. G., J. A. Bittikofer, H. L. Pardue, and D. W. Margerum: *Anal. Chem.*, 42:1340 (1970).
43. Savitzky, A., and M. J. E. Golay: *Anal. Chem.*, 36:1627 (1964).
44. Toren, E. C., A. A. Effert, A. E. Sherry, and G. P. Hicks, *Clin. Chem.*, 16:215 (1970).
45. Klauminzer, G. K.: *Appl. Opt.*, 9:2183 (1970).
46. Cohen, J. S., R. I. Shrager, M. McNeel, and A. N. Schechter: *Biochem. Biophys. Res. Commun.*, 40:144 (1970).
47. Ohno, K., and J. Sohma: *Chem. Instrum.*, 2:121 (1969).
48. Birnan, A., A. Shoshani, and P. A. Montano: *Nucl. Instrum. Methods*, 89:21 (1970).
49. Goodman, R. H.: *Moessbauer Eff. Methodol.*, 3:163 (1967).
50. Kanzig, H., U. P. Wild, and S. P. Perone: *Chem. Instrum.*, in press, 1972.
51. Taft, H. D., and P. J. Martin: **Mezhdunar**. *Konf. Fiz, Vys. Energ.*, 12th Dubna, USSR, 1964, 2:390 (1966); *Phys. Abstr. Sci. Abstr.*, 70:449 (1967).
52. Wilburn, N. P., and L. D. Coffin: *AEC Rep. BNWL-CC-700*, Battelle N.W. Lab., Richland, Wash., 1966.
53. Der-Mateosin, E.: *Nucl. Instrum. Methods*, 73:77 (1969).
54. Pagdan, I. M., G. J. Pearson, and V. N. Beck: *IEEE Trans. Nucl. Sci.*, 17:211 (1970).
55. Hamilton, W. C.: *AEC Rep. BNL-10238*, Brookhaven Nat. Lab., Upton, N.Y., 1966.
56. Friedes, J. L., R. Sutter, H. Palensky, G. Bennett, G. Igo, W. D. Simpson, R. L. Stearns, and D. M. Corley: *Nucl. Instrum. Methods*, 54:1 (1967).
57. Eiben, B., H. Faissner, M. Holder, J. Konig, K. Krisor, and H. Umbach: *Nucl. Instrum. Methods*, 73:83 (1969).
58. Ford, W. T., P. A. Pirdue, R. S. Remmal, A. J. Smith, and P. A. Souder: *Nucl. Instrum. Methods*, 87:241 (1970).
59. Ashkenazi, Y., and I. Carmi: *Nucl. Instrum. Methods*, 89:125 (1970).
60. Hiebert, J. C., C. C. Hamilton, T. H. Sathre, and R. C. Rogers: *Nucl. Instrum. Methods*, 86:45 (1970).
61. Alderson, P. R., and N. Dawson: *Nucl. Instrum. Methods*, 86:35 (1970).
62. Emmerich, W. D., P. Elzer, and A. Hofmann: *Nucl. Instrum. Methods*, 81:277 (1970).
63. Merchez, F., J. Pouxe, and M. Tournier: *Nucl. Instrum. Methods*, 81:173 (1970).
64. Bevington, P. R.: *IBM J. Res. Dev.*, 13:119 (1969).
65. Fujiwara, S., Y. Umezawa, and T. Kugo: *Bunseki Kagaku*, 19:1119 (1970).
66. Scott, C. D., D. D. Chilcote, and W. W. Pitt: *Clin. Chem.*, 16:637 (1970).
67. Westerberg, A. W.: *Anal. Chem.*, 41:1770 (1969).
68. Hancock, H. A., Jr., L. A. Dahm, and J. F. Muldoon: *J. Chromatogr. Sci.*, 8:57 (1970).
69. Gutknecht, W. F., and S. P. Perone: *Anal. Chem.*, 42:906 (1970).
70. Moore, R. H., and R. K. Zeigler, "The Solution of the General Least Squares Problem with Special Reference to High-speed Computers," *LA*-2367, U.S. Government Printing Office, Washington, D.C., 1959.

71. Mandel, John: "The Statistical Analysis of Experimental Data," Interscience, New York, 1964.
72. Willis, B. G., W. H. Woodruff, J. R. Frysinger, D. W. Margerum, and H. L. Pardue: *Anal. Chem.*, 42:1350 (1970).
73. Low, M. J. D.: *Anal. Chem.*, 41:97A (1969).
74. Cooley, J. W., and J. W. Tukey: *Math. Comput.*, 19(90):297 (1965).
75. Kiss, A. Z.: *Hewlett-Packard J.*, 21:10 (1970).
76. Horlick, G., and H. V. Malmstadt: *Anal. Chem.*, 42: 1361 (1970).
77. Ridyard, J. N. A.: *Colloq. Int. CNRS*, 161:62 (1967).
78. Farrar, T. C.: *Anal. Chem.*, 42:109A (1970).
79. Klein, P. D., and B. A. Kunze-Falkner: *Anal. Chem.*, 37:1245 (1965).
80. Anderson, A. H., T. C. Gibb, and A. B. Littlewood: *Anal. Chem.*, 42:434 (1970); *J. Chromatog. Sci.*, 8:640 (1970).
81. Luenberger, D. G., and U. E. Dennis: *Anal. Chem.*, 38:715 (1966).
82. Keller, W. D., T. R. Lusebrink, and C. H. Sederholm: *J. Chem. Phys.*, 44:782 (1966).
83. Bauder, A., and R. J. Meyers: *J. Mol. Spectros.*, 27:110 (1968).
84. Stone, H.: *J. Opt. Soc. Am.*, 52:998 (1962).
85. Papousek, D., and J. Pliva: *Collect. Czech. Chem. Commun.*, 30:3007 (1965).
86. Pitha, J., and R. N. Jones: *Can. J. Chem.*, 44:3031 (1966).
87. Aczel, T., D. E. Allan, J. H. Harding, and E. A. Knipp: *Anal. Chem.*, 42:341 (1970).
88. Morrey, J. R.: *Anal. Chem.*, 40:905 (1968).
89. Margoshes, M., and S. D. Rasberry: *Anal. Chem.*, 41:1163 (1969).
90. Erley, D. S.: *Anal. Chem.*, 40:894 (1968).
91. Lytle, F. E.: *Anal. Chem.*, 42:355 (1970).
92. Lytle, F. E., and T. L. Brazie: *Anal. Chem.*, 42:1532 (1970).
93. Jurs, P. C.: *Anal. Chem.*, 43:364 (1971).
94. Jurs, P. C., B. R. Kowalski, and T. L. Isenhour: *Anal. Chem.*, 41:21 (1969).
95. Jurs, P. C., B. R. Kowalski, T. L. Isenhour, and C. N. Reilley: *Anal. Chem.*, 41:690 (1969).
96. Kowalski, B. R., P. C. Jurs, T. L. Isenhour, and C. N. Reilley: *Anal. Chem.*, 41:1945 (1969).
97. Jurs, P. C., B. R. Kowalski, T. L. Isenhour, and C. N. Reilley: *Anal. Chem.*, 41:1949 (1969).
98. Kowalski, B. R., P. C. Jurs, T. L. Isenhour, and C. N. Reilley: *Anal. Chem.*, 41:695 (1969).
99. Wangen, L. E., and T. L. Isenhour: *Anal. Chem.*, 42:737 (1970).
100. Sybrandt, L. B., and S. P. Perone: *Anal. Chem.*, 43:382 (1971).
101. A wide variety of computerized instrument systems for the chemistry laboratory are commercially available. A partial list of manufacturers and types of computerized systems is given here:

Angstrom-West, Div. of Angstrom, Inc., Belleville, Mich. (emission spectrometer system)
Applied Research Laboratory, Sunland, Calif. (optical emission and x-ray fluorescence spectrometers)
Baird-Atomic, Bedford, Mass. (emission spectrometer)
Beckman Instruments, Fullerton, Calif (Fourier transform spectrometer)
Datacomp, Inc., Pittsburgh, Pa. (emission spectrometer)
Digilab, Inc., Cambridge, Mass. (Fourier transform spectrometer, GC/IR spectrometer, pulsed nmr)
Digital Equipment Corp., Maynard, Mass. (gas chromatography and several others—see application note 101, DEC, Maynard, Mass.)

E. I. DuPont de Nemours & Co., Inc., Instrument Products Div., Monrovia, Calif. (GC/MS/data system)

Electronic Associates, Inc., PACE Div., W. Long Branch, N.J. (quadrupole MS, gas chromatography)

Finnigan Corp., Sunnyvale, Calif. (quadrupole MS)

Hewlett-Packard, Palo Alto, Calif. (mass spectrometry, gas chromatography)

IBM Corp., Data Processing Div., White Plains, N.Y. (gas chromatography)

Japan Electron Optics Lab, Co., Tokyo, Japan (GC/MS system, pulsed nmr)

NMR Specialties, New Kensington, Pa. (pulsed/swept nmr)

Perkin-Elmer Corp., Norwalk, Conn. (gas chromatography, mass spectrometry)

Picker-Nuclear Corp., White Plains, N.Y. (mass spectrometry, x-ray diffractometer)

Siemens Corp., Iselin, N.J. (x-ray spectrometer)

Varian Associates, Palo Alto, Calif. (gas chromatography, nmr, mass spectrometry, EPR, spectrophotometry)

102. Blaivas, M. A.: *Technicon Symp.*, 2d, *N.Y. London*, 1965:452 (1966).
103. Abernathy, M. H., G. T. Bentley, D. Gartelmann, P. Gray, J. A. Owen, and G. D. Quan-Sing: *Clin. Chim. Acta*, 30:463 (1970); *Biochem. J.*, 121:4PA (1971).
104. Ottaway, J. H.: *Biochem. J.*, 121:1PA (1971).
105. Gambino, S. R.: *Anal. Chem.*, 43:20A (1971).
106. Ramaley, L., and G. S. Wilson: *Anal. Chem.*, 42:606 (1970).
107. Lauer, G., and R. A. Osteryoung: *Anal. Chem.*, 40:30A (1968).
108. James, G. E., and H. L. Pardue: *Anal. Chem.*, 41:1618 (1969).
109. Hicks, G. P., A. A. Eggert, and E. C. Toren, Jr.: *Anal. Chem.*, 42:729 (1970).
110. Reynolds, W. E., V. A. Bacon, J. C. Bridges, T. C. Coburn, B. Halpern, J. Lederberg, E. C. Levinthal, E. Steed, and R. B. Tucker: *Anal. Chem.*, 42:1122 (1970).
111. Stacy, R. W., and B. Waxman: "Computers in Biomedical Research," vol. II, pp. 35-66, Academic, New York, 1965.
112. Housemann, J., and F. W. Hafner: *J. Phys. E.*, 4:46 (1971).
113. Jones, N. A., R. D. Friesen, and J. W. Pyper: *Rev. Sci. Instrum.*, 41:1828 (1970).
114. Bonelli, E. J.: *Am. Lab.*, 3(2):27 (1971).
115. Lagergren, C. R., and J. J. Stoffels: *Int. J. Mass Spectrom. Ion Phys.*, 3:429 (1970).
116. Stephens, F. B., F. Jakob, L. P. Rigdon, and J. E. Harrar: *Anal. Chem.*, 42:764 (1970).
117. Ramaley, L.: *Chem. Instrum.*, 2:415 (1970).
118. Keller, H. E., and R. A. Osteryoung: *Anal. Chem.*, 43:342 (1971).
119. Anderson, R. E.: Lawrence Livermore Lab., Livermore, preprint, *UCRL*-72093, December 1969; see also preprint, *UCRL*-70638.
120. Johnson, B., T. Kuga, and H. M. Gladney: *IBM J. Res. Dev.*, 13:36 (1969).
121. Grant, P. M.: *IBM J. Res. Dev.*, 13:15 (1969).
122. Hannon, D. M.: *IBM J. Res. Dev.*, 13:79 (1969).
123. Bridwell, L., W. E. Maddox, L. M. Beyer, and R. C. Etherton: *Appl. Opt.*, 9:929 (1970).
124. Satoh, S., K. Kushida, and T. Nishida: *Bunseki Kagaku*, 19:1110 (1970).
125. Schoolery, J. N., and L. H. Smithson: *J. Am. Oil Chem. Soc.*, 47:153 (1970).
126. Cooper, H. R., and R. L. Vaughn: *Trans. AIME*, 244:295 (1969).
127. Boyle, W. G., and W. Sunderland: *Anal. Chem.*, 42:1403 (1970).
128. Kirschner, G. L., and S. P. Perone: Symposium on Photochemical Methods, 9th national meeting, Soc. Appl. Spectrosc., New Orleans, October 1970; *Anal. Chem.*, 44:443 (1972).
129. Perone, S. P.: "Computers in Chemistry and Instrumentation" vol. 2, T. Mattson, H. B. Mark, Jr., and T. MacDonald (eds.), Marcel Dekker, New York, 1972.
130. Cole, H., and Y. Okaya: *Lab. Manage.*, November 1965.

131. Cole, H.: "Proceedings of the IBM Scientific Computing Symposium on Computer-aided Experimentation," p. 1, IBM Corp., White Plains, N.Y., 1966.
132. Busing, W. R., R. D. Ellison, H. A. Levy, S. P. King, and R. T. Roseberry: *AEC Rep. ORNL*-4143, January 1968.
133. Desper, C. R.: "Advances in X-ray Analysis," vol. 12, p. 404, Plenum, New York, 1969.
134. Fueg, G., E. Sanchez, H. Gasparoux, and S. Flandrois: *C. R. Acad. Sci.*, series B, 271:784 (1970).
135. Freeman, H. C., J. M. Guss, C. F. Nackolds, R. Page, and A. Webster: *Acta Crystallogr.*, sect. A, 26:149 (1970).
136. Jitaka, Y.: *Kogyo Kagaku Zasshi*, 71:14 (1968).
137. Burke, M. F., and R. G. Thurman: *J. Chromatogr. Sci.*, 8:39 (1970).
138. Thurman, R. G., K. A. Mueller, and M. F. Burke: *J. Chromatogr. Sci.*, 9:77 (1971).
139. Perone, S. P., D. O. Jones, and W. F. Gutknecht: *Anal. Chem.*, 41:1154 (1969).
140. Perone, S. P., and J. Zipper: unpublished results.
141. Deming, S. N., and H. L. Pardue: *Anal. Chem.*, 43:192 (1971).
142. Frazer, J. W., A. M. Kray, L. R. Carlson, M. R. Bertoglio, and S. P. Perone: *Anal. Chem.*, 43:1479 (1971).
143. Perone, S. P., J. W. Frazer, and A. Kray: *Anal. Chem.*, 43:1485 (1971).
144. Jones, D. O., and S. P. Perone: *Anal. Chem.*, 42:1151 (1970).
145. Frazer, J. W.: *Anal. Chem.*, 40:26A (1968).
146. Frazer, J. W.: *Chem. Instrum.*, 2:271 (1970).
147. Westerberg, A. W.: *Anal. Chem.*, 41:1595 (1969).
148. Westerberg, A. W.: *Anal. Chem.*, 41:1770 (1969).
149. Hancock, H. A., Jr., L. A. Dahm, and J. F. Muldoon: *J. Chromatogr. Sci.*, 8:57 (1970).
150. Murray, G. L., and B. E. F. Macefield: *Nucl. Instrum. Methods*, 51:229 (1967).
151. Fryklund, J. W.: *IBM J. Res. Dev.*, 13:75 (1969).
152. Gladney, H. M.: *J. Comp. Phys.*, 2:255 (1968).
153. Sederholm, C. H., P. J. Friedl, and T. R. Lusebrink: paper presented at Pittsburgh Conf. Anal. Chem. Appl. Spectros., March 1968.
154. Wiederhold, G.: "Proceedings of IBM Scientific Computing Symposium on Computers in Chemistry," p. 249, IBM Data Processing Division, White Plains, N.Y., 1969: H. M. Gladney, ibid., p. 157 (and references therein).
155. Scherer, J. R., and S. Kint: *Appl. Opt.*, 9:1615 (1970).
156. Ziegler, E., D. Henneberg, and G. Schomburg: *Anal. Chem.*, 42:51A (1970).
157. Shapiro, M., and A. Schultz: *Anal. Chem.*, 43:398 (1971).
158. Briggs, P. P.: *Control Eng.*, 14:75 (1967).
159. Felton, H. R., H. A. Hancock, and J. L. Knupp, Jr.: *Instrum. Control Syst.*, 40:83 (1967).
160. Raymond, A. J., D. M. G. Lawrey, and T. J. Mayer: *J. Chromatogr. Sci.*, 8:1 (1970).
161. Tivin, F.: *J. Chromatogr. Sci.*, 8:13 (1970).
162. Wilson, W. O., and J. G. W. Price: *J. Chromatogr. Sci.*, 8:31 (1970).
163. Baumann, F., D. L. Wallace, and L. G. Brenden: "Interfacing Chromatographs to the Computer," Varian Aerograph, Walnut Creek, Calif., 1969.
164. Gill, J. M.: *J. Chromatogr. Sci.*, 7:731 (1969).
165. Klopfenstein, C. E.: paper presented at Pittsburgh Conf. Anal. Chem. Appl. Spectrosc., March 1971.
166. Sederholm, C. H.: paper presented at Pittsburgh Conf. Anal. Chem. Appl. Spectrosc., March 1971.
167. Anderson, R. E.: *J. Chromatogr. Sci.*, 7:725 (1969).
168. Perone, S. P., and J. F. Eagleston: *J. Chem. Educ.*, 48:317 (1971).

169. Childs, C. W., P. S. Hallman, and D. D. Perrin: *Talanta*, 16:629 (1969).
170. Smith, D. E.: *J. AOAC*, 52:206 (1969).
171. Kuzel, N. R., H. E. Roudebush, and C. E. Stevenson: *J. Pharm. Sci.*, 58:381 (1969).
172. Venkataraghavan, R., R. J. Klimowski, and F. W. McLafferty: *Acc. Chem. Res.*, 3:158 (1970).
173. Edwards, R. A.: *Lab. Manage.*, September 1965; ibid., October 1965.
174. Swalen, J.: "Proceedings of IBM Scientific Computing Symposium on Computers in Chemistry," p. 9, IBM Data Processing Division, White Plains, N.Y., 1969.
175. Sederholm, C.: ibid., p. 35.
176. Perone, S. P.: *J. Chromatog. Sci.*, 7:714 (1969).
177. Hobbs, L. C.: *Datamation*, March 1969.
178. *Sci. Res.*, May 26, 1969; *Sci. Res.*, Oct. 4, 1969, p. 21.
179. Krugers, J. F.: *Chem. Weekbl.*, 66(51):26 (1970).
180. Wiberg, K. B.: *J. Chem. Educ.*, 47:113 (1970).
181. Minami, S.: *Bunseki Kagaku*, 19:721 (1970).
182. French, M.: *EEE*, 18:62 (1970).
183. Carter, W. C.: *Comput. Decis.*, November 1970, p. 17.
184. Liljenzin, J. O.: *Kem. Tidskr.*, 82(7):14 (1970).
185. Perone, S. P.: *J. Chem. Educ.*, 47:105 (1970).
186. Uttal, W. R.: "Real-time Computers," Harper & Row, New York, 1968.
187. Hoeschele, D. F., Jr.: "Analog-to-digital/Digital-to-analog Conversion Techniques," Wiley, New York, 1968.
188. Malmstadt, H. V., and C. G. Enke: "Digital Electronics for Scientists," Benjamin, New York, 1969.
189. Malmstadt, H. V., and C. G. Enke: "Computer Logic," Benjamin, New York, 1970.
190. Fisher, D. J.: *Chem. Instrum.*, 2:3 (1969).
191. Lorenz, L. J., R. A. Culp, and L. B. Rogers: *Anal. Chem.*, 42:979 (1970).
192. Deming, S. N., and H. L. Pardue: *Anal. Chem.*, 42:1466 (1970).
193. Glenn, T. H., and S. P. Cram: *J. Chromatogr. Sci.*, 8:46 (1970).
194. Enke, C. G.: *Anal. Chem.*, 43:69A (1971).
195. Anderson, N. G.: *Science*, 166:317 (1969).
196. Hatcher, D. W., and N. G. Anderson: *Am. J. Clin. Pathol.*, 52:645 (1969).
197. Kelley, M. T., and J. M. Jansen: *Clin. Chem.*, 17:701 (1971).
198. Maclin, E.: *Clin. Chem.*, 17:707 (1971).
199. Tiffany, T. O., G. F. Johnson, and M. E. Chilcote: *Clin. Chem.*, 17:715 (1971).
200. Ottaway, J. H.: *Biochem. J.*, 121:1PA (1971).
201. Fay, C., and S. Hayes: *Biochem. J.*, 121:1PB (1971).
202. Abernathy, M. H., G. T. Bentley, D. Gartelmann, P. Gray, J. A. Owen, and G. D. Quan-Sing, *Biochem. J.*, 121:4PA (1971).
203. Holmes, W. F., W. H. Holland, and J. A. Parker: *Anal. Chem.*, 43:1806 (1971).
204. Butterer, F., J. Roboz, L. Sarkozi, A. Ruhig, and P. Bacchin, *Clin. Chem.*, 17:789 (1971).
205. Plattner, J. R., and S. P. Markey: *Org. Mass Spectrom.*, 5:463 (1971).
206. Klimowski, R. J., R. Venkataraghavan, F. W. McLafferty, and E. B. Delany: *Org. Mass Spectrom., suppl.*, 4:17 (1970).
207. Oshima, S., M. Mitooka, T. Nishishita, and S. Nakoshi: *Sekiya Gakkai Shi*, 13:963 (1970).
208. Craven, D. A., E. S. Everett, and M. Rubel: *J. Chromatogr. Sci.*, 9:541 (1971).
209. Alber, L. L., M. W. Overton, and D. E. Smith: *J. Assoc. Off. Anal. Chem.*, 54:620 (1971).
210. Burke, M. F., and D. G. Ackerman: *Anal. Chem.*, 43:573 (1971).
211. Braun, G.: *J. Appl. Polym. Sci.*, 15:2321 (1971).

212. Chilcote, D. D., and J. E. Mrochek: *Clin. Chem.*, **17**:751 (1971).
213. Schmid, E. D., G. Berthold, H. Berthold, and B. Brosa: *Ber. Bunsenges. Phys. Chem.*, **75**:149 (1971).
214. Vandlen, R. L., and A. Tulinsky: *Acta Crystallogr.*, sect. B, **27**:437 (1971).
215. Andrinova, M. E., I. A. Kudryashov, and D. M. Kheiker: *Kristallografiya*, **16**:54 (1971).
216. Bridwell, L., L. M. Beyer, W. E. Maddox, and R. C. Etherton: *Nucl. Instrum. Methods*, **90**:187 (1970).
217. Lawless, J. L., and T. L. Anna: *IEEE Trans. Nucl. Sci.*, **18**:327 (1971).
218. Woods, R., H. K. Jennings, M. W. Collins, D. G. Harder, and D. E. McMillan: *J. Vac. Sci. Technol.*, **8**:352 (1971).
219. Schmitt, M., J. Henry, and Y. Flamaut: *Nucl. Instrum. Methods*, **91**:321 (1971).
220. Smith, D. H., R. W. Olsen, F. C. Walls, and A. L. Burlingame: *Anal. Chem.*, **43**:1796 (1971).
221. Housemann, J., and F. W. Hafner: *J. Phys. E.*, **4**(1):46 (1971).
222. McLafferty, F. W., R. Venkataraghavan, J. E. Coutant, and B. G. Giessner: *Anal. Chem.*, **43**:967 (1971).
223. Swingle, R. S., and L. B. Rogers: *Anal. Chem.*, **43**:810 (1971).
224. Cushley, R. J., D. R. Anderson, and S. R. Lipsky: *Anal. Chem.*, **43**:1281 (1971).
225. Lawrence, J., and D. M. Mohilner: *J. Electrochem. Soc.*, **118**:1596 (1971).
226. Mueller, K. A., and M. F. Burke: *Anal. Chem.*, **43**:641 (1971).
227. Johnson, S. A., W. M. Fairbank, and A. L. Schawlow: *Appl. Opt.*, **10**:2259 (1971).
228. Abbott, R. H.: *Biochem. J.*, **121**:3PB (1971).
229. Frueh, P. U., J. T. Clerc, and W. Simon: *Helv. Chim. Acta*, **54**:1445 (1971).
230. Chesler, S. N., and S. P. Cram: *Anal. Chem.*, **43**:1922 (1971).
231. Herlan, A.: *Fresenius' Z. Anal. Chem.*, **253**:1 (1971).
232. Molodenkova, I. D., and I. F. Kovalev: *Izv. Vyssh. Uchebn. Zaved., Fiz.*, **14**:157 (1971).
233. Johnson, K. J.: *J. Chem. Educ.*, **47**:819 (1970).
234. Jurs, P. C.: *Anal. Chem.*, **43**:1812 (1971).
235. Isenhour, T., and P. Jurs: *Anal. Chem.*, **43**:20A (1971).
236. Wangen, L. E., N. M. Frew, and T. L. Isenhour: *Anal. Chem.*, **43**:845 (1971).
237. Wangen, L. E., N. M. Frew, T. L. Isenhour, and P. C. Jurs: *Appl. Spectrosc.*, **25**:203 (1971).
238. Sybrandt, L. B., and S. P. Perone: *Anal. Chem.*, in press, 1972.
239. Hertz, H. S., R. A. Hites, and K. Biemann: *Anal. Chem.*, **43**:681 (1971).
240. Hertz, H. S., D. A. Evans, and K. Biemann: *Org. Mass Spectrom.*, suppl., **4**:453 (1970).
241. Arnett, E. M.: *Science*, **170**:1370 (1970).
242. Lytle, F. E.: *Anal. Chem.*, **43**:1334 (1971).
243. Wangen, L. E., W. S. Woodward, and T. L. Isenhour: *Anal. Chem.*, **43**:1605 (1971).
244. Stahl, D.: *Chimia*, **25**:149 (1971).
245. Van't-Klooster, H. A.: *Chem. Weekbl.*, **67**:9 (1971).
246. Levy, F.: *Chim. Anal. (Paris)*, **53**:402 (1971).
247. Brown, G. A., D. A. Nelson, and W. Ritzert: *Am. Lab.*, March 1971, p. 15.
248. Margoshes, M.: *Anal. Chem.*, **43**:101A (1971).
249. Nicoud, J. D.: *Chimia*, **25**:141 (1971).
250. Spragg, S. P.: *Biochem. J.*, **121**:2PB (1971).
251. Goeremann, T.: *Chimia*, **25**:144 (1971).
252. Brown, J. E.: *J. Am. Oil Chem. Soc.*, **48**:185 (1971).
253. Munson, A. W., and E. L. Schneider: *J. Am. Oil Chem. Soc.*, **48**:220 (1971).
254. Schneider, E. L., and A. W. Munson: *J. Am. Oil Chem. Soc.*, **48**:217 (1971).

13
Alternative Minicomputer Systems and Interface Design

In the preceding chapters describing on-line computer operations, a specific computer system was assumed in all discussions. This was necessary in order to make the discussions realistic and meaningful. The particular computer system used as a model for all previous discussions was the H.P. 2100 family of computers [1-3]. As pointed out earlier, however, the reader should be able to translate the material presented to the particular computer system which he is using.

To facilitate the reader in translating the concepts and specific discussions in this book to other computer systems, we have included in this chapter a discussion of the fundamental differences which exist between available minicomputer systems. In addition, a detailed listing of some software comparisons for various commercially available minicomputer systems is included in Appendix D. In this chapter, as well as in Appendix D, we have limited specific references to only a few of the many computer systems available. (The reader should see Ref. 4 for a more extensive and detailed discussion of a variety of computer systems.) The systems selected for specific discussion here are among those which have been most widely implemented in the laboratory. The characteristics of these systems are representative of the wide variety of minicomputer systems that are available. The specific systems considered in the following discussion and in Appendix D include the Digital Equipment Corp. (DEC) PDP-8, the DEC PDP-11, the Varian 620, and the Data General NOVA. (See Refs. 5 to 8.)

13-1 GENERAL CHARACTERISTICS OF MINICOMPUTER SYSTEMS

It should be pointed out here, first of all, that the discussions in this section and in succeeding sections are important to the reader who is interested in a general-purpose laboratory computer system. If the scientist is interested in purchasing a *turn-key* system for a particular job such as mass spectrometry, gas chromatography, Fourier transform spectroscopy, etc., then he need not be concerned with the detailed characteristics of the computer hardware. In this case, the scientist should only be interested in the overall characteristics of the turn-key system and its capability for doing the specified job.

A list of the general characteristics of minicomputer systems is provided in Table 13-1. Within the categories outlined in Table 13-1, the various commercially available minicomputers can differ widely in their characteristics. When all the minicomputer features are taken together, they determine the applicability of the system to the particular problem. It is difficult to say which of these features will be of greatest importance to the user. However, as long as the computer is to be used for on-line laboratory applications, among the most important features for evaluation of the system will be the I/O structure, the overall processor architecture, and the characteristics of the machine language (particularly with respect to I/O instructions). In this chapter, we will focus our attention primarily on these three characteristics.

13-2 PROCESSOR ARCHITECTURE

We will consider here three fundamentally different kinds of processor architecture: *CPU-oriented architecture, register-oriented architecture,* and

Table 13-1   General minicomputer characteristics

1. Word size
2. Processor architecture
3. I/O structure
4. Machine-language features
5. Speed (memory cycle and execution)
6. Hardware accessories included or available (EAU, DMA, priority interrupt, read-only memory, etc.)
7. Standard peripherals available
8. Standard software available

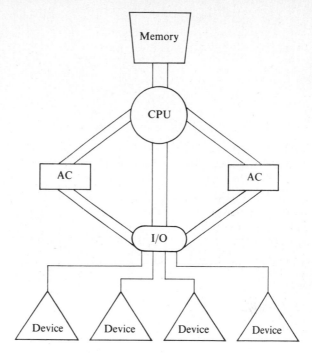

Fig. 13-1 CPU-oriented architecture.

device-oriented architecture. Each of these configurations includes a central processing unit (CPU), which carries out functions similar to those described earlier. However, in each of the three different structures, the orientation of the CPU with respect to the other computer hardware may be considerably different.

Figures 13-1 to 13-3 show diagrammatically the fundamental differences in computer architecture. In the CPU-oriented architecture the computer system usually has a limited number of accumulators (ACs) (one or two) which can be used for arithmetic and/or logical operations. Normal I/O operations will proceed through these ACs also. The CPU dominates all machine operations. All logical, arithmetic, and information-transfer functions proceed under its control. Moreover, because of the limited number of general-purpose registers (ACs), many references to memory are required to accomplish a given function. For example, in the simplest case where a single AC is available, the program frequently must save the AC's contents at appropriate intermediate points in a processing program. (The CPU-oriented architecture is representative of the type of computer system discussed in the rest of this book, namely the H.P. 2100 series. Other computers which fall in this category are the DEC PDP-8 and Varian 620.)

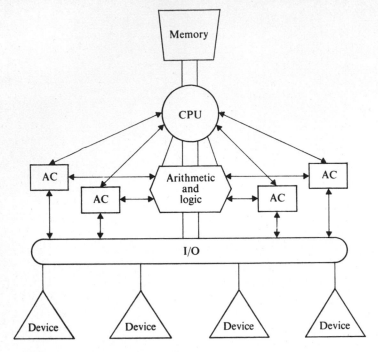

Fig. 13-2   Register-oriented architecture.

Figure 13-2 represents a register-oriented architecture. In this case, at least two general-purpose registers are included. Typical examples of this kind of architecture are seen in the NOVA or PDP-11 computers. The NOVA has four general-purpose registers, and the PDP-11 has eight. Aside from the availability of a larger number of working registers, this architecture provides for direct lines of communication among registers and the accomplishment of arithmetic and logical operations during register-to-register transfers. In this fashion, a minimum of memory reference operations is required to carry out computational functions. Any of the general-purpose registers may be used for specific functions, such as acting as index registers for memory reference instructions. (This is also possible in the CPU-oriented architecture with multiple registers.) I/O functions normally will proceed through the working registers. However, some register-oriented computers, like the PDP-11, provide for a different pathway for I/O functions. These computer systems fall into the third category of device-oriented systems.

In a device-oriented computer architecture (Fig. 13-3), every device which can accept or produce binary information is hung on a common bus. Normally, the CPU has control of the bus and determines what operations are to be performed, the direction of information transfer, and the *source* and *destination*

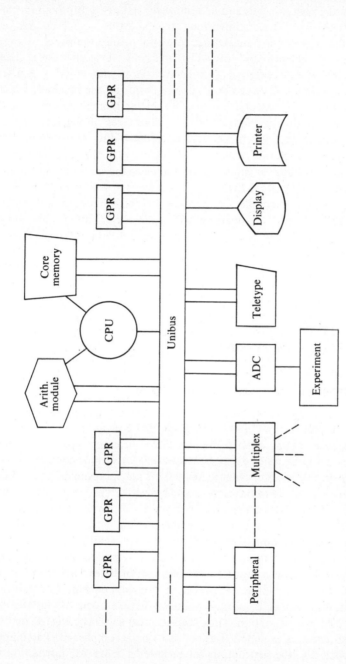

Fig. 13-3 Device-oriented computer configuration.

of information transfers. However, any device, except memory, can establish control over the common bus and determine the nature of information transfers which will take place. The distinguishing characteristics of such an architecture should be immediately apparent. Direct device-to-memory transfers, for example, can take place independent of CPU functions. That is, direct lines of communication among external peripheral devices and working registers or memory can be established without passing through the channels which are necessary in the CPU-oriented or register-oriented architectures. Thus, the device-oriented architecture retains all the advantages of the register-oriented architecture but adds to it the capability of providing more direct lines of communication among all devices associated with the computer system. Moreover, the possibility of imposing arithmetic or logical functions on the data transfers among devices other than working registers opens up a whole new range of possibilities. For example, information transfer between a register and memory might include a superimposed addition or subtraction function. Arithmetic operations can be performed directly on data obtained in memory. Data transfers from data-acquisition devices to memory could include superimposed arithmetic or logical operations. In general, then, each device on the bus can be accessed as readily as core memory. Moreover, the external devices can be considered as auxiliary general-purpose registers. This is true also for each location in memory, except memory can never exercise control over the common bus. Data transfers to or from memory must always be executed under the control of some other device.

13-3   MACHINE-LANGUAGE COMPARISONS

Obviously, the specific characteristics of the machine language will be very much dependent on the overall architecture of the computer system. Thus, the machine-language characteristics for the three different types of architecture described here will be quite different. However, for various computers within a given category, the machine-language functions are very similar. For example, it is almost trivially simple to translate from Hewlett-Packard assembly language to the PDP-8 or Varian 620 languages, or vice versa. (See Sec. D-1.) Despite the fact that the mnemonics are somewhat different, the fundamental CPU functions are very similar and the fundamental algorithms are readily transferable. Some instructions available on one computer system may not be available on another. For example, the Varian computer instruction set includes a Subtract instruction and a Decrement instruction. However, it does not include Compare or ISZ instructions. The PDP-8 computer does not have a Load AC instruction but utilizes an Add AC instruction. Thus, the AC must always be cleared first before an Add instruction is executed if a load is to be accomplished. To compensate for this deficiency, the instruction set includes a Store AC instruction which automatically clears the AC.

In the register-oriented computer system, the outstanding feature of the instruction set is that arithmetic **and logical** instructions involve register-to-register operations which are essentially Move instructions. These instructions include specification of a source, a destination, an operation, and a test on the resultant operation. (See Sec. D-3.) Any or all of these functions may be performed during the execution of the Move instruction. For device-oriented computer systems, the Move instruction is utilized also, except that the definition of source and destination can be expanded to include memory locations and peripheral devices. Unquestionably, the features of an assembly-language program to carry out a given computational function will look considerably different for a register-oriented machine compared to a CPU-oriented machine. The same fundamental algorithms for handling binary-coded data will hold, but the machine-level operations for implementing these algorithms will be considerably different for the three different architectures.

It is very difficult to establish whether one type of computer architecture is significantly advantageous over another type of architecture. This kind of evaluation can only be made with a very specific kind of application in mind. For example, if the majority of the programming effort will involve data-handling and data processing functions, the programmer may prefer to do most of the programming in a high-level language, as suggested in Chap. 11. In this case, the specific characteristics of the machine language are important only in so far as the computer manufacturer has provided a Compiler or Interpreter which efficiently utilizes that machine language. The user can compare the effectiveness of two different computer systems by executing identical high-level programs and determining which program executes most rapidly, with the most efficient utilization of memory and with adequate retention of numerical significance. However, only a limited aspect of the computer's capability is evaluated.

Regardless of whether the user utilizes high-level languages or assembly language for his programming effort, he must be concerned with the I/O structure and language of his machine if he is to use it for on-line applications. Sec. 13-4 focuses on this aspect of computer systems.

13-4   COMPARISON OF I/O STRUCTURES AND INSTRUCTIONS

One general observation should be made in comparing the I/O considerations for various computer architectures. Although the channeling of information may be considerably different because of the computer architecture, the fundamental aspects of the machine language for I/O operations are surprisingly similar from one computer system to another. (See Appendix D.) This is because the I/O functions can be reduced to the fundamental operations of *data transfers in and out, control commands from the computer, checking the status of external devices, and device selection.* The fundamental aspects of these functions were discussed in Chaps. 6 and 9. Thus, if we consider only program-controlled data transfers to and from peripheral devices, the machine-language considerations for the CPU-oriented and register-oriented computer architectures are very similar. For the device-oriented architecture, the fundamental difference is that the data need not be channeled through a working register. Moreover, interrupt-controlled data transfers can involve considerably different functions than for the CPU-oriented computer system. For example, with a device-oriented system, the generation of an interrupt by an external device will allow that device to exercise control over the common bus and transfer the data directly to the appropriate register or memory location. This kind of function is very similar to the direct-memory-access (DMA) function described in Chap. 9. For the device-oriented system, however, this is a standard feature which is inherent in the architecture.

There are some fundamental differences in I/O structures which may exist even within systems that have similar overall architecture. For many computer systems, we can consider at least three different I/O structures. One of these is the multichannel, buffered, hardware priority interrupt structure inherent in the Hewlett-Packard computer system used as a model in this text. Another structure is the party-line structure representative of the Varian 620, DEC PDP-8, and Data General NOVA computer systems. In the party-line structure, a single I/O line is available, and all external peripheral devices hang in parallel on this common line, or bus. This bus provides control lines, device-selection lines, interrupt, and a status-check line. The device-selection decoding logic is provided by the device rather than the computer hardware. For example, Fig. 13-4 shows schematically the party-line I/O structure for a PDP-8 computer. The I/O bus must provide at least 23 lines. (In fact, the PDP-8/e OMNIBUS[1] provides a total of 96 lines, which include power and logic-level voltages, ground, timing pulses, etc. [5].) Twelve lines are for I/O to the 12-bit AC, and 9 lines are required for the output of control bits and a select code. In addition, interrupt and skip lines are required. The 9-bit output code includes 6 bits for device selection and 3 bits to generate control functions. All devices hang in parallel on the AC bus, and data are strobed in or out of these devices from or to the AC by the proper decoding of control commands to the device-selector hardware associated with each device. Status checking is provided through the skip line, which can be activated through the device-selection hardware by feeding back a command bit to the computer. The interrupt line can be set by any of the devices in parallel, but the computer will not know a priori which device has caused the interrupt. The computer must execute a prearranged sequence of status-check software operations to determine which of the devices on the party-line has caused the interrupt. If several devices are capable of interrupting the system, then the interrupt service program may be quite lengthy and involved. The priority of

---

[1] OMNIBUS is a registered trademark of Digital Equipment Corp., Maynard, Mass.

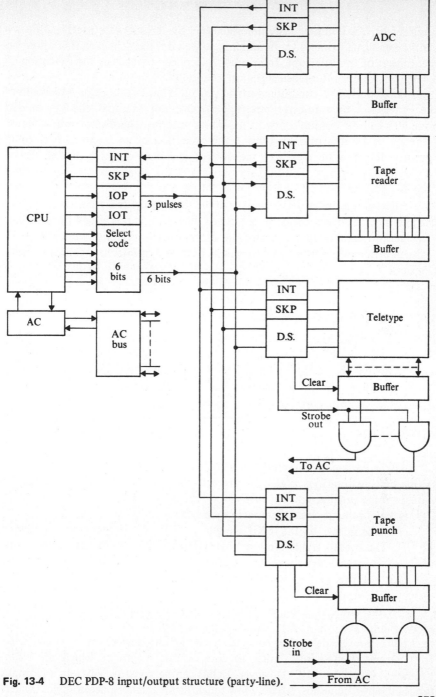

Fig. 13-4   DEC PDP-8 input/output structure (party-line).

353

interrupt service will obviously be established by the software routine. Priority can be established simply by determining which device is checked first upon the recognition of the interrupt. Interrupts are serviced by causing the execution of a JMS (jump subroutine) instruction to location 0.

Some party-line computer systems provide a hardware option which allows the conversion to a priority interrupt system. For example, the Varian 620 computer can be configured with a priority interrupt hardware option which responds to an interrupt by causing the execution of an instruction at some specified location in memory; the address specified depends on the device causing the interrupt. This is very similar to the interrupt service function described for the multichannel system in Chaps. 6 and 9.

The third type of I/O structure is the device-oriented architecture where each peripheral device can be in control of the common bus as described earlier. This allows each external device to communicate directly with computer registers or memory, either with program-controlled or interrupt-controlled data transfers.

13-5   MICROPROGRAMMABLE COMPUTERS

Regardless of the computer architecture, as represented by Figs. 13-1 to 13-3, the types of computer systems discussed thus far all share one common feature. They have all been described as fixed-instruction general-purpose (FIGP) machines. They are general-purpose in that the instruction set and overall architecture are not designed for any specific task but can be adapted to a wide variety of applications. They are fixed-instruction devices because the machine instructions are hard-wired into the CPU and are not generally alterable by the user.

Other common characteristics include the fact that the speed of execution of machine instructions is tied to the core memory cycle time and generally requires some integral number of memory cycles for completion. Moreover, the CPU is the dominant *control device* in each system.

An alternative to the FIGP system is the *microprogrammed computer* [2-4,9]. The overall architecture of this type of system can be analogous to any of the three considered above. One possible configuration is shown in Fig. 13-5. The heart of this system is the *control memory*. This is usually a *read-only-memory* (ROM) device.

As the name implies, an ROM is a memory module with an unalterable array of binary words [3,9]. The contents of each location can be fetched and read but cannot be deleted or changed. Such memory devices can be constructed from magnetic materials similar to core or disc memories which are write-protected. (See Chap. 10.) However, ROMs can be constructed from solid-state semiconductor materials, which allow much faster memory access time. In fact, access times on the order of 30 to 100 nsec are possible.

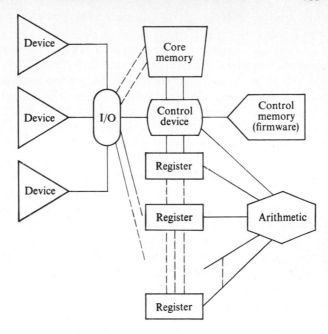

Fig. 13-5 Typical microprogrammable computer configuration.

It is the availability of high-speed ROMs which makes microprogrammable computers feasible and advantageous. With reference to Fig. 13-5, the general philosophy of the microprogrammable computer is to provide a *control device* capable of executing a limited, but very fundamental, set of *microinstructions—* such as logical AND, OR, 1-bit rotate, 1-bit shift, add, clear, complement, logical tests, I/O, register-register and register-memory transfers, etc. These hard-wired machine instructions (microinstructions) are normally more limited and fundamental than the instruction set provided with a FIGP computer. To generate more powerful instructions, the ROM is made to contain various subroutines which generate appropriate sequences of the microinstructions to accomplish a given algorithm. Because the ROM has such rapid access time and because several semiconductor (high-speed) working registers are usually provided as *scratch-pad* memory, these ROM subroutines can be executed rapidly. Thus, the speed of execution of many instructions can be divorced from the core memory cycle time of the computer. (In fact, in some designs the control memory can operate completely independently of the normal CPU core memory operations.)

Very sophisticated and complex algorithms can be implemented by the ROM subroutines (called *firmware*), and each can be called by a macro instruction defined by the user for his assembly-language programming. These firmware subroutines can be executed very rapidly but may be limited ultimately by the number of references to slow core memory required for completion.

The microprogrammable computer functions are similar to the FIGP computer functions, except that the user may *define his own machine-language instruction set.* That is, the nature of the programming language used for his machine will depend on the firmware subroutines he has specified. Thus, he may choose to have macro instructions—such as MPY, SQRT, LOG, STRTADC, DACOUT, etc.—which call specific firmware subroutines to accomplish indicated operations. This flexibility is possible because of the reasonable cost of producing an infinite variety of specified ROMs.

The advantages of microprogramming should be obvious. One is that a computer can be tailor-made, as specified by the user, for specific computational or control operations. Another is that the macro instruction set can be changed by replacing the control memory. Also, it should be obvious that the macro instruction set can provide a very high-level programming language—which is essentially the machine language as far as the programmer is concerned. Thus, "machine-language" instruction sets defined by BASIC, FORTRAN, ALGOL, etc., are possible.

Examples of microprogrammable computers include the H.P. 2100, Digital Scientific META 4, and Microdata MICRO 800. Some pertinent characteristics of each of these machines are given in Table 13-2. One of these machines, the H.P. 2100, is one which we have used as a model in this text; although no mention had been made previously of its microprogramming characteristics. In

Table 13-2   Some characteristics of typical microprogrammable computers[a]

Characteristics	H.P. 2100	Digital Scientific META 4	Microdata MICRO 800
ROM access time	196 nsec	35 nsec	220 nsec
ROM capacity	up to 1,024 (24-bit)	up to 4,096 (16-bit)	up to 1,024 (16-bit)
Scratch-pad registers	6 (16-bit)	up to 256 (16-bit)	16 (8-bit)
Core memory size	up to 32K (16-bit)	up to 458K (18-bit)	up to 32K (8-bit)
Core memory cycle time	980 nsec	900 nsec	1,100 nsec

[a]Taken from manufacturers' literature supplied by:

Hewlett Packard, Cupertino, Calif.
Digital Scientific Corp., San Diego, Calif.
Microdata Corp., Santa Ana, Calif.

fact, the H.P. 2100 was designed to be a completely flexible, microprogrammable computer. However, to provide compatibility with the earlier Hewlett-Packard models (2116, 2115, 2114), one 256-word ROM module was programmed to implement a machine-language instruction set identical to the earlier models, including integer-multiply, integer-divide, etc., previously hard-wired in an EAU option. An additional 256-word ROM module provides complete floating-point arithmetic firmware (with features noted in Chap. 5). This allows the user an additional 512 words of ROM to use for customized firmware. For more details on the H.P. 2100, see Refs. 2 and 3; for a good general discussion of microprogrammable computers, see Ref. 9.

13-6   FUNDAMENTAL PRINCIPLES FOR INTERFACE DESIGN

It seems appropriate here to pull together many of the instrumental concepts discussed throughout the book relating to the design of interfacing between computer and experimental systems.

Scope of the interface   It should be recognized, first of all, that the interface includes everything from the transducer or transducers used in the laboratory instrumentation to the computer software designed to communicate with that experiment. Everything in between contributes to the interface. This includes the computer I/O hardware, the signal-handling devices of the analog instrumentation, translation elements, signal transmission devices, and connections to and from the computer. It should be emphasized here that it is unrealistic to consider interface design without recognizing all the components which are part of the interface, starting with the chemical instrumentation itself.

Interface design   In this section, we will try to provide some general guidelines for interface design. The various considerations are summarized in Table 13-3. The first of these steps involves a definition of what the outside world sees at the computer terminals. As described earlier, the general approach will be similar for various computers. That is, each computer system must provide for the output of command bits as well as status checking and input and output of binary-coded data. Details may differ from one computer to the next. For example, with the PDP-8 computer, the device-selection code is output to the outside world and must be decoded by the external device. In addition, the PDP-8 outputs three command bit lines, where these command bits are sequential pulses appearing at different points in time after initiation of the instruction. If the user desires to check the status of the particular device, he must use one of these output pulses to feed back to the skip line. This can be accomplished by the hardware shown in Fig. 13-6. The user composes the instruction word such that the particular line externally wired to check the status of the device is pulsed by the instruction. Then the instruction is inserted

Table 13-3   General interface design considerations

---

1. CPU I/O structure
    a. Primary (I/O bus features: command, status check, data in/out)
    b. Secondary (standard interface devices: buffer registers, sense switches, DACs, ADCs, clock, multiplexer, T/H, logic devices, etc.)

2. Laboratory instrumentation characteristics
    a. Output features (transducer, noise, range, location, bandwidth, etc.)
    b. Control requirements (start/stop, time-base synchronization, experimental control parameters, etc.)

3. Processing and data handling
    a. Passive (data logging)
    b. Active (real-time control)
    c. Dedicated or time-sharing

---

into the program as a Skip-on-flag instruction. Thus, if the external device status is correct, the output pulse will be fed back to the skip line when the instruction is executed, and the computer will skip the next sequential instruction. By comparison with the discussion of I/O hardware for the Hewlett-Packard computer described in Chaps. 6 and 9, the reader can see the contrast in functions. Moreover, one must recognize that a clear understanding of the specific I/O structure of the computer is necessary to generate appropriate experimental interfacing.

In addition to the primary features of the computer's I/O structure, the user must consider the characteristics of any standard interface devices incorporated into the system. For example, a computer system might include an interfaced internal time-base generator or a real-time clock. Other standard interface devices might include I/O buffer registers, sense switches, multiplexers, digital-to-analog converters (DACs), analog-to-digital converters (ADCs), etc. These and other devices may be included as an integral part of a purchased computer system, and it is necessary for the user to understand the required input characteristics, the specified output characteristics, and the nature of the programming required to properly operate the standard interface devices. Most importantly, the user must be aware of the limitations of applicability of the primary and secondary I/O devices. For example, he must be aware of whether digital data input are buffered or unbuffered at the computer interface. He must know whether the computer, the external device, or both determine whether a true flag bit will cause an interrupt or not. With regard to the secondary devices, the user must be aware of the analog range limitation of the ADC or DAC devices, of how the real-time clock is synchronized with external events, of how the computer determines the status of sense switches, etc.

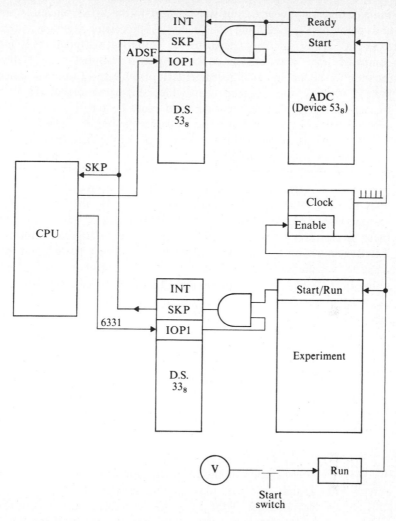

Fig. 13-6 Typical communication lines for on-line DEC PDP-8.

The second major function in interface design is to define the characteristics of the laboratory instrumentation which the user desires to place on-line to the computer. One aspect of the laboratory instrumentation that must be considered is the output characteristics. For example, one needs to know what the inherent features of the transducer element are. Does it produce a current or a voltage signal related to the fundamental phenomenon of interest? Perhaps a discontinuous output is obtained such as in the discrete pulse output of a photomultiplier when it is used for photon counting, or in the pulse-train output of the quartz thermometer. Also to be considered are the characteristics of the fundamental transducer output with respect to the signal frequency, amplitude, and noise background, and the analog preprocessing functions desired. For example, the signal may need to be amplified, differentiated, integrated, compared, etc., with analog devices. Another essential consideration is the location of the fundamental instrumentation with respect to the data-acquisition system, that is, the distance over which analog or digital signals must be transmitted. This distance consideration must be coupled with the inherent characteristics of the signal in order to design the proper transmission elements. The user must also determine the overall nature of the experiment and associated signals with which the time base must be synchronized. The user must be aware of what inherent experimental characteristics are accessible for synchronization. For example: Can the experiment be initiated at a precise moment in time or does the experiment initiate spontaneously? Finally, the required digitization features must be established. That is, the bandwidth of the experimental signal must be established in order to define an adequate data-acquisition frequency. The dynamic range of the analog elements producing the experimental signal must be evaluated in order to establish the precision of the ADC required. The noise-rejection features of the interface hardware or data-acquisition software can be determined only when the noise level of the experimental signal has been established.

The other major aspect of laboratory instrumentation to be considered for interface design includes the control requirements. For example: Exactly how is an experiment initiated? Does it require a manual-mechanical operation, such as mixing, or can an electrical signal be used? Also, how is the experiment terminated? What mechanical or electronic control functions are required? Do these control functions involve the establishment of voltage, current, or waveform frequency values? What precision and accuracy of controls are desired?

A third and very important consideration in overall interface design involves the determination of the specific processing and data-handling requirements. For example, the user must determine whether his on-line application will involve dedicated use of the computer or perhaps time sharing with several different devices or instruments. Furthermore, the design of the interface may be considerably different, depending on whether the program will involve simple data-logging operations or will involve real-time interaction with the experiment. To be more specific, if the application involves a time-sharing operation, the user may very well choose to design his data acquisition around the interrupt system. If the application requires a considerable amount of real-time computer processing and interaction with the experiment, the user may need to design the interfacing to minimize the amount of real-time bookkeeping achieved by the computer. For example, a DMA channel may be used, or the hardware interface might be designed to include certain bookkeeping or logical functions which might ordinarily be executed by the computer in real time.

Obviously, the preceding discussion on general considerations for interface design leaves much to be desired with regard to specific systems. However, to do an adequate job of describing interface considerations for the wide variety of possible applications is beyond the scope of this text. It is hoped that the inexperienced reader will be able to use the preceding discussion as a guide for approaching his interface design problems. The specific and detailed discussions included in Chaps. 6-9 should be helpful in elucidating the intricacies of specific interface hardware and software.

BIBLIOGRAPHY

1. "A Pocket Guide for Hewlett-Packard Computers," Hewlett-Packard, Cupertino, Calif.
2. Coury, F. F.: *Hewlett-Packard J.*, October 1971, p. 2.
3. Leis, C. T.: *Hewlett-Packard J.*, October 1971, p. 4.
4. Bell, G. G., and A. Newell: "Computer Structures: Readings and Examples," McGraw-Hill, New York, 1971.
5. "Small Computer Handbook," Digital Equipment Corp., Maynard, Mass., 1972.
6. "PDP-11 Handbook," Digital Equipment Corp., Maynard, Mass.
7. "620/f Computer Handbook," Varian Data Machines, Irvine, Calif.
8. "How to Use the NOVA Computers," Data General Corp., Southboro, Mass.
9. "Microprogramming Handbook," Microdata Corp., Santa Ana, Calif., 1971.

appendix A

A-1 ASSEMBLY-LANGUAGE MACHINE INSTRUCTIONS FOR H.P. 2100 FAMILY

(Cycle times: model 2116, 1.6 μsec; models 2115 and 2114, 2.0 μsec; model 2100, 0.98 μsec)

MEMORY REFERENCE

[All require 2 cycles except JMP, requiring 1 cycle, and ISZ, requiring 2.25 cycles or 3 cycles (2100 only).]

Jump and increment-skip

ISZ	m [,I]	$(m) + 1 \rightarrow m$; then if $(m) = 0$, execute P + 2; otherwise execute P + 1.[1]
JMP	m [,I]	Jump to m; $m \rightarrow$ P.
JSB	m [,I]	Jump subroutine to m: P + 1 $\rightarrow$ m; $m + 1 \rightarrow$ P.

[1] P is the program counter register which normally contains the address of the instruction being executed.

Add, load and store

ADA	m [,I]	$(m) + (A) \to A$.
ADB	m [,I]	$(m) + (B) \to B$.
LDA	m [,I]	$(m) \to A$.
LDB	m [,I]	$(m) \to B$.
STA	m [,I]	$(A) \to m$.
STB	m [,I]	$(B) \to m$.

Logical

AND	m [,I]	$(m) \cdot (A) \to A$.
XOR	m [,I]	$(m) \oplus (A) \to A$.
IOR	m [,I]	$(m) + (A) \to A$.
CPA	m [,I]	If $(m) \neq (A)$, execute $P + 2$; otherwise execute $P + 1$.
CPB	m [,I]	If $(m) \neq (B)$, execute $P + 2$; otherwise execute $P + 1$.

REGISTER REFERENCE

(All require one cycle on models 2116, 2115, 2114; two cycles are required on model 2100.)

Shift-rotate

CLE	$0 \to E$.
ALS	Shift (A) left 1 bit; $0 \to A_0$; A_{15} is unaltered.
BLS	Shift (B) left 1 bit, $0 \to B_0$; B_{15} is unaltered.
ARS	Shift (A) right 1 bit; $(A_{15}) \to A_{14}$.
BRS	Shift (B) right 1 bit; $(B_{15}) \to B_{14}$.
RAL	Rotate (A) left 1 bit.
RBL	Rotate (B) left 1 bit.
RAR	Rotate (A) right 1 bit.
RBR	Rotate (B) right 1 bit.
ALR	Shift (A) left 1 bit; $0 \to A_{15}$.
BLR	Shift (B) left 1 bit; $0 \to B_{15}$.
ERA	Rotate E and A right 1 bit.
ERB	Rotate E and B right 1 bit.
ELA	Rotate E and A left 1 bit.
ELB	Rotate E and B left 1 bit.
ALF	Rotate A left 4 bits.
BLF	Rotate B left 4 bits.
SLA	If $(A_0) = 0$, execute $P + 2$; otherwise execute $P + 1$.
SLB	If $(B_0) = 0$, execute $P + 2$; otherwise execute $P + 1$.

Shift-rotate instructions can be combined as follows

ALS			ALS
ARS			ARS
RAL			RAL
RAR	[,CLE]	[,SLA]	RAR
ALR			ALR
ALF			ALF
ERA			ERA
ELA			ELA

BLS			BLS
BRS			BRS
RBL			RBL
RBR	[,CLE]	[,SLB]	RBR
BLR			BLR
BLF			BLF
ERB			ERB
ELB			ELB

No-operation

NOP　　　　Execute $P + 1$.

Alter-skip

CLA	$0\text{'s} \rightarrow A$.
CLB	$0\text{'s} \rightarrow B$.
CMA	$(\overline{A}) \rightarrow A$.
CMB	$(\overline{B}) \rightarrow B$.
CCA	$1\text{'s} \rightarrow A$.
CCB	$1\text{'s} \rightarrow B$.
CLE	$0 \rightarrow E$.
CME	$(\overline{E}) \rightarrow E$.
CCE	$1 \rightarrow E$.
SEZ	If $(E) = 0$, execute $P + 2$; otherwise execute $P + 1$.
SSA	If $(A_{15}) = 0$, execute $P + 2$; otherwise execute $P + 1$.
SSB	If $(B_{15}) = 0$, execute $P + 2$; otherwise execute $P + 1$.
INA	$(A) + 1 \rightarrow A$.
INB	$(B) + 1 \rightarrow B$.
SZA	If $(A) = 0$, execute $P + 2$; otherwise execute $P + 1$.
SZB	If $(B) = 0$, execute $P + 2$; otherwise execute $P + 1$.
SLA	If $(A_0) = 0$, execute $P + 2$; otherwise execute $P + 1$.
SLB	If $(B_0) = 0$, execute $P + 2$; otherwise execute $P + 1$.
RSS	Reverse sense of skip instructions; if no skip instructions precede, execute $P + 2$.

Alter-skip instructions can be combined as follows

```
CLA            CLE
CMA    [,SEZ]  CME    [,SSA] [,SLA] [,INA] [,SZA] [,RSS]
CCA            CCE


CLB            CLE
CMB    [,SEZ]  CME    [,SSB] [,SLB] [,INB] [,SZB] [,RSS]
CCB            CCE
```

INPUT/OUTPUT, OVERFLOW, AND HALT

(All require one cycle on models 2116, 2115, 2114; two cycles are required on model 2100.)

Input/output

STC sc[1] [,C][2] Set control bit; enable transfer of one
                    element of data between device and buffer.

CLC sc [,C] Clear control bit; if sc = 0, clear all
              control bits.

LIA sc [,C] $(buffer_{sc}) \rightarrow A$.

LIB sc [,C] $(buffer_{sc}) \rightarrow B$.

MIA sc [,C] $(buffer_{sc}) \vee (A) \rightarrow A$.

MIB sc [,C] $(buffer_{sc}) \vee (B) \rightarrow B$.

OTA sc [,C] $(A) \rightarrow buffer_{sc}$.

OTB sc [,C] $(B) \rightarrow buffer_{sc}$.

STF sc Set flag bit; if sc = 0, enable interrupt
         system; sc = 1 sets overflow bit.

CLF sc Clear flag bit; if sc = 0, disable interrupt
         system; if sc = 1, clear overflow bit.

SFC sc If (flag bit) = 0, execute P + 2; otherwise
         execute P + 1; if sc = 1, test overflow bit

SFS sc If (flag bit) = 1, execute P + 2; otherwise
         execute P + 1; if sc = 1, test overflow bit.

Overflow

CLO $0 \rightarrow$ overflow bit.

STO $1 \rightarrow$ overflow bit.

SOC   If (overflow bit) = 0, execute P + 2; otherwise
          execute P + 1

SOS   If (overflow bit) = 1, execute P + 2; otherwise
          execute P + 1

[1] sc = select code = I/O channel number.
[2] ,C = execute CLF simultaneously.

Halt

HLT   sc [,C]   Halt computer.

EXTENDED ARITHMETIC UNIT

(Requires EAU version of Assembler or Extender Assembler.)

MPY m [,I]   (A) $\times$ (m) $\rightarrow$ $(B_{\pm msb}$ and $A_{lsb})$.

DIV m [,I] $(B_{\pm msb}$ and $A_{lsb})/(m)$ $\rightarrow$ A; remainder $\rightarrow$ B.

DLD m [,I] (m) and $(m+1)$ $\rightarrow$ A and B.

DST m [,I]   (A) and (B) $\rightarrow m$ and $(m+1)$.

(DLD and DST require 5 cycles; MPY, 12 cycles; DIV, 13 cycles on models 2116, 2115, 2114. For model 2100, respective times are 5.9, 12.7, and 16.7 μsec.)

ASR b Arithmetically shift (B and A) right b bits;
                 B_{15} is extended.

ASL b Arithmetically shift (B and A) left b bits;
                 B_{15} is unaltered; 0's to A_{lsb}.

RRR b Rotate (B and A) right b bits.

RRL b Rotate (B and A) left b bits.

LSR b Logically shift (B and A) right b bits; 0's
                 to B_{msb}.

LSL b Logically shift (B and A) left b bits; 0's
                 to A_{lsb}.

SWP   (A) $\rightarrow$ B; (B) $\rightarrow$ A.

[Extended shift-rotate instructions require two to five cycles (or 2.9 to 7.8 μsec on model 2100) depending on the number of shift-rotate operations.]

A-2   ASCII CODE[1]

| | Octal code | | | Octal code | |
Character	First byte	Second byte	Character	First byte	Second byte
A	040400	000101	H	044000	000110
B	041000	000102	I	044400	000111
C	041400	000103	J	045000	000112
D	042000	000104	K	045400	000113
E	042400	000105	L	046000	000114
F	043000	000106	M	046400	000115
G	043400	000107	N	047000	000116

[1] Seven-bit code.

A-2 (continued)

Character	Octal code First byte	Octal code Second byte	Character	Octal code First byte	Octal code Second byte
O	047400	000117	>	037000	000076
P	050000	000120	?	037400	000077
Q	050400	000121	@	040000	000100
R	051000	000122	[	055400	000133
S	051400	000123		056000	000134
T	052000	000124	]	056400	000135
U	052400	000125	↑	057000	000136
V	053000	000126	←	057400	000137
W	053400	000127	ACK	036000	000174
X	054000	000130	1	036400	000175
Y	054400	000131	ESC	037000	000176
Z	055000	000132	DEL	037400	000177
			NULL	000000	000000
0	030000	000060	SUM	000400	000001
1	030400	000061	EOA	001000	000002
2	031000	000062	EOM	001400	000003
3	031400	000063	EOT	002000	000004
4	032000	000064	WRU	002400	000005
5	032400	000065	RU	003000	000006
6	033000	000066	BELL	003400	000007
7	033400	000067	RE_0	004000	000010
8	034000	000070	HT/SK	004400	000011
9	034400	000071	LF	005000	000012
			VTAB	005400	000013
"space"	020000	000040	FF	006000	000014
!	020400	000041	CR	006400	000015
..	021000	000042	SO	007000	000016
#	021400	000043	SI	074000	000017
$	022000	000044	DC_0	010000	000020
%	022400	000045	DC_1	010400	000021
&	023000	000046	DC_2	011000	000022
'	023400	000047	DC_3	011400	000023
(	024000	000050	DC_4	012000	000024
)	024400	000051	ERR	012400	000025
*	025000	000052	SYNC	013000	000026
+	025400	000053	LEM	013400	000027
,	026000	000054	S_0	014000	000030
−	026400	000055	S_1	014400	000031
	027000	000056	S_2	015000	000032
/	027400	000057	S_3	015400	000033
:	035000	000072	S_4	016000	000034
;	035400	000073	S_5	016400	000035
<	036000	000074	S_6	017000	000036
=	036400	000075	S_7	017400	000037

A-3 MATHEMATICAL EQUIVALENTS

2^n		n	2^{-n}						
	1	0	1.0						
	2	1	0.5						
	4	2	0.25						
	8	3	0.125						
	16	4	0.0625						
	32	5	0.03125						
	64	6	0.01562	5					
	128	7	0.00781	25					
	256	8	0.00390	625					
	512	9	0.00195	3125					
	1024	10	0.00097	65625					
	2048	11	0.00048	82812	5				
	4096	12	0.00024	41406	25				
	8192	13	0.00012	20703	125				
	16384	14	0.00006	10351	5625				
	32768	15	0.00003	05175	78125				
	65536	16	0.00001	52587	89062	5			
1	31072	17	0.00000	76293	94531	25			
2	62144	18	0.00000	38146	97265	625			
5	24288	19	0.00000	19073	48632	8125			
10	48576	20	0.00000	09536	74316	40625			
20	97152	21	0.00000	04768	37158	20312	5		
41	94304	22	0.00000	02384	18579	10156	25		
83	88608	23	0.00000	01192	09289	55078	125		
167	77216	24	0.00000	00596	04644	77539	0625		
335	54432	25	0.00000	00298	02322	38769	53125		
671	08864	26	0.00000	00149	01161	19384	76562	5	
1342	17728	27	0.00000	00074	50580	59692	38281	25	
2684	35456	28	0.00000	00037	25290	29846	19140	625	
5368	70912	29	0.00000	00018	62645	14923	09570	3125	
10737	41824	30	0.00000	00009	31322	57461	54785	15625	
21474	83648	31	0.00000	00004	65661	28730	77392	57812	5
42949	67296	32	0.00000	00002	32830	64365	38696	28906	25

A-4   NUMERICAL CONVERSION

Octal	Decimal	Decimal	Octal	2's complement
0-7	0-7	1	1	177777
10-17	8-15	10	12	177766
20-27	16-23	20	24	177754
30-37	24-31	40	50	177730
40-47	32-39	100	144	177634
50-57	40-47	200	310	177470
60-67	48-55	500	764	177014
70-77	56-63	1,000	1750	176030
100	64	2,000	3720	174040
200	128	5,000	11610	166170
400	256	10,000	23420	154360
1000	512	20,000	47040	130740
2000	1,024	32,768		100000
4000	2,048			
10000	4,096			
20000	8,192			
40000	16,384			
77777	32,767			

appendix B
Digital Logic Electronics

It is the purpose of Appendix B to describe the operation of the more common bipolar transistor integrated-circuit logic types, to compare their characteristics, and to present some common techniques used to satisfactorily wire up logic systems using integrated-circuit devices.

Integrated-circuit logic devices can be divided into two major categories: The first consists of those devices which use conventional bipolar transistors. The second category consists of those devices using metal-oxide-silicon field-effect transistors, which are often designated MOS logic. Logic systems using bipolar transistors can further be subdivided into two types: In one the transistor driving stages use *saturated* switching techniques, and in the other they use *current-mode* or *unsaturated* switching techniques. Saturated transistor logic consists of direct-coupled transistor logic (DCTL), resistor-transistor logic (RTL), diode-transistor logic (DTL), transistor-transistor logic (T²L), high-level logic (HLL), and some others. Current-mode logic fits within a classification called emitter-coupled logic (ECL). The differences between types are generally based upon modifications of the standard ECL design to provide increasingly greater switching speeds.

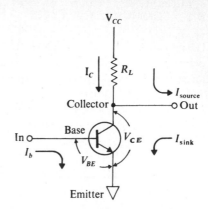

Fig. B-1 NPN transistor common-emitter configuration.

B-1   Saturated Bipolar Transistor Switching

The basis of saturated transistor logic is the use of a transistor as a saturated switch. This is often called the *common-emitter circuit configuration*. It is presented in Fig. B-1 for an *NPN* transistor. The transistor is a three-terminal device with *base, collector,* and *emitter* terminals. For normal operation of an *NPN* transistor, the collector must be positive with respect to the emitter terminal. The transistor operates as follows: When the voltage applied between the base and the emitter exceeds a threshold or breakdown value of about 0.8 V, the transistor turns on in saturation. That is, the transistor impedance becomes very low, and the collector current through the transistor is determined by the magnitude of the load resistor R_L. The voltage drop $i_C R_L$ is essentially the total power-supply voltage. The output voltage V_{CE} goes low to the saturation voltage of the transistor. The collector-emitter saturation voltage V_{CE} is typically about 0.1 V for a silicon unit. When the base-emitter voltage V_{BE} goes significantly below the threshold value, the transistor turns off, creating a high impedance between the collector and emitter terminals. When the transistor is turned off, its impedance is usually much higher than the magnitude of R_L. Therefore, the output voltage V_{CE} is essentially the value of V_{cc}. The collector current through the divider string consisting of resistor R_L and the transistor is then limited by the impedance of the transistor. The transistor thus acts as a switch with two states, on and off, producing two logic states, low and high, respectively. The logic levels are voltage levels, and the logic swing is determined by the magnitude of the power-supply voltage V_{CC}.

   The common-emitter transistor configuration presented in Fig. B-1 can be used in two modes for digital logic gates. Current can either flow out of the output terminal from V_{cc} across R_L or, when the transistor is turned on, into the output terminal to ground. When current flows out of the output terminal, the common-emitter output stage is said to be *current sourcing*. When current flows into the output terminal, it is said to be *current sinking*. Both types of output-current conditions are used in saturated transistor logic.

B-2   COMMON LOGIC TYPES

Direct-coupled transistor logic (DCTL)   A DCTL two-input NOR gate is presented in Fig. B-2. The voltages driving the inputs are applied directly to the bases of transistors Q_1 and Q_2, which are used to perform the logical, inversion, and driving operations.

The DCTL NOR gate presented in Fig. B-2 operates as follows: When both inputs A and B are low, that is open, grounded, or connected to a DCTL low output, both transistors Q_1 and Q_2 are turned off, and the output voltage is high. However, when either inputs A or B or both are high, that is connected to V_{cc} or to a DCTL high output, the corresponding transistor(s) is turned on, and the output voltage is low.

The successful operation of DCTL gates depends upon the consistency of V_{BE} in the transistors being used in the gates. If V_{BE} values of the transistors are not identical when one output is driving several inputs, the input transistor with the lower V_{BE} holds the output voltage at a breakdown value below that needed by the other inputs and "hogs" the current from the driving stage. The *current-hogging* characteristics of transistors with the lower V_{BE} limit drastically the successful use of DCTL logic. It is only presented here as an introduction into the first popular integrated-circuit logic type, RTL. Notice in Fig. B-2 that the DCTL driving stage supplies current to turn on the driven stage. DCTL is a current-sourcing type of saturated logic.

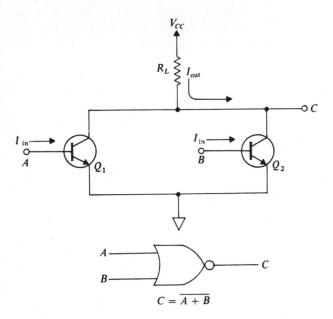

Fig. B-2   Direct-coupled transistor logic (DCTL).

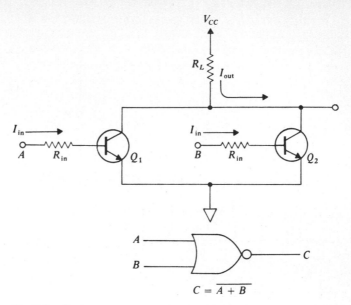

Fig. B-3   Resistor-transistor logic (RTL).

Resistor-transistor logic (RTL)   RTL is derived directly from DCTL. The RTL circuit presented in Fig. B-3 is essentially the same as the DCTL circuit presented in Fig. B-2 except that small base resistors have been added to the base terminals of each transistor. The small base resistors, which are on the order of a few hundred ohms, increase the input impedance of the transistors and eliminate the current-hogging problem that occurs with DCTL. The addition of the base resistors makes RTL an easy-to-manufacture, reliable logic system. Like DCTL, RTL is current-sourcing logic. The RTL gate presented in Fig. B-3 operates in essentially the same way as the DCTL gate presented in Fig. B-2. RTL typically uses a power-supply voltage of +3.6 V, producing ideal logic levels of 0 and +3.6 V.

Diode-transistor logic (DTL)   A typical integrated-circuit DTL two-input NAND gate is presented in Fig. B-4. The logic is performed by the input diodes D_1 and D_2 which form an AND gate. The DTL gate operates as follows: When both of the diode inputs A and B are high, the node point shown in the diagram is the high. Transistor Q_1 and diode D_3 act as a level shifter and turn on the driver for transistor Q_2. When the node point is high, the base terminal of the output transistor Q_2 is above the switching threshold. Transistor Q_2 is turned on, and the output is low. When either input A or input B or both are low, the node point is low, the base terminal of transistor Q_2 is below threshold, and it is turned off. In this condition, the output is high. Notice that DTL is current-sinking logic. That is, current flows out of the inputs into the outputs in

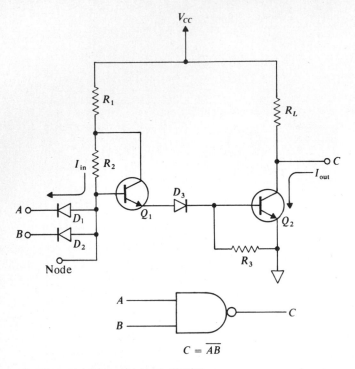

Fig. B-4   Diode-transistor logic (DTL).

the logical low state. DTL typically uses a power-supply voltage of +5 V, producing ideal logic levels of 0 and +5 V.

Some DTL logic gates have circuit diagrams that differ from the one presented in Fig. B-4. The differences are primarily in the output structure to provide higher sink currents and drive capabilities for power and buffer gates.

Other circuit modifications have been made to provide what is called *low-power* DTL logic. Modifications are made so that the power dissipation per gate is less than that for the type of DTL presented here. The user sacrifices circuit speed to obtain low-power operation. Low-power devices generally run slower with longer gate propagation delays than do high-power devices. However, they require much less operating current.

TRANSISTOR-TRANSISTOR LOGIC (T^2L)

A basic circuit for a T^2L NAND gate is presented in Fig. B-5. T^2L is basically a modified form of DTL that has been designed for higher-speed operation with higher drive-current capabilities.

The typical two-input T^2L NAND gate presented in Fig. B-5 operates as follows: The input diodes for the DTL gate have been replaced by an input

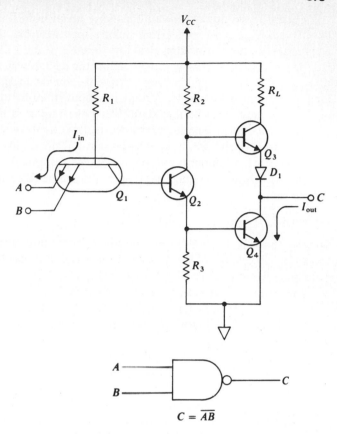

$$C = \overline{AB}$$

Fig. B-5   Transistor-transistor logic (TTL, T^2L).

transistor with multiple emitters. The multiple-emitter transistor allows faster input switching. When both inputs A and B of Q_1 are high, the collector of Q_1 and the base of Q_2 are high. Q_2 is turned on, which turns on Q_4. When Q_4 is turned on, the output is low. If, however, either input A or input B or both of Q_1 are low, the collector of Q_1 and the base of Q_2 are low. Q_2 is turned off, and as a result Q_4 is turned off. In this condition, Q_3 is turned on. The output of the gate is high, and its impedance to power supply is quite low because the size of R_L is quite low—much lower than for DTL.

Notice that the output structure in the T^2L gate is different from that in the DTL gate. The use of two transistors, one switching to power supply and the other switching to ground, is called a *totem-pole* output. It is used to provide higher switching speeds and lower output impedances, resulting in higher capacitive drive capabilities in saturated bipolar transistor digital logic. The stage, preceding the output stage, consisting of Q_2, R_2, and R_3, is called the *phase-splitter* stage. It is used to turn on either Q_3 or Q_4. Notice also that Q_3

and Q_4 are never turned on simultaneously under normal circumstances. T^2L, like DTL, is a form of current-sinking logic. It, like DTL, uses a +5-V power-supply voltage providing ideal logic levels of 0 and +5 V.

As a matter of review, bipolar transistor digital logic can be divided into three groups: current-sinking logic, current-sourcing logic, and current-mode logic. In current-sinking logic, current flows out of an input and into an output when an output is in its low state. When the output is in its high state, no current except for a leakage current flows. In current-sourcing logic, when the output is in its high state, it provides current to drive inputs. When the output is in its low state, no currents except for leakage currents flow between gates. In *current-mode* logic, presented next, an output can either source or sink current in both states.

EMITTER-COUPLED LOGIC (ECL)

So far, the digital logic types considered here have used saturated bipolar transistor switching. Current-mode logic does not use saturated transistor switching. Instead, it uses small logic voltage-level changes to produce current steering within the device. Current-mode logic is commonly called ECL. The circuit diagram for an ECL two-input OR/NOR gate is presented in Fig. B-6.

Unlike the previously discussed logic systems which use positive power-supply voltages and logic levels, ECL uses a negative power supply voltage V_{EE} of around −5.2 V. The logical low state is usually −1.55 V, and the logical high state is then −0.75 V. A fixed bias voltage, which a user may or may not have to supply depending on type and manufacturer, is applied to the base input of Q_3 in Fig. B-6. The bias voltage is usually about −1.15 V.

The ECL gate presented in Fig. B-6 operates as follows: If a logical low is applied to the inputs A and B, Q_1 and Q_2 are turned off. The current through R_3 is supplied by the bias transistor Q_3 through R_2. The size of R_2 is selected so that a voltage drop of about −0.8 V occurs across it. This level is applied to the base of Q_5. A characteristic base-emitter voltage drop V_{BE} of about 0.75 V occurs across Q_5. It adds to the −0.8-V base voltage applied, causing the D or OR output to be at −1.55 V, a logical low. Since no current flows through the input transistors Q_1 and Q_2 as they are turned off, the base of transistor Q_4 is at essentially 0 V or ground. The C or NOR output is one V_{BE} below ground or at about −0.75 V, a logical high. When a logical high is applied to either or both of input A or B, Q_1 or Q_2 or both pass current. Current is switched from R_2 to R_1 causing a −0.8-V drop across R_1. Q_4 then has −0.8 V applied to its base. The NOR or C output is one V_{BE} below −0.8 V, or −1.55 V. Since the current has been switched from R_2 to R_1, the collector of transistor Q_3 and the base of transistor Q_5 are essentially at ground. The D or OR output of the gate is at one V_{BE} below ground or −0.75 V.

The logic levels occurring in ECL are a function of the characteristic base-emitter junction voltages V_{BE} of the transistors involved. The logic is

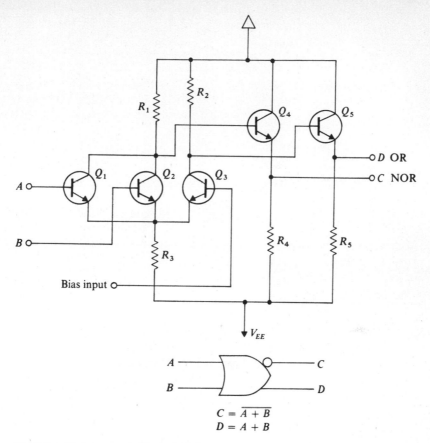

Fig. B-6   Emitter-coupled logic (ECL).

performed by current steering at the input stage consisting of Q_1, Q_2, Q_3, R_1, R_2, and R_3. The output stages are in the common-collector or emitter-follower configuration where the transistor emitter voltages follow one V_{BE} below the base voltages. The transistors are not allowed to saturate, reducing charge-storage time and increasing switching speed. In fact, ECL logic has been developed to produce very high-speed digital systems.

B-3 OPERATING PARAMETERS AND SELECTION CONSIDERATIONS

Having presented simple working descriptions of some of the different types of bipolar transistor logic, we can now discuss their operating characteristics. They include logic-level definitions, operating speed, power dissipation, noise characteristics, and gate driving capabilities.

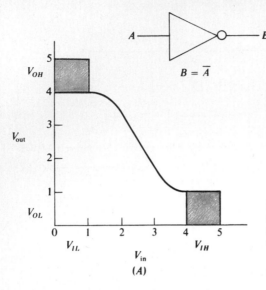

Fig. B-7 Transfer characteristics for a typical inverter gate (V_{OH} = high output voltage; V_{OL} = low output voltage; V_{IH} = high input voltage; V_{IL} = low input voltage; V_{out} = output voltage; V_{in} = input voltage).

A typical transfer-characteristic curve for a hypothetical inverter gate is presented in Fig. B-7. Notice that when the input is low, the output is high. The logic swing for the hypothetical gate is 5 V. From the shaded area in the transfer curve, one can see that the logical 0 or low state for both the input and output covers an area from 0 to 1 V. A logical high state covers an area from 4 to 5 V. In between the two shaded operating areas exists an area where the logical states are not clearly defined. In other words, from 1 to 4 V is an undefined area where logical switching occurs. Since the logic levels for the hypothetical inverter are defined over voltage ranges of 1 V, the inverter described by the transfer-characteristic curve in Fig. B-7 is said to have a dc noise immunity of 1 V.

The speed of logic devices is generally described by gate propagation delays and binary-element or flip-flop maximum operating frequencies. Gate propagation delays can be understood by considering Fig. B-8, where the response curves for a typical inverter gate are presented. The *propagation delay* is defined as the time it takes for the output of a gate to respond after a step function has been presented to the input. The higher the speed of a logic system, the shorter will be its propagation delay.

Maximum operating frequencies are the maximum signal frequencies that the flip-flops can follow. Since flip-flops are essentially constructed from logic gates, the toggle frequency is a function of the propagation delay of the gates. It is also a function of the configuration of the gates making up the flip-flop. Common logic systems have toggle frequencies from less than 1 MHz up to the order of 350 MHz.

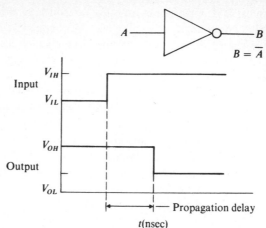

Fig. B-8 Propagation delay for a
typical inverter gate.

Dc noise immunity was described above and defines the voltage range over which logical levels are defined. However, in addition to dc noise, one can also have ac noise. Generally speaking, one can consider ac noise in terms of the amplitude and duration of the transient signal. The susceptibility of a particular type of logic is a function of the dc noise immunity, or *noise threshold*, and the width or time duration of the transient ac noise spike. Noise signals above the threshold value must exist for a minimum period of time in order for the gate output to respond. Higher-speed logic systems are susceptible to noise spikes of shorter time duration because they are designed to run with shorter-duration input signals and gate propagation delays.

The susceptibility of a given logic system to dc and ac noise can be divided into internal and external noise margins. The *external* noise margin is the susceptibility of a given logic system to noise introduced from the outside world. The noise may be a result of power-supply line transients, capacitive or inductive pickup of transient signals, ground loops, etc. The *internal* noise margin is the susceptibility of a logic device to internally generated noise. Many logic devices generate internal noise signals that travel on power-supply and ground lines as well as on logic interconnection lines.

The gate drive capability or *fan-out* of a particular type of logic may be important in selecting a logic family. It is defined as the number of inputs to other logic devices that the outputs of a given type of logic can drive. There is a wide variation in fan-out capabilities among the different logic systems. In addition, the ability to drive high capacitive loads without significant signal distortion is an important drive characteristic.

The temperature characteristics of logic devices are important. Generally speaking, integrated-circuit manufacturers produce functionally equivalent logic families that will operate over more than one temperature range. For example, a typical DTL logic family may be divided into two different type number series. One series might be designed for commercial applications over narrow temperature ranges with its operating characteristics specified from 0 to 75°C. Another might be designed for use over wide temperature ranges, and its operating characteristics will be specified from −55 to +125°C. For narrow temperature-range applications, the commercial temperature-range devices are generally satisfactory and considerably cheaper than devices for use over extended temperature ranges.

It is also important to know what the characteristics of a given logic system are when used at temperatures widely different from 25°C. The effects on logic levels, ac and dc noise immunity, fan-out, etc., all need to be considered. These characteristics can generally be found in the detailed literature produced by the manufacturers and will not be discussed here.

B-4   COMPARISON OF BIPOLAR TRANSISTOR LOGIC TYPES

Now that some of the parameters used to compare digital logic systems have been discussed, the various types of logic can be considered in detail. Some of the operating parameters for RTL, DTL, TTL, and ECL are presented in Table B-1. RTL can be divided into two categories: conventional and low-power RTL (LPRTL). Both are moderately fast with typical gate propagation delays of 15 nsec for standard RTL and 30 nsec for LPRTL. Typical binary-element clock frequencies are 8 and 3 MHz, respectively. Internal and external noise margins are fair with a typical dc value of 0.4 V. The fan-out of a typical RTL basic gate is limited to about 5 and to about 25 for large fan-out drivers. RTL fan-out is generally limited because logic swing and noise immunity are heavily dependent upon the number of inputs being driven by a given output. This can be explained easily with reference to Fig. B-3. RTL logic is current-sourcing. When an input is connected to an output, current flows through the load resistor R_L and the input resistor and transistor to ground. As more inputs are connected, increasing amounts of current are drawn from the output resulting in higher voltage drops across R_L. The high-level output voltage is reduced, causing the gate to become more susceptible to noise by reducing the dc noise immunity. The dependency of logic swing and noise immunity on fan-out is typical for all types of current-sourcing logic.

Wired collector gating is possible with RTL and increases its flexibility. The outputs of more than one logic device can be tied together to produce positive logic AND gates. The number of outputs that can be tied together is limited by the characteristics of the gates and is usually specified by manufacturers.

DTL characteristics are also presented in Table B-1. The average DTL gate propagation delay is somewhat longer than that for RTL. Typical binary-element

Table B-1 Comparison of bipolar transistor digital logic types

Logic type	Typical average gate propagation delay (nsec)	Typical binary-element clock frequency (MHz)	Typical power-supply voltage	Logic levels V_L	V_H	Typical dc noise margin (V)	Typical internal noise margin[a]	Typical external noise margin[a]	Wired logic capability
Resistor-transistor (RTL)	15	8	3.6	0	3.6	0.4	F	F	Yes
Low-power resistor-transistor (LPRTL)	30	3	3.6	0	3.6	0.4	F	F	Yes
Diode-transistor (DTL)	25	12	5	0	5	1.0	F	G	Yes
Low-power diode-transistor (LPDTL)	55	1	5	0	5	1.0	F	G	Yes
High-level diode-transistor (HLDTL)	120	3	30[b]	[c]	[c]		G	Ex	Yes[d]
Transistor-transistor (T²L)	10	20	5	0	5	1.0	G	G	No[d]
Low-power transistor-transistor (LPT²L)	25	10	5	0	5	1.0	G	G	No[d]
High-speed transistor-transistor (HST²L)	6	50	5	0	5	1.0	G	G	Yes
Emitter-coupled (ECL)	5	60	-5.2	-1.55	-0.75	0.4	Ex	G	Yes
High-speed emitter-coupled (HSECL)	1	350 (maximum)	-5.2	-1.55	-0.75	0.4	Ex	G	Yes

[a]F = fair; G = good; Ex = excellent.
[b]Varies among manufacturers.
[c]Depends on power-supply voltage V_{cc}.
[d]Some MSI devices allow for wired collector outputs.

clock frequencies are about the same as for RTL. Again, as in RTL, there is a low-power series of DTL called LPDTL. The typical gate propagation for LPDTL is somewhat longer than that for standard DTL. Typical binary clock frequencies are considerably lower. Notice that DTL typically uses a 5-V rather than a 3.6-V power supply. Higher logic swings and greater noise immunity result from the higher power-supply voltage. Typical dc noise margins are 1 V. The susceptibility of DTL to external noise is quite favorable as a result of the larger typical dc noise margin. The typical internal noise margin is only fair because of the relatively high output impedance of the gates. R_L is typically equal to or greater than 2K Ω. DTL can produce the wired AND function.

The noise immunity of DTL is less dependent on fan-out than it is for RTL. The reason for this can be seen by again referring to Fig. B-4. Since DTL is current-sinking logic, adding more inputs to a given output will not increase the voltage drop across the load resistor. Instead, when the output transistor is turned on, the output current passes through it to ground. It will maintain its saturation voltage level of about 0.1 to 0.2 V over a fairly wide range of currents. In other words, the impedance of the transistor to ground is not purely resistive. The result is that the typical dc noise margins hold up better over varying load situations than they do for RTL.

High-level DTL (HLDTL) is available in integrated-circuit form for those situations where very large amounts of noise immunity are needed. HLDTL has long gate propagation delays and low typical binary-element clock frequencies. Power-supply voltages can range from 8 to 10 V to above 30 V with corresponding logic swings. The dc noise margins range from 2 V at the lower power-supply voltages to upwards of 8 V at the higher power-supply voltages. Because of the larger logic swings, both internal and external noise margins are very good. Like standard DTL, many HLDTL systems can perform the wired-collector positive-logic AND function.

The characteristics of T^2L, LPT^2L, and HST^2L are also presented in Table B-1. The typical average gate propagation delays are shorter in T^2L than in either RTL or DTL. In fact, HST^2L has a gate propagation delay of about 6 nsec. Typical binary-element clock frequencies are 20, 10, and 50 MHz for T^2L, LPT^2L, and HST^2L, respectively. The 5-V power-supply voltage used for T^2L is the same as that for DTL. The logic levels are again about the same for the two logic lines. Typical dc noise margins are 1 V in T^2L, the same as in DTL. However, internal and external noise margins are both good for T^2L; while typical internal noise margins are only fair in DTL. This is a result of the lower output impedance of the T^2L gate. The lower output impedance can be understood with reference to Figs. B-4 and B-5. The output configuration of the T^2L gate has a lower impedance to the power supply when the output is high than does the DTL gate with a passive load resistor. As a result, low-energy high-frequency noise spikes are more readily shunted off to the power supply in T^2L than in DTL.

There is another important characteristic of T^2L that has not been described in the table. T^2L tends to generate more noise along power-supply and ground lines than either RTL or DTL. As a result, more precautions are often needed when wiring up T^2L systems. The noise is a current spike and results from the following: Normally, a T^2L gate sits with its output high or low. When it changes state, one output transistor turns on, while the other turns off. During the transition, there is a time when both transistors Q_3 and Q_4 (Fig. B-5) are partially turned on, creating a low impedance from power supply to ground. The result is a fairly high-magnitude current pulse on the power-supply and ground interconnecting wires.

T^2L cannot perform the wired collector functions that DTL and RTL can. If two T^2L outputs are connected together and if on one of the outputs the upper transistor is turned on (high state) and on the other output the lower transistor is turned on (low state), a low-impedance current path is produced from power supply to ground, which is likely to burn out the output transistors of the T^2L gates. However, some complex T^2L integrated circuits can have their outputs switched off to a logical "nothing" so that wired logic can be implemented. In other words, transistors Q_3 and Q_4 in Fig. B-5 would both be turned off. It allows T^2L to be used in common data-bus systems.

Finally, ECL characteristics are presented in Table B-1. ECL and HSECL are considerably faster than other logic types. Typical dc noise margins are about 0.4 V. The typical internal noise margin is excellent for ECL because the current-steering approach does not produce large surge currents. The immunity to external noise is poorer for ECL than for most saturated logic types. This is because of the narrow logic swing and resulting small dc noise margin and because of the extremely high speed of ECL, which makes it susceptible to very short noise spikes. Wired output logic functions can be performed using ECL.

B-5 SELECTION OF A LOGIC FAMILY

How does a scientist choose a particular logic line for his use? Obviously, no logic line is superior to all the others. The user must consider his total system requirements, including such factors as speed, noise environment, cost availability of complex logic functions, ease of use, etc. Some considerations will dictate the use of one logic system over another. If a worker requires extremely high-speed logic, then his selection will obviously have to be either ECL or T^2L. If the logic must be used in an extremely high-noise environment, HLDTL should be selected. The choice of a logic type to use for a variety of general applications is somewhat different. Not too long ago, RTL logic was used extensively. It was more easily available, had more complex functions, and was cheaper than other types. This is no longer true today. In recent years most of the development work has been in logic areas other than RTL. It is the authors' feeling that one should not choose RTL for new systems because of its poorer noise immunity, lack of a more complete line of complex functions, and lack of a cost advantage.

DTL has some strong advantages and disadvantages. It is a current-sinking type of logic and is electrically compatible with T^2L in many situations. In some situations, however, DTL will not successfully drive a T^2L device because of the longer output rise time. DTL logic does not generate large amounts of noise on power-supply and ground lines, as does T^2L. In addition, the passive pullup output structure of DTL allows the use of wired collector logic functions, which can be extremely helpful in logic design. Unfortunately, not many complex logic functions exist in DTL.

T^2L has one very strong advantage. Most new development technology in the area of medium-scale integrated circuits (MSI)—which is the insertion of many logical elements in a complex device to perform complex counting, arithmetic, and logical operations—has been directed toward T^2L. The availability of MSI functions is a great advantage in implementing logic designs. It means that fairly complex instrumental systems can be built easily and cheaply by workers with a minimal electronics background.

B-6   WIRING AND INTERCONNECTION CONSIDERATIONS

The logic systems requiring the most care in wiring and interconnection are those that run the fastest and have the shortest gate propagation delays. Obviously, ECL with gate propagation delays on the order of 1 nsec is going to be somewhat more difficult to wire up than the slower logic. This is because the time interval for a logic pulse to move along a wire may be longer than a gate propagation delay resulting in timing and synchronization problems. The same sort of situation exists for high-speed T^2L. In addition to the timing and synchronization problems, which always occur in logic design, the actual wiring techniques used will often determine the success or failure of the system. Some of the practical approaches to use in wiring logic systems are presented below.

The power supply used to drive digital logic should be well regulated, have a fast transient response, and maintain ripple to less than 5 percent. The power-supply interconnection wires should be as large as possible to reduce impedances and the resulting voltage drops that occur upon current surges, particularly when using T^2L. If possible, a ground plane is extremely desirable on logic cards. It is especially necessary when a large number of packages are included on one board.

Logic packages should be decoupled at least every five or six packages with at least 0.1 μfarad ceramic or solid tantalum capacitors. *Decoupling* refers to the connection of a capacitor between the power supply and ground immediately adjacent to the integrated-circuit power-supply connection. The capacitors are used to supply large instantaneous current demands and to shunt high-frequency power-supply noise to ground. The authors routinely decouple every logic package. As a result, noise problems seldom exist on interconnection lines. For one-of-a-kind systems, the reduction in troubleshooting time is well worth the cost of decoupling each package. In addition, all circuit-board power-supply connections should be decoupled at the board entrance with at least a 2 μfarad solid tantalum capacitor.

Unused inputs of logic gates should never be left open. They should be tied through resistors equal in value to the collector pullup resistors, but greater than or equal to 1K Ω, to either power supply or ground. Unused inputs to a particular gate may be tied to a driven input if they do not exceed the fan-out requirements of the driving gate and still perform the desired logic. If noise pulses persist to cause erratic operation, small capacitors can be connected from the inputs of gates and other logic devices to ground in an attempt to shunt off short noise spikes and synchronize logic pulses. When using expandable gates, the gate expanders must be as close as possible to the gate being expanded to avoid capacitive loading of the expander nodes and to maintain switching speeds.

When using flip-flops, care must be taken that the proper timing and synchronization described by the manufacturer's data sheet are maintained. For example, care must be taken that data inputs of master-slave JK flip-flops do not change while clock pulses are transferring the data.

When interconnecting printed circuit boards, twisted-wire pairs or coaxial cables should be used for signal wires exceeding about 1 ft in length. The authors have found that the best line impedance to use when driving cables is about 100 Ω. High impedances tend to increase cross talk between wires; while lower impedances become difficult to drive. When using twisted pairs, number 26 or 28 wire with thin insulation, twisted about 30 turns per foot, appears to work well. In order to improve rise times, resistive pullup at the receiving ends of long cables can be used. These resistors should be on the order of 1K Ω for common logic types. When using gates as line drivers and receivers, only one gate input and output should be used per line. The excessive delays that occur down long lines make synchronization of several drivers and receivers difficult. In addition, multiple gate inputs and outputs on a single line cause erroneous signals as a result of line reflections and excessive driver loading. Flip-flops make poor line drivers because the flip-flop output states can be changed from noise by what is called *collector triggering*. It is absolutely necessary to decouple line drivers and receivers.

Care must be taken that ground loops do not occur in the system. Choose a single ground point close to the power supply, and tie all system grounds to this point, not to multiple points. Ground loops tend to cause system grounds to have different voltage levels which reduce the dc noise margin of the system as a whole. In addition, the use of many ground connections produces unnecessary impedances in the ground line. The use of extremely large ground buses and extremely large power-supply leads, preferably rectangular in shape, to reduce lead inductances is strongly urged.

BIBLIOGRAPHY

Much of the information presented here comparing logic types comes from Ref. 1. A basic understanding of digital electronics can be obtained from any of Refs. 2 to 9. References 2 to 5 are for beginners; while the remainder provide more detailed discussions. In addition, Ref. 9 provides much useful working information concerning integrated-circuit digital logic.

1. Meyer, C. S., D. K. Lynn, and D. J. Hamilton: "Analysis and Design of Integrated Circuits," McGraw-Hill, New York, 1968.
2. Malmstadt, H. V., and C. G. Enke: "Digital Electronics for Scientists," Benjamin, New York, 1969.
3. Malmstadt, H. V., C. G. Enke, and E. C. Toren, Jr.: "Electronics for Scientists," Benjamin, New York, 1963.
4. Hunter, D. M.: "Introduction to Electronics," Holt, New York, 1964.
5. Hibberd, R. G.: "Solid-state Electronics," McGraw-Hill, New York, 1968.
6. Miller, J. (ed.): "Transistor Circuit Design," McGraw-Hill, New York, 1963.
7. Delhom, Louis A.: "Design and Application of Transistor Switching Circuits," McGraw-Hill, New York, 1968.
8. Malvino, A. P.: "Transistor Circuit Approximations," McGraw-Hill, New York, 1968.
9. Morris, R. L., and J. R. Miller: "Designing with TTL Integrated Circuits," McGraw-Hill, New York, 1971.

appendix C
Boolean Algebra

C-1 INTRODUCTION

Boolean algebra provides a symbolic method of studying logical operations. Its use for the description of modern electronic logic systems is based on the fact that modern digital electronic systems employ signal levels consisting usually of two states. Since Boolean algebra is essentially binary, it is well suited to the description of these systems.

While conventional algebra manipulates numerical quantities and arithmetic relations, Boolean algebra manipulates connectives and truth values [1]. *Connectives* can be defined as those elements which connect the parts of a logical function or expression together, such as AND and OR statements. The manipulation of truth values rather than numerical quantities implies that Boolean algebra is a *calculus of validity* [1]. In other words, in Boolean algebra a statement is either true or false. Because Boolean algebra manipulates connectives and truth values, some of the terms used with conventional algebra must be redefined. It is the purpose of Appendix C to list, define, and discuss briefly some of the basic terms and postulates of Boolean algebra.

C-2 CHARACTERISTICS OF BOOLEAN ALGEBRA

Boolean algebra differs from conventional algebra in the following ways [2]:

1. The terms in a Boolean expression have no coefficients as in ordinary algebra. For example, for the variable A, in ordinary algebra it can have a coefficient such as the number 4 to make the term $4A$. In Boolean algebra, no such coefficients exist.
2. Each letter in a Boolean expression designates which one of only two possible events exists. The value assigned to the letter for one event is 1 and for the other event is 0. In other words, the two numbers used in the binary number system are also used in Boolean algebra. We can use a letter in a Boolean expression to tell whether a switch is open or closed, a voltage is high or low, or a statement is true or false. All these conditions can be represented by only two states, one of which is assigned the number 1 and the other of which is assigned the number 0. This is summarized in the case of Table C-1 if we let the switch be denoted by the letter A.

Table C-1 Possible Boolean states

Case	A	Switch condition
1	0	Open
2	1	Closed

3. Like each letter in a Boolean function, a Boolean function can only state which of the two possible states exist. It can only tell if a switch is open or closed, a voltage is high or low, or a statement is true or false. As is indicated above, these conditions are indicated by the numbers 1 and 0.
4. In Boolean algebra, no subtraction, division, square root, etc., operators exist, as in conventional algebra. All operations are performed using Boolean algebra operators.
5. For the logical description of digital computers and control logic, the subset of Boolean operators most often used includes the primitive operators OR, AND, and NOT. If the two possible values of a Boolean algebraic variable are assigned the values of the binary number system 0 and 1, then the three primitive operators are analogous to addition, multiplication, and complementation, respectively.

C-3 PRIMITIVE OPERATORS

The inclusive OR operator   The inclusive OR operator is designated literally by OR and in equations is given the symbol "+" [2]. As a result, a Boolean function in which A or B or both are equal to C is written as

$$C = A + B$$

The OR function given above states that C is true only if A or B or both are true and is read C equals A or B.

If a table is constructed which describes all possible combinations of a number of variables A, B, C, ... and defines the corresponding values for X, where $X = f(A, B, C, ...)$, the information can be condensed by a Boolean function [3]. Such a table is called a *truth table* and is very useful in summarizing Boolean functions. The truth table for the OR function given above is presented in Table C-2. If we define a logical 1 as true and a logical 0 as false, then when variable A is true, $A = 1$, but for a false condition, $A = 0$.

Table C-2   Truth table for the inclusive OR function

$C = A + B$

Case	A	B	C
1	0	0	0
2	0	1	1
3	1	0	1
4	1	1	1

Note that in Table C-2, for cases 1, 2, and 3, the value of C is the same as that for normal addition in ordinary algebra. However, the result for case 4 is different. This is a result of the *idempotency* law and will be discussed in the next section.

THE AND OPERATOR

The AND operation is designated literally by AND and in equations is given the symbol "·" or is denoted by placing the variables to be operated on adjacent to each other [2]. As a result, a Boolean function in which A and B equals C is written in an equation as

$$C = A \cdot B$$

or

$$C = AB$$

In this text, the latter of these two conventions is adopted where the symbol is not needed for clarity.

The AND function states that C is true only if A and B are both true. The truth table defining this relationship is presented in Table C-3. Notice that the

Table C-3   Truth table for the AND function

$C = AB$

Case	A	B	C
1	0	0	0
2	0	1	0
3	1	0	0
4	1	1	1

AND function gives the same numerical result as ordinary multiplication in conventional algebra in all four cases.

The NOT operator   The complement or negation operation is designated literally by NOT and in equations by placing a bar or a prime over the variables [2]. For example, the NOT function of a single variable A can be written as $\overline{A}$ or A'. The truth table defining the NOT operation is presented in Table C-4. Note that this is the same as complementing a variable in ordinary algebra.

Table C-4   Truth table for the NOT function

Case	A	$\overline{A}$
1	0	1
2	1	0

C-4   FUNDAMENTAL THEOREMS OF BOOLEAN ALGEBRA

From the above primitive operators and the following fundamental theorems, most Boolean algebra operations can be performed. The theorems are enumerated below [3]:

1 Uniqueness
    a. The element 1 is unique.
    b. The element 0 is unique.
2. Complementation
    a. $A + \overline{A} = 1$
    b. $A\overline{A} = 0$
3. Involution (double negative)
    a. $\overline{\overline{A}} = A$

4. Adsorption
   a. $A + AB = A$
   b. $A(A + B) = A$
5. Idempotency
   a. $A + A = A$
   b. $AA = A$
6. Intersection
   a. $A + 1 = 1$
   b. $A \cdot 0 = 0$
7. Union
   a. $A + 0 = A$
   b. $A \cdot 1 = A$
8. Commutative
   a. $AB = BA$
   b. $A + B = B + A$
9. Associative
   a. $A(BC) = (AB)C$
   b. $A + (B + C) = (A + B) + C$
10. Distributive
   a. $A + BC = (A + B)(A + C)$
   b. $A(B + C) = AB + AC$
11. De Morgan's theorem
   a. $(A + B)' = \overline{A} \cdot \overline{B}$
   b. $(AB)' = \overline{A} + \overline{B}$

The theorems just enumerated are discussed below. No formal proofs are given, but the unique elements 0 and 1 are substituted into some of the theorems to illustrate their validity.

Theorem 1 states that the elements 0 and 1 are unique. This is a small but necessary property for the proof of subsequent theorems.

Theorem 2 states that any variable ORed with its complement always results in a true function while any element ANDed with its complement results in a false function. These results are summarized in Table C-5.

Table C-5   Truth table for theorem 2, complementation

Case	A	$\overline{A}$	$A + \overline{A} = 1$	$A\overline{A} = 0$
1	1	0	$1 + 0 = 1$	$1 \cdot 0 = 0$
2	0	1	$0 + 1 = 1$	$0 \cdot 1 = 0$

Theorem 3 states that the double complement of any variable is equal to that variable. This theorem is obvious from the definition of the NOT function. Some results of using theorem 3 and the NOT function are illustrated in Table C-6.

Table C-6   Truth table for theorem 3, involution

$$\overline{\overline{A}} = A$$

Case	A	$\overline{A}$	$\overline{\overline{A}}$	$\overline{\overline{\overline{A}}}$
1	1	0	1	0
2	0	1	0	1

Theorem 4 states that in each of two cases, a variable can be adsorbed by another variable. The results of this are summarized in Table C-7.

Table C-7   Truth table for theorem 4, adsorption

$(a)\ A + AB = A; (b)\ A(A + B) = A$

Case	A	B	$A + AB = A$	$A(A + B) = A$
1	0	0	$0 + 0 \cdot 0 = 0$	$0(0 + 0) = 0$
2	0	1	$0 + 0 \cdot 1 = 0$	$0(0 + 1) = 0$
3	1	0	$1 + 1 \cdot 0 = 1$	$1(1 + 0) = 1$
4	1	1	$1 + 1 \cdot 1 = 1$	$1(1 + 1) = 1$

Theorem 5 states that any variable ORed or ANDed with itself is equal to the variable. The results of this are summarized in Table C-8.

Table C-8   Truth table for theorem 5, idempotency

$(a)\ A + A = A; (b)\ AA = A$

Case	A	$A + A = A$	$AA = A$
1	0	$0 + 0 = 0$	$0 \cdot 0 = 0$
2	1	$1 + 1 = 1$	$1 \cdot 1 = 1$

Theorem 6 states that any variable ORed with the unique element 1 results in a 1 and any variable ANDed with the unique variable 0 results in a 0. This is illustrated in Table C-9.

Table C-9   Truth table for theorem 6, intersection

(a) $A + 1 = 1$; (b) $A + 0 = 0$

Case	A	$A + 1 = 1$	$A \cdot 0 = 0$
1	0	$0 + 1 = 1$	$0 \cdot 0 = 0$
2	1	$1 + 1 = 1$	$1 \cdot 0 = 0$

Theorem 7 states that any variable ORed with the unique element 0 or any variable ANDed with the unique element 1 results in the value of that variable. This is illustrated in Table C-10.

Table C-10   Truth table for theorem 7, union

(a) $A + 0 = A$; (b) $A \cdot 1 = A$

Case	A	$A + 0 = A$	$A \cdot 1 = A$
1	0	$0 + 0 = 0$	$0 \cdot 1 = 0$
2	1	$1 + 0 = 1$	$1 \cdot 1 = 1$

Theorems 8 and 9, the commutative and associative laws, are the same as for ordinary algebra. It is left to the reader to verify these theorems. Theorem 10, the distributive law, states that A OR B AND C is equal to A OR B AND A OR C, and that A ANDed to B OR C is equal to A AND B OR A AND C. The results are summarized in Table C-11. The reader should work these out.

De Morgan's theorem, theorem 11, can be used to complement complex Boolean functions. Because of its importance, it will be discussed in more detail

Table C-11   Truth table for theorem 10, distribution

(a) $A + BC = (A + B)(A + C)$; (b) $A(B + C) = AB + AC$

Case	A	B	C	$A + BC = (A + B)(A + C) = D$	$A(B + C) = AB + AC = D$
1	0	0	0	$0 + 0 \cdot 0 = (0 + 0)(0 + 0) = 0$	$0(0 + 1) = 0 \cdot 0 + 0 \cdot 1 = 0$
2	0	0	1	$0 + 0 \cdot 1 = (0 + 0)(0 + 1) = 0$	$0(0 + 1) = 0 \cdot 0 + 0 \cdot 1 = 0$
3	0	1	0	$0 + 1 \cdot 0 = (0 + 1)(0 + 0) = 0$	$0(1 + 0) = 0 \cdot 1 + 0 \cdot 0 = 0$
4	0	1	1	$0 + 1 \cdot 1 = (0 + 1)(0 + 1) = 1$	$0(1 + 1) = 0 \cdot 1 + 0 \cdot 1 = 0$
5	1	0	0	$1 + 0 \cdot 0 = (1 + 0)(1 + 0) = 1$	$1(0 + 0) = 1 \cdot 0 + 1 \cdot 0 = 0$
6	1	0	1	$1 + 0 \cdot 1 = (1 + 0)(1 + 1) = 1$	$1(0 + 1) = 1 \cdot 0 + 1 \cdot 1 = 1$
7	1	1	0	$1 + 1 \cdot 0 = (1 + 1)(1 + 0) = 1$	$1(1 + 0) = 1 \cdot 1 + 1 \cdot 0 = 1$
8	1	1	1	$1 + 1 \cdot 1 = (1 + 1)(1 + 1) = 1$	$1(1 + 1) = 1 \cdot 1 + 1 \cdot 1 = 1$

than those above. In order to accomplish this, the principle of *duality* and the definition of *literal* must be understood.

The principle of duality states that given a Boolean expression its dual is found by interchanging all occurrences of "+" and "·" and of 1 and 0 [3]. For example, Table C-12 lists a series of expressions and the corresponding duals. It is important to remember that if a theorem is valid, its dual is also valid.

Table C-12   Illustrations of the principle of duality

Expression	Dual
$A \cdot 1$	$A + 0$
$\overline{A} + B$	$\overline{A}B$
$A \cdot 0 + B \cdot 1$	$(A + 1)(B + 0)$
$\overline{A}B + C$	$(\overline{A} + B)C$
$A(B + \overline{C}) + D$	$(A + B\overline{C})D$

In taking the dual of an expression, particular care must be taken in cases where implied parentheses exist. For example, a case such as this is presented in Table C-13.

Table C-13   Principle of duality as used with implied parentheses

Expression	Dual Correct	Incorrect
$\overline{A} + BC$	$\overline{A}(B + C)$	$\overline{A}B + C$

The expression $\overline{A} + BC$ has implied parentheses around BC such that $\overline{A} + (BC)$ is the implied expression. Taking the dual of this expression leads to the correct answer. The dual can be checked by writing a truth table for the expression and its complement as will be shown below.

The second term, literal, defines any single variable or its complement. For example, A, $\overline{A}$, B and $\overline{B}$ are literals since they are only single variables or their complements; while $\overline{A} \cdot \overline{B}$, AB, $\overline{A}B$, and $A\overline{B}$ are not literals since each is an expression consisting of two variables.

With the terms dual and literal defined, we can now return to De Morgan's theorem and generalize it. *Shannon's generalized De Morgan's theorem states that the complement of any Boolean expression or function can be found by complementing all literals in the dual expression.* For example, to complement the expression

$$A + \overline{B}$$

first the dual is written down:

$$A\overline{B}$$

Then all literals are complemented giving

$$\overline{A}B$$

Therefore,

$$(A + \overline{B})' = \overline{A}B$$

This can easily be checked by inserting a 1 for A and B and a 0 for A and B. The results are tabulated in Table C-14.

Table C-14   Checking results of using De Morgan's theorem

Expression	Complement
$A + \overline{B} = 1 + 0 = 1$	$\overline{A}B = 0 \cdot 1 = 0$

Some examples of the use of De Morgan's theorem are presented in Table C-15. The reader should work through these examples. It was stated above that an incorrect dual can result from a misinterpretation of implied parentheses.

Table C-15   Use of De Morgan's theorem

Procedure: complement all literals in the dual expression; substitute 0's and 1's to check.

1. Given: $A\overline{B} + \overline{A}C$
   Dual: $(A + \overline{B})(\overline{A} + C)$
   Complement: $(\overline{A} + B)(A + \overline{C})$
   Check: $A\overline{B} + \overline{A}C = 1 \cdot 0 + 0 \cdot 1 = 0$
   $(\overline{A} + B)(A + \overline{C}) = (0 + 1)(1 + 0) = 1$
2. $[A\overline{B}(C + D)]' = \overline{A} + B + \overline{CD}$
3. $[A(B + \overline{C}D)]' = (\overline{A} + \overline{B})(\overline{A} + C + \overline{D})$

Because the validity of a complement can be checked easily by constructing a truth table, one can determine if the correct dual has been taken. Using the example from Table C-13, the truth table given in Table C-16 can be

Table C-16   Checking the complement of an expression

				Expression	Complement	
					Correct	Incorrect
Case	A	B	C	$\overline{A} + BC$	$A(\overline{B} + \overline{C})$	$A\overline{B} + \overline{C}$
1	0	0	0	1	0	1
2	0	0	1	1	0	0
3	0	1	0	1	0	1
4	0	1	1	1	0	0
5	1	0	0	0	1	1
6	1	0	1	0	1	1
7	1	1	0	0	1	1
8	1	1	1	1	0	0

constructed. Notice that the incorrect dual expression leads to an incorrect complement.

C-5 MORE LOGICAL OPERATORS

At this point, the reader may wonder why De Morgan's theorem has been discussed in such detail. The reason is that many modern electronic logic systems do not generate AND or OR functions directly but rather generate AND-NOT or NAND functions and OR-NOT or NOR functions. It is important for the reader to be able to think in terms of the NAND and NOR functions. To arrive at the NAND or NOR functions, one must merely complement the corresponding AND or OR functions. For example, the inclusive OR function was defined as

$$C = A + B$$

To get the NOR function, the complement of the OR function is generated such that

$$C = (A + B)' = \overline{A} \cdot \overline{B}$$

The truth table for the NOR function is presented in Table C-17.

In a similar manner, the NAND function can be generated from the AND function, yielding the defining equation and truth table presented in Table C-18.

Finally, two more logical operations will be presented here. The first is the exclusive OR function denoted by the symbol $\oplus$ and written in sentences as XOR. Because of the use of this symbol, it is often called the *ring-sum*.

An exclusive OR function of two variables is written as

$$C = A \oplus B$$

Table C-17   Truth table for the NOR function

$C = (A + B)' = \overline{A} \cdot \overline{B}$

Case	A	B	C
1	0	0	1
2	0	1	0
3	1	0	0
4	1	1	0

Table C-18   Truth table for the NAND function

$C = (AB)' = \overline{A} + \overline{B}$

Case	A	B	C
1	0	0	1
2	0	1	1
3	1	0	1
4	1	1	0

For this case, C is true if A or B is true but not if both are true. It can be defined in terms of the basic OR concept as

$$A \oplus B = A\overline{B} + \overline{A}B$$

The exclusive OR is the basic building block for binary addition. The defining truth table is presented in Table C-19.

The second of these two functions is the *coincidence* function denoted by the symbol $\odot$. Because of the symbol, it is often called the *ring-dot*. In terms of the basic OR function, the ring-dot is defined as

$$A \odot B = AB + \overline{A}\overline{B}$$

The defining truth table is presented in Table C-20. The coincidence function is useful in cases where two variables must be detected as equal whether they are logical 1's or logical 0's.

C-6   POSITIVE AND NEGATIVE LOGIC

So far in this appendix (C), we have defined a logical true as a logical 1 and a logical false as a logical 0. This is often called *positive logic*. Notice that we can

Table C-19   Truth table for the XOR function

$C = A\overline{B} + \overline{A}B = A + B$

Case	A	B	C
1	0	0	0
2	0	1	1
3	1	0	1
4	1	1	0

Table C-20   Truth table for the coincidence function

$C = AB + \overline{A} \cdot \overline{B} = A \cdot B$

Case	A	B	C
1	0	0	1
2	0	1	0
3	1	0	0
4	1	1	1

Table C-21   Positive and negative logic

Column	1	2	3	4	5
Case	A	B	C	C	C
1	0	0	0	F	T
2	0	1	1	T	F
3	1	0	1	T	F
4	1	1	1	T	F

also define a logical true as a logical 0 and a logical false as a logical 1. This, then, is *negative logic*. The only difference between positive and negative logic is the way in which we define a logical true or a logical false. It has nothing to do with the voltage levels or polarities used in the electronic devices with which we will be implementing the logic we are developing here. It is the purpose of this section to point out some of the characteristics of changing to and from negative and positive logic.

We previously defined the inclusive OR operation as

$$C = A + B$$

We did this in terms of positive logic and generated columns 1 to 3 of the truth table presented in Table C-21. Column 4 of Table C-21 shows that for positive logic we get a true in all cases except for the case where both A and B are 0. This, then, is the truth table for an OR function. Notice that we can restate the above sentence in a different manner if we redefine our logic as negative logic. Then we can say that we get a false output in all cases except that in which both A and B are 0, or true. This is illustrated in column 5 of Table C-21. What we have shown is that an OR function for positive logic becomes an AND function for negative logic. In a like manner, an OR function for negative logic is an AND function for positive logic. In fact, in going from positive to negative logic or vice versa, OR functions become AND functions, AND functions become OR functions, NOR functions become NAND functions, NAND functions become NOR functions, and exclusive OR functions become coincidence functions. The reader should work through the truth tables of those functions to convince himself of the validity of these relationships.

The concept of positive and negative logic just presented is extremely important in the development of *efficient* logic systems. That is, a minimization of the number of gates required can be achieved by using both positive and negative logic. When one is using electronics to implement logic, a vast reduction in the amount of components needed saves money and increases the speed of the overall logic systems by reducing signal delay times.

C-7 BOOLEAN FUNCTIONS FROM TRUTH TABLES

So far we have considered generating truth tables from Boolean functions. However, often the reverse operation is necessary, that of generating a Boolean function from a truth table. This is particularly true for complex logic functions and devices. Often manufacturers' literature gives only the truth table for these devices. We will consider in this section, with the use of examples, the generation of Boolean functions from truth tables.

Consider the following truth table:

Case	A	B	X
1	0	0	1
2	0	1	0
3	1	0	0
4	1	1	1

Notice that $X = 1$ when

$\quad A = 0 \qquad$ AND $\qquad B = 0$

$\qquad\qquad$ OR

$\quad A = 1 \qquad$ AND $\qquad B = 1$

This can be rewritten in the form of $X = 1$ when

$\quad \overline{A} = 1 \qquad$ AND $\qquad \overline{B} = 1$

$\qquad\qquad$ OR

$\quad A = 1 \qquad$ AND $\qquad B = 1$

Thus, X is true when $\overline{A}$ AND $\overline{B} = 1$ OR A AND $B = 1$. This can be written in an equation as

$$X = \overline{A}\overline{B} + AB$$

What is the name of this function?

Notice that a Boolean expression can be generated from this truth table in another manner. Since X is true when $X = 1$, one can solve for $\overline{X}$, or $X = 0$. In doing this, $X = 0$ when

$\quad A = 0 \qquad$ AND $\qquad B = 1$

$\qquad\qquad$ OR

$\quad A = 1 \qquad$ AND $\qquad B = 0$

which can be rewritten as

$A = 0$ AND $\bar{B} = 0$

OR

$\bar{A} = 0$ AND $B = 0$

Notice that in this case the terms were rewritten to equal 0 because analysis is for the case where $X = 0$, while in the previous example, the terms were written to equal 1. This then leads to the following function:

$$\bar{X} = A\bar{B} + \bar{A}B$$

If the function for $X = 1$ or X is wanted, this function can be inverted using De Morgan's theorem, which gives

$$X = (\bar{A} + B)(A + \bar{B}) = \bar{A}A + \bar{A}\bar{B} + BA + B\bar{B} = \bar{A}\bar{B} + BA$$

which is the same as before.

This same technique can be applied to cases where more variables exist. For example, consider the following truth table:

Case	A	B	C	X
1	0	0	0	0
2	0	0	1	0
3	0	1	0	0
4	0	1	1	1
5	1	0	0	0
6	1	0	1	1
7	1	1	0	1
8	1	1	1	1

We find that $X = 1$ when

$A = 0$ AND $B = 1$ AND $C = 1$

OR

$A = 1$ AND $B = 0$ AND $C = 1$

OR

$A = 1$ AND $B = 1$ AND $C = 0$

OR

$A = 1$ AND $B = 1$ AND $C = 1$

This can be rewritten as $X = 1$ when

$\overline{A} = 1$ AND $B = 1$ AND $C = 1$

OR

$A = 1$ AND $\overline{B} = 1$ AND $C = 1$

OR

$A = 1$ AND $B = 1$ AND $\overline{C} = 1$

OR

$A = 1$ AND $B = 1$ AND $C = 1$

This leads to the function

$$X = \overline{A}BC + A\overline{B}C + AB\overline{C} + ABC$$

which can be simplified using theorems $5a, 8b, 10b$, and $2a$ to

$$X = AB + AC + BC$$

C-8 THE USE OF THEOREMS AND OPERATORS

Often one needs to simplify Boolean functions to perform efficient logic design. It is the purpose of this section to present some examples of this operation based on the operators and theorems presented in the preceding pages. Several examples are presented below in the form of proofs for some identities [3]; the reader should work through each of these in detail.

Example C-1   Prove the following identities:

1. $AB + CD = (A + C)(A + D)(B + C)(B + D)$

Proof	Theorem used
$AB + CD = (AB) + CD$	
$= (AB + C)(AB + D)$	$10a$
$= (C + AB)(D + AB)$	$8a$
$= (C + A)(C + B)(D + A)(D + B)$	$10a$
$= (A + C)(B + C)(A + D)(B + D)$	$8b$
$= (A + C)(A + D)(B + C)(B + D)$	$8a$

2. $(A + B)(C + D) = (AC + BC + AD + BD)$

Proof	Theorem used
$(A + B)(C + D) = (A + B)C + (A + B)D$	$10b$
$= C(A + B) + D(A + B)$	$8a$
$= CA + CB + DA + DB$	$10b$
$= AC + BC + AD + BD$	$8a$

3. $(A + B\overline{C} + C)\overline{C} = AB\overline{C} + A\overline{B}\overline{C} + \overline{A}B\overline{C}$

	Proof	Theorem used
$(A + B\overline{C} + C)\overline{C}$	$= A\overline{C} + B\overline{C}\overline{C} + C\overline{C}$	10b
	$= A\overline{C} + B\overline{C} + C\overline{C}$	5b
	$= A\overline{C} + B\overline{C} + 0$	2b
	$= A\overline{C} + B\overline{C}$	7a
	$= A(B + \overline{B})\overline{C} + (A + \overline{A})B\overline{C}$	2a and 7b
	$= AB\overline{C} + A\overline{B}\overline{C} + AB\overline{C} + \overline{A}B\overline{C}$	10b
	$= AB\overline{C} + A\overline{B}\overline{C} + \overline{A}B\overline{C}$	5a

4. $(A + B + C + \dots)' = \overline{A}\,\overline{B}\,\overline{C}\dots$

Proof

This identity is used to extend De Morgan's theorem, number 11a, to many variables.

$ABC \dots$ dual
$\overline{A}\,\overline{B}\,\overline{C}\dots$ literals complemented

$\therefore (A + B + C \dots)' = \overline{A}\,\overline{B}\,\overline{C}\dots$

5. $A\overline{B} + B = A + B$

	Proof	Theorem used
$A\overline{B} + B$	$= A\overline{B} + (A + \overline{A})B$	2a and 7b
	$= A\overline{B} + AB + \overline{A}B$	10b
	$= A\overline{B} + AB + \overline{A}B + AB$	5a
	$= A(\overline{B} + B) + (\overline{A} + A)B$	10b
	$= A(B + \overline{B}) + (A + \overline{A})B$	8a
	$= A + B$	2a and 7b

While working the above proofs is a good exercise, now and then one has a more complicated Boolean expression that can be simplified. Several examples of these are presented below [3].

Example C-2 Simplify the following Boolean functions:

1. $(A + \overline{B})B$

	Simplification	Theorem used
$(A + \overline{B})B$	$= AB + \overline{B}B$	10b
	$= AB + 0$	2b
	$= AB$	7a

2. $AB + A\overline{B}$

	Simplification	Theorem used
$AB + A\overline{B}$	$= A(B + \overline{B})$	10b
	$= A \cdot 1$	2a
	$= A$	7b

3. $AB + \overline{A}B + A\overline{B}$

	Simplification	*Theorem used*
$AB + \overline{A}B + A\overline{B}$	$= AB + \overline{A}B + A\overline{B} + AB$	5a
	$= B(A + \overline{A}) + A(B + \overline{B})$	10b
	$= B + A$	2a and 7b

4. $(A + B)(A + \overline{B})$

	Simplification	*Theorem used*
$(A + B)(A + \overline{B})$	$= AA + BA + A\overline{B} + B\overline{B}$	10b
	$= A + BA + A\overline{B} + B\overline{B}$	5b
	$= A + BA + A\overline{B} + 0$	2b
	$= A + BA + A\overline{B}$	7a
	$= A + A(B + \overline{B})$	10b
	$= A + A$	2a and 7b
	$= A$	5a

EXERCISES

C-1. How many variables are used in Boolean algebra?

C-2. Boolean algebra has been described as a calculus of validity. Explain why this is true.

C-3. How does Boolean algebra differ from ordinary algebra?

C-4. Define the term "truth table," and give an example using the inclusive OR function.

C-5. Is the inclusive OR function the same as ordinary addition in ordinary algebra? If it is not, why not? Does the binary number system have anything to do with this?

C-6. Construct truth tables to verify the commutative and associative laws for Boolean algebra.

C-7. Why is De Morgan's theorem so important to Boolean algebra?

C-8. Find the dual of the following expressions:

(a) $A + 1$

(b) $AB + CD$

(c) $A(BC + D)$

(d) $(1 + 1) \cdot 1$

(e) $(A\overline{B} + \overline{A}B)C + D\overline{E}$

(f) $(\overline{A}\overline{B} + AB)(A\overline{B} + \overline{A}B) + C(D + \overline{C}D) + E$

C-9. Find the complements of the expressions given in Exercise C-8.

C-10. How does the NAND operation differ from the AND operation? Construct truth tables for each using three variables, and compare these tables.

C-11. What exactly is the ring-sum operation? What is its purpose?

C-12. What is the difference between positive and negative logic?

C-13. Given the following function in positive logic,

$D = A(A\overline{B} + \overline{A}B) + C$

what is the corresponding function in negative logic?

C-14. Prove the following identities [1] :

(a) $AB + \overline{A}C = AB + \overline{A}C + BC$

(b) $A\overline{B} + \overline{A}B = (AB + \overline{A}\overline{B})'$

(c) $AB\overline{C} + \overline{A}BC + \overline{A}\overline{B}C = \overline{A}B + \overline{A}C$

(d) $AB + BC + AC = (A + B)(B + C)(A + C)$

(e) $B\overline{C} + \overline{A}B + \overline{A}C = B\overline{C} + \overline{A}C$

C-15. Simplify the following Boolean expressions [6] :

(a) $\overline{A}B + AB + \overline{A}\overline{B}$

(b) $(A\overline{B} + \overline{A}B)'$

(c) $A + \overline{A}B$

(d) $(AB + C)A$

BIBLIOGRAPHY

1. Maley, Gerald A., and John Earle: "The Logic Design of Transistor Digital Computers," Prentice-Hall, Englewood Cliffs, N.J., 1963.
2. Braun, Edward L.: "Digital Computer Design," Academic, New York, 1963.
3. Chu, Yaohan: "Digital Computer Design Fundamentals," McGraw-Hill, New York, 1962.
4. Gillie, Angelo C.: "Binary Arithmetic and Boolean Algebra," McGraw-Hill, New York, 1962.
5. Hoernes, Gerhard E., and Melvin F. Heilweil: "Introduction to Boolean Algebra and Logic Design," McGraw-Hill, New York, 1964.
6. Nahselsky, Louis: "Digital Computer Theory," Wiley, New York, 1966.

appendix D

D-1 COMPARISON OF ASSEMBLY-LANGUAGE INSTRUCTIONS; H.P. 2100 FAMILY, VARIAN 620, DEC PDP-8

a. MEMORY REFERENCE

H.P. 2100 family (A and B accumulators, 16 bits)		Varian 620 [Two accumulators (A,B), 16 bits; one index register, X]			DEC PDP-8 (One accumulator, 12 bits)		
LDA	m [,I]	LDA	m	(LDAE[h]) (LDAI[i])	–		
LDB	m [,I]	LDB	m	(LDBE[h]) (LDBI[i])	–		
–		LDX	m	(LDXE[h]) (LDXI[i])	–		
STA	m [,I]	STA	m	(STAE[h]) (STAI[i])	[c]DCA	m [I	m]
STB	m [,I]	STB	m	(STBE[h]) (STBI[i])	–		
–		STX	m	(STXE[h]) (STXI[i])	–		
ADA	m [,I]	[d]ADD	m	(ADDE[h]) (ADDI[i])	TAD	m [I	m]
		[d]SUB	m	(SUBE[h]) (SUBI[i])	–		
ADB	m [,I]	–					
[a]CPA	m [,I]	–					
[a]CPB	m [,I]	–					
JMP	m [,I]	JMP[h]			JMP	m [I	m]
JSB	m [,I]	JMPM[h]			JMS	m [I	m]
–		[e]XEC[h]			–		
ISZ	m [,I]	–			ISZ	m [I	m]
–		[f]JXY, JXYM, XXY					
–		[g]INR	m	(INRE[h]) (INRI[i])	–		
[b]AND	m [,I]	[b]ANA	m	(ANAE[h]) (ANAI[i])	AND	m [I	m]
[b]IOR	m [,I]	[b]ORA	m	(ORAE[h]) (ORAI[i])	–		
[b]XOR	m [,I]	[b]ERA	m	(ERAE[h]) (ERAI[i])	–		

[a] Compare A or B with the contents of location m. Skip next instruction if *different*.
[b] Works with the A register only.
[c] DCA m results in clearing the accumulator (AC) also.
[d] Only uses A register.
[e] XEC instruction causes the execution of the instruction at the memory location referred to, followed by the return to the next instruction following XEC.
[f] Branch—if specific conditions are met. XY specifies conditions. For example, XY = AP, branch if A register is positive; XY = BZ, branch if B = 0.
[g] INR m *only increments* the contents of location m.
[h] Two-word instruction. The second word contains the memory reference address.
[i] Immediate instruction. Memory reference is always the next sequential location.

b.   REGISTER-ALTER INSTRUCTIONS

H.P. 2100 family	Varian 620	DEC PDP-8
[c]NOP	NOP	NOP
CLA	TZA	CLA
CLB	TZB	–
–	TZX	–
CMA	CPA	CMA
CMB	CPB	–
–	CPX	–
CCA	–	STA
CCB	–	–
INA	IAR	IAC
INB	IBR	–
–	IXR	–
[a]CLE	–	CLL
[a]CCE	–	STL
–	–	[b]GLK
CMA, INA	–	CIA
HLT	HLT	HLT
LDB 0	TAB	–
LDA 1	TBA	–
–	TBX	–
–	TAX	–
–	TXA	–
–	TXB	–
–	[d]DAR	–
–	[d]DBR	–
–	[d]DXR	–
–	[e]AOFA	–
–	[e]AOFB	–
–	[e]AOFX	–
–	[f]SOFA	–
–	[f]SOFB	–
–	[f]SOFX	–

[a] E register in Hewlett-Packard corresponds to Link register in PDP-8, except that it does not complement on a carry like the Link.

[b] Get link—set AC to value of Link.

[c] NOP does nothing except use up one machine cycle.

[d] Decrement register contents.

[e] Add *overflow* bit to register.

[f] Subtract *overflow* bit from register.

c. TEST INSTRUCTIONS[c]

H.P. 2100 family	Varian 620	DEC PDP-8	
SSA	[a]JAP; JAPM; XAP	SPA	
SSA, RSS	JAN; JANM; XAN	SMA	
SSB	−	−	
SSB, RSS	−	−	
SZA	JAZ; JAZM; XAZ	SZA	
SZA, RSS	−	SNA	
SZB	JBZ; JBZM; XBZ	−	
SZB, RSS	−	−	
−	JXZ; JXZM; XXZ	−	
SEZ	−	SZL	
SEZ, RSS	JOF; JOFM; XOF	SNL	
SLA	−	−	
SLA, RSS	−	−	
SLB	−	−	
SLB, RSS	−	−	
RSS	JMP; JMPM; XEC	SKP	
−	[b]JSSN; JSNM; XSN	−	
(Some combinations)	−	−	
SEZ, SSA	−	SPA	SZL
SEZ, SSA, RSS	−	SMA	SNL
− −	−	SPA	CLA
− −	−	SMA	CLA
CLA, SEZ, CLE	−	−	−

[a] JAP = JMP to m if A is positive. JAPM = JSB to m if A is positive. XAP = Execute instruction at m if A is positive. *Note*: These are memory reference instructions, *not* skip instructions. Subsequent instructions are interpreted likewise.

[b] JSSN = JMP if sense switch N is set. N = 1 to 3.

[c] For other possible combinations, see H.P. 2116 and DEC PDP-8 manuals.

d. SHIFT-ROTATE INSTRUCTIONS

H.P. 2100 family	Varian 620		DEC PDP-8
—	[a]LRLA	n	—
—	LRLB	n	—
RAL	LRLA	1^h	—
RBL	LRLB	1^h	—
RAR	LRLA	15^h	—
RBR	LRLB	15^h	—
RAR, RAR	LRLA	14^h	—
RBR, RBR	LRLB	14^h	—
RAL, RAL	LRLA	2^h	—
RBL, RBL	LRLB	2^h	—
ALF	LRLA	4^h	—
BLF	LRLB	4^h	—
ALF, ALF	LRLA	8^h	—
BLF, BLF	LRLB	8^h	—
ERA	—		[b]RAR
ERB	—		—
ELA	—		[b]RAL
ELB	—		—
ALS	ASLA	1^h	—
BLS	ASLB	1^h	—
ARS	ASRA	1^h	—
BRS	ASRB	1^h	—
ALR	—		—
BLR	—		—
ERA, ERA	—		RTR
ELA, ELA	—		RTL
CLE, ELA	—		CLL RAL
CLE, ERA	—		CLL RAR
—	[c]LSRA	n^h	—
—	[c]LSRB	n^h	—
—	[d]LLSR	n^h	—
—	[e]LLRL	n^h	—
—	[f]LASR	n^h	—
—	[g]LASL	n^h	—

[a] Logical rotation of A register n places, n can be from 0 to 31 places. Thus, $RAL_{(H.P.)} \equiv LRLA\ 1_{(620)}$; RAL, $RAL_{(H.P.)} \equiv LRLA\ 2_{(620)}$; $ALF_{(H.P.)} \equiv LRLA\ 4_{(620)}$.

[b] Note that the PDP-8 RAR and RAL instructions rotate the AC *with the Link.*

[c] Logical shift right n places. Zeros→MSB; LSB→lost.

[d] *Long* logical shift right n places. Both A and B registers shifted. $A_0 \rightarrow B_{15}$: $B_0 \rightarrow$ lost; $0 \rightarrow A_{15}$.

[e] Long logical rotate left n places; combined rotation of A and B registers.

[f] Long arithmetic shift right n places. A_{15} extended right; $A_0 \rightarrow B_{14}$; $B_0 \rightarrow$ lost. B_{15} unaltered.

[g] Long arithmetic shift left. $0 \rightarrow B_0$: $B_{14} \rightarrow A_0$; B_{15}, A_{15} unaltered.

[h] Number of shifts can be 0 to 31.

e. INPUT/OUTPUT

H.P. 2100 family		Varian 620		DEC PDP-8
[h]STC	sc	[a]EXC	YXX	[a]$6YXX_8$
[g]STC	sc, C	[i]EXC	YXX	$6YXX_8$
CLC	sc	EXC	YXX	$6YXX_8$
CLC	sc, C	EXC	YXX	$6YXX_8$
LIA	sc	CIA	XX	$6YXX_8$
LIA	sc, C	–		$6YXX_8$
LIB	sc	CIB	XX	–
LIB	sc, C	–		–
MIA	sc	INA	XX	–
MIA	sc, C	–		–
MIB	sc	INB	XX	–
MIB	sc, C	–		–
OTA	sc	OAR	XX	$6YXX_8$
OTA	sc, C	–		$6YXX_8$
OTB		OBR	XX	–
OTB	sc, C	–		–
STF	sc	EXC	YXX	$6YXX_8$
CLF	sc	EXC	YXX	$6YXX_8$
SFS	sc	SEN	YXX	$6YXX_8$
SFC	sc	SEN	YXX	$6YXX_8$
LIA	1	–		LAS
MIA	1	–		OSR
–		[b]IME	XX	–
–		[c]OME	XX	–
–		[d]INAB	XX	–
–		[e]CIAB	XX	–
–		[f]OAB	XX	–

[a] Y is a value from 0 to 7 and determines the specific function of the instruction at the device. For example, Y = 0, start; y = 1, clear flag, etc.

[b] Input to memory; two-word instructions. The second word contains the memory address.

[c] Output memory; two-word instruction. The second word contains the memory address.

[d] Inclusive OR input to both A and B registers.

[e] Load input word into both A and B.

[f] Output inclusive OR of A and B.

[g] [,C] following I/O instruction causes CLF to be executed simultaneous with I/O instruction.

[h] sc = device's channel number.

[i] XX = value from 0 to 77_8 for device-selection logic.

f. SOME COMPARISONS OF I/O INSTRUCTIONS

H.P. 2100 family		*Varian 620*		*DEC PDP-8*	
STF	0	[c]EXC	240	ION	(6001_8)
CLF	0	EXC	440	IOF	(6002_8)
[a]LIA	TTY, C	[d]CIA	00	KRB	(6036_8)
[a]SFS	TTY	SEN	100	KSF	(6031_8)
[a]SFS	TTY	SEN	100	TSF	(6041_8)
[a]OTA	TTY	OAR	00	TLS	(6046_8)
STC	TTY, C	EXC	300	–	
[b]SFS	A/D	SEN	YXX	ADSF	(6531_8)
[b]STC	A/D, C	EXC	YXX	ADVD	(6532_8)
[b]LIA	A/D, C	CIA	XX	ADRB	(6534_8)

[a] TTY = channel number for TTYP, e.g., 12_8.
[b] A/D = channel number for ADC, e.g., 13_8.
[c] 40 = device code for interrupt control.
[d] 00 = device code for Teletype printer.

D-2 COMPARISON OF SAMPLE PROGRAMS FOR TWO DIFFERENT ASSEMBLY LANGUAGES

(H.P. 2100 family, DEC PDP-8)

a. CLEAR BLOCK OF CORE MEMORY

H.P. 2100 family			*DEC PDP-8*		
ASMB,A,B,L,T			*50		
	ORG	50B		CLA	
	CLA		LOOP,	DCA	I PBLCK
LOOP	STA	. BLCK, I		ISZ	PBLCK
	ISZ	. BLCK		ISZ	CNTR
	ISZ	CNTR		JMP	LOOP
	JMP	LOOP		HLT	
	HLT	77B	DECIMAL		
CNTR	DEC	−33	CNTR,	−33	
. BLCK	DEF	BLOCK	PBLCK,	BLOCK	
	ORG	140B	OCTAL		
BLOCK	OCT	0	*140		
	END		BLOCK,	0	
			$		

b. BLOCK SUM PROGRAM

H.P. 2100 family			DEC PDP-8		
ASMB,A,B,L,T			*1000		
	ORG	1000B	START,	CLA	
START	LDA	PDTST		TAD	PDTST
	STA	PDATA		DCA	PDATA
	LIA	1		LAS	
	CMA, INA			CIA	
	STA	DCNTR		DCA	DCNTR
	CLA			TAD	I PDATA
	ADA	PDATA,I		ISZ	PDATA
	ISZ	PDATA		ISZ	DCNTR
	ISZ	DCNTR		JMP	.−3
	JMP	*−3		DCA	SUM
	STA	SUM		HLT	
	HLT	77B		JMP	START
	JMP	START	PDTST,	DATA	
PDTST	DEF	DATA	PDATA,	0	
PDATA	OCT	0	DCNTR,	0	
DCNTR	OCT	0	SUM,	0	
SUM	OCT	0	*2000		
			DATA,	0	
	ORG	2000B	$		
DATA	OCT	0			
	END				

c. DOUBLE-PRECISION ADDITION

H.P. 2100 family (using B accumulator)			DEC PDP-8		
ASMB,A,B,L,T			*1000		
	ORG	1000B		CLA	
	CLB			DCA	MSH
	CLA,CLE		DLUPE,	CLL	
DLUPE	ADA	.DTA,I		TAD	I PDATA
	SEZ,CLE			SZL	
	INB			ISZ	MSH
	ISZ	.DTA		ISZ	PDATA
	ISZ	CNTR		ISZ	CNTR
	JMP	DLUPE		JMP	DLUPE
	STA	LSH		DCA	LSH
	STB	MSH		HLT	
	HLT	77B	MSH,	0	
MSH	OCT	0	LSH,	0	
LSH	OCT	0	PDATA,	DATA	
.DTA	DEF	DATA	CNTR,	−N	
CNTR	DEC	−N	*2000		
			DATA,	0	
	ORG	2000B	$		
DATA	OCT	0			
	END				

d. PROGRAM-CONTROLLED DATA TRANSFERS

1 Subroutine to print one character

H.P. 2100 family

```
        •
        •
        •
TYPE    NOP                /ENTER W/CHAR TO TYPE IN "A"
        LDB  DOPP          /GET TTYP FUNCTION IN "B"
        OTB  TTY           /TELL TTYP WHAT TO DO ("PRINT")
        AND  MASK          /MASK THRU LOWER 8 BITS OF "A"
        OTA  TTY           /OUTPUT ASCII CHAR TO TTYP
        STC  TTY,C         /TELL TTYP TO PRINT. CLR FLG.
        SFS  TTY           /WAIT FOR FLAG
        JMP  *−1
        CLF  TTY           /CLR FLAG BEFORE RETRN
        JMP  TYPE,I        /RETRN MAIN PROGRM
MASK    OCT  377
TTY     EQU  12B           /TTYP IS ON CHANNEL 12B
DOPP    OCT  130000        /FUNCTION CODE: DATA OUT PRINT
                             AND PUNCH
        •
        •
        •
```

DEC PDP-8

```
        •
        •
        •
TYPE,   0                  /ENTER ROUT   W/ASCII CHAR IN AC
        TLS                /OUTPUT CHAR   CLR TTYP FLG.   PRINT
                             AND/OR PUNCH.
        TSF                /WAIT FOR TTYP FLAG
        JMP  .−1
        JMP  I   TYPE      /RETRN MAIN PROGRM
        •
        •
        •
```

---

2 Subroutine to print six-digit (or four-digit) octal equivalent of 16-bit (or 12-bit) binary value

---

H.P. 2100 family

```
                    •
                    •
                    •
T6DG      NOP              ENTER WITH BINARY # IN "A"
          RAL              /GET BIT 15 INTO BIT 0
          STA   1          /STORE PRTLY ROTTD WRD IN "B"
                             ("B" = LOC'N #1)
          AND   MSK1       /MASK THRU BIT 0
          ADA   N260       /CNVRT TO ASCII-CODED CHAR
          JSB   PTYPE,I    /GO TO SUBROUT TO PRINT ONE CHARACTR
LOOP      LDA   1          /GET PRTLY ROTTD WRD FROM "B"
          ALF, RAR         /ROTATE NEXT 3 BITS INTO BITS 0-2
          STA   1          /STORE PRTLY ROTTD WRD
          AND   MSK7       /MASK THRU BITS 0-2
          ADA   N260       /CNVRT TO ASCII-CODED CHARCTR
          JSB   PTYPE,I    /GO TO SUBROUT TO PRINT ONE CHRCTR
          ISZ   CNT5       /CNVRTD AND PRNTD ALL 16 BITS?
          JMP   LOOP       /NO
          LDA   CT5ST      /YES-RESET FOR RE-ENTRY
          STA   CNT5
          JMP   T6DG,I     /RETRN MAIN PROGRM
CT5ST     OCT   -5
CNT5      OCT   -5
N260      OCT   260
MSK1      OCT   1
MSK7      OCT   7
                    •
                    •
                    •
```

DEC PDP-8

```
                    •
                    •
                    •
TYP4DG    0                /ENTR W/ 12-BIT BINARY DATUM IN AC
          CLL              /CLR LINK FOR ROT'N OPERATION
          RAL              /INITIALIZE LEFT ROT'N INTO LINK
          DCA   TEMP       /SAVE PRTTLY ROT'D WRD
          TAD   M4
          DCA   CNTR       /INITIALIZE COUNTER
TYPLP,    TAD   TEMP       /GET PRTLLY ROT'D DATUM
          RTL              /ROTATE UPPER 3 BITS OF ORIG WRD
                            INTO LOWEST 3 BITS
          RAL
          DCA   TEMP       /SAVE
```

2 *(continued)*

DEC PDP-8

```
          TAD   TEMP
          AND   MSK7          /MASK THRU ONLY LOWER 3 BITS
          TAD   N260          /ADD 260 TO MAKE ASCII-CODED NUMERIC
                                 CHARCTR
          JMS  I   PTYPE      /GO TO TYPE SUBROUT
          ISZ   CNTR          /TYPD 4 DIGS?
          JMP   TYPLP         /NO–GET NXT OCTAL DIG
          JMP  I   TYP4DG
M4,       –4
MSK7,     7
N260,     260
TEMP,     0
CNTR,     0
$
```

---

3   Data acquisition with external clock control of ADC

H.P. 2100 family

```
           •
           •
           •
          CLF   0             /TURN OFF INTERRPT SYSTEM
          STC   13B,C         /SET CONTROL BIT (ENCODE) ON ADC CHNNL
                                 (CHNNL 13B); STRT CLOK
WAIT      SFS   13B           /WAIT FOR ADC FLG FOR EACH CNVRSN
          JMP   *–1
          LIA   13B,C         /GET DIGITAL VALUE WHEN READY, CLR FLG
          STA   PBFFR,I
          ISZ   PBFFR
          ISZ   DCNTR         /ALL DATA TAKEN?
          JMP   WAIT          /NO–WAIT FOR NXT CONVRSN
          CLC   13B           /YES–STOP EXPT AND/OR CLOK TO STOP
                                 DATA ACQUISITION
           •
           •
           •
```

3 *(continued)*

DEC PDP-8

```
IOT1=6331
ADSF=6531
ADRB=6534

*200
```

```
          CLA
          •                          /INITIALIZE
          •
          •
          IOF                        /DISABLE INTRRUPT SYSTEM
          IOT1                       /STRT EXPT AND CLOK
          CLA
DATIN,    ADSF                       /WAIT FOR ADC CONVRSNS
          JMP   .-1
          ADRB                       /READ ADC BFFR
          DCA  I    PBFFR
          ISZ   PBFFR
          ISZ   DCNTR                /ALL DATA IN?
          JMP   DATIN                /NO
          IOT1                       /YES—DISABLE CLOK, STOP EXPT
          JMP   PRCSS                /GO TO PRCSSNG ROUT
          •
          •
          •
```

---

e. INTERRUPT-CONTROLLED I/O

H.P. 2100 family

```
          •
          •
          •
          ORG  10B
          JSB   PRDSV,I      /STORAGE OF INTRRPT SERVICE INSTRUCTIONS
                                FOR DEVICES ON CHNNLS #10 THRU 13.
          JSB   PPNCH,I
          JSB   PTYSV,I
          •
          •
          •
PRDSV     DEF  RDSVC
PPNCH     DEF  PNSVC
PTYSV     DEF  TYSVC
```

e. *(continued)*

H.P. 2100 family

```
              •
              •
              •
          ORG  2000B
STRT      NOP                  /MAIN PROGRM
              •
              •
              •
          CLF  TTY             /CLR FLG ON TTYP AND ALL OTHER DEVICES
                                  TO BE USED
          STF  0               /TURN ON INTRRPT SYSTEM
              •
              •
              •
          LDA  CHAR,I          /GET ASCII CHRCTR TO PRINT
          LDB  DOPP            /GET TTYP AND FUNCTION CODE
          OTB  TTY             /TELL TTYP WHAT TO DO ("PRINT")
          OTA  TTY             /OUTPUT ASCII CHRCTR FOR PRINT
          STC  TTY
          JMP  MAIN            /GO TO MAIN PROGRAM & WAIT FOR INTRRPT
              •
              •
              •
TYSVC     NOP                  /TTYP INTRRPT SERVICE ROUT
          STA  ASAV1           /SAVE "A" REGISTER
          STB  BSAV1           /SAVE "B" REGISTER
          CLA
          ELA                  /GET "E" INTO "A"
          STA  ESAV1
          ISZ  CHAR            /SET UP TO GET NXT CHRCTR FOR PRINT
          ISZ  DCNTR           /ALL DATA PRINTD?
          RSS
          JMP  EXIT1           /YES
          LDA  CHAR,I          /NO–GET NXT CHAR TO PRINT
          LDB  DOPP
          OTB  TTY
          OTA  TTY
          STC  TTY             /PRINT
          LDA  ESAV1           /RESTORE WRKNG REGS
          ERA
          LDA  ASAV1
          LDB  BSAV1
          CLF  TTY
          JMP  TYSVC,I         /RETRN MAIN PROGRM
```

e. *(continued)*

DEC PDP-8

```
*0

INRPT,     0
           JMS  I  2
PSRVC,     SRVC

*200
             .
             .
             .
(Main program)
             .
             .
             .
           ION
             .
             .
             .

SERVC,     0
           IOF
           DCA  ACSAV      /SAVE WRKNG REGISTERS
           GLK
           DCA  LNKSAV
           KSF             /KYBRD INTRRPT?
           SKP             /NO
           JMP  KYSVC      /YES-GO TO KYBRD SRVC ROUT
           TSF             /PRNTR INTRRPT?
           SKP             /NO
           JMP  PRSVC      /YES-GO TO PRNTR SRVC ROUT
           ADSF            /ADC INTRRPT?
           SKP             /NO
           JMP  ADSVC      /YES-GO TO ADC SRVC ROUT
             .
             .
             .
(Restore working registers)
           ION
           JMP  I  0       /RETRN TO MAIN PROGRM
```

---

[1] *Note*: The interrupt service routine should clear the device flag causing the interrupt, otherwise another interrupt will occur when ION is executed again.

D-3　INSTRUCTION SETS FOR DATA GENERAL NOVA AND DIGITAL EQUIPMENT CORP. PDP-11

The following instruction sets (for the Data General NOVA and the Digital Equipment Corp. PDP-11) are included to provide the reader with a handy comparison with those of the other systems studied in this text. However, because the architecture and hardware features of these computers are quite different and have not been discussed in detail here, the reader is referred to the manufacturers' literature for the additional information required to interpret properly the various instructions listed below. (See "How to Use the NOVA Computers," Data General Corp., Southboro, Mass.; "PDP-11 Handbook," Digital Equipment Corp., Maynard, Mass.) A general description of these machines is provided in Chap. 13.

a　INSTRUCTION SET, DATA GENERAL NOVA

1　Derivation of the instruction mnemonics

$\left. \begin{array}{l} \text{LoaD} \\ \text{STore} \end{array} \right\}$ Accumulator

$\left. \begin{array}{l} \text{Increment} \\ \text{Decrement} \end{array} \right\}$ and Skip if Zero

JuMP

Jump to SubRoutine

$\left. \begin{array}{l} \text{COMplement} \\ \text{NEGate} \\ \text{MOVe} \\ \text{INCrement} \\ \text{ADd Complement} \\ \text{SUBtract} \\ \text{ADD} \\ \text{AND} \end{array} \right\}$ $\begin{array}{l} \text{for carry bit} \\ \text{base value use} \end{array}$ $\left\{ \begin{array}{l} \text{current carry} \\ \text{Zero} \\ \text{One} \\ \text{Complement of current carry} \end{array} \right\}$ $\left\{ \begin{array}{l} \text{shift Left} \\ \text{shift Right} \\ \text{Swap bytes} \end{array} \right\}$

$\left\{ \begin{array}{l} \text{SKiP} \\ \\ \text{Skip} \end{array} \right.$ $\left. \begin{array}{l} \text{on Zero} \\ \text{on Nonzero} \\ \text{if Either is Zero} \\ \text{if Both are Nonzero} \end{array} \right\}$ $\left\{ \begin{array}{l} \text{Carry} \\ \text{Result} \end{array} \right.$

No I/O transfer

Data $\left\{ \begin{array}{l} \text{In} \\ \text{Out} \end{array} \right\}$ $\left\{ \begin{array}{l} \text{A} \\ \text{B} \\ \text{C} \end{array} \right\}$ buffer — and $\left\{ \begin{array}{l} \text{Start} \\ \text{Clear} \\ \text{special Pulse} \end{array} \right.$

SKiP if $\left\{ \begin{array}{l} \text{Busy} \\ \text{Done} \end{array} \right\}$ is $\left\{ \begin{array}{l} \text{Nonzero} \\ \text{Zero} \end{array} \right.$

READ Switches

I/O ReSeT

HALT

INTerrupt Acknowledge

MaSK Out

INTerrupt ENable

INTerrupt DiSable

MULtiply

DIVide

2   Instruction mnemonics (NOVA)

Alphabetic Listing

Notation: ACS = source AC; ACD = destination AC; PC = program counter.

ADC   Add the complement of ACS to ACD; use Carry as base for carry bit.

ADCC   Add the complement of ACS to ACD; use complement of Carry as base for carry bit.

ADCCL   Add the complement of ACS to ACD; use complement of Carry as base for carry bit; rotate left.

ADCCR   Add the complement of ACS to ACD; use complement of Carry as base for carry bit; rotate right.

ADCCS   Add the complement of ACS to ACD; use complement of Carry as base for carry bit; swap halves of result.

ADCL   Add the complement of ACS to ACD; use Carry as base for carry bit; rotate left.

ADCO   Add the complement of ACS to ACD; use 1 as base for carry bit.

ADCOL   Add the complement of ACS to ACD; use 1 as base for carry bit; rotate left.

ADCOR   Add the complement of ACS to ACD; use 1 as base for carry bit; rotate right.

ADCOS   Add the complement of ACS to ACD; use 1 as base for carry bit; swap halves of result.

ADCR   Add the complement of ACS to ACD; use Carry as base for carry bit; rotate right.

ADCS   Add the complement of ACS to ACD; use Carry as base for carry bit; swap halves of result.

ADCZ   Add the complement of ACS to ACD; use 0 as base for carry bit.

ADCZL   Add the complement of ACS to ACD; use 0 as base for carry bit; rotate left.

ADCZR   Add the complement of ACS to ACD; use 0 as base for carry bit; rotate right.

ADCZS   Add the complement of ACS to ACD; use 0 as base for carry bit; swap halves of result.

ADD   Add ACS to ACD; use Carry as base for carry bit.

ADDC   Add ACS to ACD; use complement of Carry as base for carry bit.

ADDCL　Add ACS to ACD; use complement of Carry as base for carry bit; rotate left.

ADDCR　Add ACS to ACD; use complement of Carry as base for carry bit; rotate right.

ADDCS　Add ACS to ACD; use complement of Carry as base for carry bit; swap halves of result.

ADDL　Add ACS to ACD; use Carry as base for carry bit; rotate left.

ADDO　Add ACS to ACD; use 1 as base for carry bit.

ADDOL　Add ACS to ACD; use 1 as base for carry bit; rotate left.

ADDOR　Add ACS to ACD; use 1 as base for carry bit; rotate right.

ADDOS　Add ACS to ACD; use 1 as base for carry bit; swap halves of result.

ADDR　Add ACS to ACD; use Carry as base for carry bit; rotate right.

ADDS　Add ACS to ACD; use Carry as base for carry bit; swap halves of result.

ADDZ　Add ACS to ACD; use 0 as base for carry bit.

ADDZL　Add ACS to ACD; use 0 as base for carry bit; rotate left.

ADDZR　Add ACS to ACD; use 0 as base for carry bit; rotate right.

ADDZS　Add ACS to ACD; use 0 as base for carry bit; swap halves of result.

AND　And ACS with ACD; use Carry as carry bit.

ANDC　And ACS with ACD; use complement of Carry as carry bit.

ANDCL　And ACS with ACD; use complement of Carry as carry bit; rotate left.

ANDCR　And ACS with ACD; use complement of Carry as carry bit; rotate right.

ANDCS　And ACS with ACD; use complement of Carry as carry bit; swap halves of result.

ANDL　And ACS with ACD; use Carry as carry bit; rotate left.

ANDO　And ACS with ACD; use 1 as carry bit.

ANDOL　And ACS with ACD; use 1 as carry bit; rotate left.

ANDOR　And ACS with ACD; use 1 as carry bit; rotate right.

ANDOS　And ACS with ACD; use 1 as carry bit; swap halves of result.

ANDR　And ACS with ACD; use Carry as carry bit; rotate right.

ANDS　And ACS with ACD; use Carry as carry bit; swap halves of result.

ANDZ　And ACS with ACD; use 0 as carry bit.

ANDZL　And ACS with ACD; use 0 as carry bit; rotate left.

ANDZR　And ACS with ACD; use 0 as carry bit; rotate right.

ANDZS　And ACS with ACD; use 0 as carry bit; swap halves of result.

COM　Place the complement of ACS in ACD; use Carry as carry bit.

COMC　Place the complement of ACS in ACD; use complement of Carry as carry bit.

COMCL　Place the complement of ACS in ACD; use complement of Carry as carry bit; rotate left.

COMCR　Place the complement of ACS in ACD; use complement of Carry as carry bit; rotate right.

COMCS   Place the complement of ACS in ACD; use complement of Carry as carry bit; swap halves of result.

COML   Place the complement of ACS in ACD; use Carry as carry bit; rotate left.

COMO   Place the complement of ACS in ACD; use 1 as carry bit.

COMOL   Place the complement of ACS in ACD; use 1 as carry bit; rotate left.

COMOR   Place the complement of ACS in ACD; use 1 as carry bit; rotate right.

COMOS   Place the complement of ACS in ACD; use 1 as carry bit; swap halves of result.

COMR   Place the complement of ACS in ACD; use Carry as carry bit; rotate right.

COMS   Place the complement of ACS in ACD; use Carry as carry bit; swap halves of result.

COMZ   Place the complement of ACS in ACD; use 0 as carry bit.

COMZL   Place the complement of ACS in ACD; use 0 as carry bit; rotate left.

COMZR   Place the complement of ACS in ACD; use 0 as carry bit; rotate right.

COMZS   Place the complement of ACS in ACD; use 0 as carry bit; swap halves of result.

DIA   Data in, A buffer to AC.

DIAC   Data in, A buffer to AC; clear device.

DIAP   Data in, A buffer to AC; send special pulse to device.

DIAS   Data in, A buffer to AC; start device.

DIB   Data in, B buffer to AC.

DIBC   Data in, B buffer to AC; clear device.

DIBP   Data in, B buffer to AC; send special pulse to device.

DIBS   Data in, B buffer to AC; start device.

DIC   Data in, C buffer to AC.

DICC   Data in, C buffer to AC; clear device.

DICP   Data in, C buffer to AC; send special pulse to device.

DICS   Data in, C buffer to AC; start device.

DIV   If overflow, set Carry. Otherwise divide AC0-AC1 by AC2. Put quotient in AC1, remainder in AC0.

DOA   Data out, AC to A buffer.

DOAC   Data out, AC to A buffer; clear device.

DOAP   Data out, AC to A buffer; send special pulse to device.

DOAS   Data out, AC to A buffer; start device.

DOB   Data out, AC to B buffer.

DOBC   Data out, AC to B buffer; clear device.

DOBP   Data out, AC to B buffer; send special pulse to device.

DOBS   Data out, AC to B buffer; start device.

DOC   Data out, AC to C buffer.

DOCC   Data out, AC to C buffer; clear device.

DOCP   Data out, AC to C buffer; send special pulse to device.

DOCS   Data out, AC to C buffer; start device.

DSZ   Decrement location E by 1, and skip if result is 0.

HALT   Halt the processor (= DOC   0, CPU).

INC   Place ACS + 1 in ACD; use Carry as base for carry bit.

INCC   Place ACS + 1 in ACD; use complement of Carry as base for carry bit.

INCCL   Place ACS + 1 in ACD; use complement of Carry as base for carry bit; rotate left.

INCCR   Place ACS + 1 in ACD; use complement of Carry as base for carry bit; rotate right.

INCCS   Place ACS + 1 in ACD; use complement of Carry as base for carry bit; swap halves of result.

INCL   Place ACS + 1 in ACD; use Carry as base for carry bit; rotate left.

INCO   Place ACS + 1 in ACD; use 1 as base for carry bit.

INCOL   Place ACS + 1 in ACD; use 1 as base for carry bit; rotate left.

INCOR   Place ACS + 1 in ACD; use 1 as base for carry bit; rotate right.

INCOS   Place ACS + 1 in ACD; use 1 as base for carry bit; swap halves of result.

INCR   Place ACS + 1 in ACD; use Carry as base for carry bit; rotate right.

INCS   Place ACS + 1 in ACD; use Carry as base for carry bit; swap halves of result.

INCZ   Place ACS + 1 in ACD; use 0 as base for carry bit.

INCZL   Place ACS + 1 in ACD; use 0 as base for carry bit; rotate left.

INCZR   Place ACS + 1 in ACD; use 0 as base for carry bit; rotate right.

INCZS   Place ACS + 1 in ACD; use 0 as base for carry bit; swap halves of result.

INTA   Acknowledge interrupt by loading code of nearest device that is requesting an interrupt into AC bits 10 to 15 (= DIB   −,CPU).

INTDS   Disable interrupt by clearing interrupt on (= NIOC   CPU).

INTEN   Enable interrupt by setting interrupt on (= NIOS   CPU).

IORST   Clear all I/O devices; clear interrupt on; reset clock to line frequency (= DICC   O,CPU).

ISZ   Increment location E by 1, and skip if result is 0.

JMP   Jump to location E (put E in PC).

JSR   Load PC + 1 in AC3, and jump to subroutine at location E (put E in PC).

LDA   Load contents of location E into AC.

MOV   Move ACS to ACD; use Carry as carry bit.

MOVC   Move ACS to ACD; use complement of Carry as carry bit.

MOVCL   Move ACS to ACD; use complement of Carry as carry bit; rotate left.

MOVCR   Move ACS to ACD; use complement of Carry as carry bit; rotate right.

MOVCS   Move ACS to ACD; use complement of Carry as carry bit; swap halves of result.

MOVL   Move ACS to ACD; use Carry as carry bit; rotate left.

MOVO   Move ACS to ACD; use 1 as carry bit.

MOVOL   Move ACS to ACD; use 1 as carry bit; rotate left.

MOVOR   Move ACS to ACD; use 1 as carry bit; rotate right.

MOVOS   Move ACS to ACD; use 1 as carry bit; swap halves of result.

MOVR   Move ACS to ACD; use Carry as carry bit; rotate right.

MOVS   Move ACS to ACD; use Carry as carry bit; swap halves of result.

MOVZ   Move ACS to ACD; use 0 as carry bit.

MOVZL   Move ACS to ACD; use 0 as carry bit; rotate left.

MOVZR   Move ACS to ACD; use 0 as carry bit; rotate right.

MOVZS   Move ACS to ACD; use 0 as carry bit; swap halves of result.

MSKO   Set up interrupt disable flags according to mask in AC (= DOB   −,CPU).

MUL   Multiply AC1 by AC2; add product to AC0; put result in AC0−AC1.

NEG   Place negative of ACS in ACD; use Carry as base for carry bit.

NEGC   Place negative of ACS in ACD; use complement of Carry as base for carry bit.

NEGCL   Place negative of ACS in ACD; use complement of Carry as base for carry bit; rotate left.

NEGCR   Place negative of ACS in ACD; use complement of Carry as base for carry bit; rotate right.

NEGCS   Place negative of ACS in ACD; use complement of Carry as base for carry bit; swap halves of result.

NEGL   Place negative of ACS in ACD; use Carry as base for carry bit; rotate left.

NEGO   Place negative of ACS in ACD; use 1 as base for carry bit.

NEGOL   Place negative of ACS in ACD; use 1 as base for carry bit; rotate left.

NEGOR   Place negative of ACS in ACD; use 1 as base for carry bit; rotate right.

NEGOS   Place negative of ACS in ACD; use 1 as base for carry bit; swap halves of result.

NEGR   Place negative of ACS in ACD; use Carry as carry bit; rotate right.

NEGS   Place negative of ACS in ACD; use Carry as carry bit; swap halves of result.

NEGZ   Place negative of ACS in ACD; use 0 as base for carry bit.

NEGZL   Place negative of ACS in ACD; use 0 as base for carry bit; rotate left.

NEGZR   Place negative of ACS in ACD; use 0 as base for carry bit; rotate right.

NEGZS   Place negative of ACS in ACD; use 0 as base for carry bit; swap halves of result.

NIO   No operation.

NIOC   Clear device.

NIOP   Send special pulse to device.

NIOS   Start device.

READS   Read console data switches into AC (= DIA   −,CPU).

SBN   Skip, if both carry and result are nonzero (skip function in an arithmetic or logical instruction).

SEZ   Skip if either carry or result is 0 (skip function in an arithmetic or logical instruction).

SKP   Skip (skip function in an arithmetic or logical instruction).

SKPBN   Skip if busy is 1.

SKPBZ   Skip if busy is 0.

SKPDN   Skip if done is 1.

SKPDZ   Skip if done is 0.

SNC   Skip if carry bit is 1 (skip function in an arithmetic or logical instruction).

SNR   Skip if result is nonzero (skip function in an arithmetic or logical instruction).

STA   Store AC in location E.

SUB   Subtract ACS from ACD; use Carry as base for carry bit.

SUBC   Subtract ACS from ACD; use complement of Carry as base for carry bit.

SUBCL   Subtract ACS from ACD; use complement of Carry as base for carry bit; rotate left.

SUBCR   Subtract ACS from ACD; use complement of Carry as base for carry bit; rotate right.

SUBCS   Subtract ACS from ACD; use complement of Carry as base for carry bit; swap halves of result.

SUBL   Subtract ACS from ACD; use Carry as base for carry bit; rotate left.

SUBO   Subtract ACS from ACD; use 1 as base for carry bit.

SUBOL   Subtract ACS from ACD; use 1 as base for carry bit; rotate left.

SUBOR   Subtract ACS from ACD; use 1 as base for carry bit; rotate right.

SUBOS   Subtract ACS from ACD; use 1 as base for carry bit; swap halves of result.

SUBR   Subtract ACS from ACD; use Carry as base for carry bit; rotate right.

SUBS   Subtract ACS from ACD; use Carry as base for carry bit; swap halves of result.

SUBZ   Subtract ACS from from ACS; use 0 as base for carry bit.

SUBZL   Subtract ACS from ACD; use 0 as base for carry bit; rotate left.

SUBZR   Subtract ACS from ACD; use 0 as base for carry bit; rotate right.

SUBZS   Subtract ACS from ACD; use 0 as base for carry bit; swap halves of result.

SZC   Skip if carry is 0 (skip function in an arithmetic or logical instruction).

SZR   Skip if result is 0 (skip function in an arithmetic or logical instruction).

b   DEC PDP-11 INSTRUCTION SET

Notation

1. src = source; dst = destination; Loc = memory address; PC = program counter; PS = program status register; C, V, Z, N = bits 0 to 3 in PS. These bits monitor results of previous instruction and are set as follows:

Z   If the result was 0
N   If the result was negative
C   If the operation resulted in a carry from the most significant bit
V   If the operation resulted in an arithmetic overflow

2. For operations:

∧   And
v   Or
~   Not
( ) Contents of
∀   XOR
↓   Is pushed onto the processor stack
↑   The contents of the top of the processor stack is popped and becomes
→   becomes

Mnemonic	Instruction operation
Double-operand group: OPR src, dst	
MOV(B)	MOVe (Byte)
	(src) → (dst)
CMP(B)	CoMPare (Byte)
	(src) − (dst)
BIT(B)	Bit Test (Byte)
	(src) ∧ (dst)
BIC(B)	Bit Clear (Byte)
	~ (src) ∧ (dst) → (dst)
BIS(B)	Bit Set (Byte)
	(src) v (dst) → (dst)
ADD	ADD
	(src) + (dst) → (dst)
SUB	SUBtract
	(dst) − (src) → (dst)
Conditional branches: Bxx Loc	
BR	BRanch (unconditionally)
	Loc → (PC)
BNE	Branch if Not Equal (0)
	Loc → (PC) if Z = 0
BEQ	Branch if EQual (0)
	Loc → (PC) if Z = 1
BGE	Branch if Greater or Equal (0)
	Loc → (PC) if N ∀ V = 0
BLT	Branch if Less Than (0)
	Loc → (PC) if N ∀ V = 1

Conditional branches: Bxx Loc   *(continued)*

BGT	Branch if Greater Than (0) Loc → (PC) if Z v (N ∀ V = 0)
BLE	Branch if Less than or Equal (0) Loc → (PC) if Z v (N ∀ V = 1)
BPL	Branch if PLus Loc → (PC) if N = 0
BMI	Branch if MInus Loc → (PC) if N = 1
BHI	Branch if HIgher Loc → (PC) if C v Z = 0
BLOS	Branch if LOwer or Same Loc → (PC) if C v Z = 1
BVC	Branch if oVerflow Clear Loc → (PC) if V = 0
BVS	Branch if oVerflow Set Loc → (PC) if V = 1
BCC (or BHIS)	Branch if Carry Clear Loc → (PC) if C = 0
BCS (or BLO)	Branch if Carry Set Loc → (PC) if C = 1

Subroutine call: JSR   reg, dst

JSR	Jump to SubRoutine (dst) → (tmp), (reg) ↓ (PC) → (reg), (tmp) → (PC)

Subroutine return: RTS   reg

RTS	ReTurn from Subroutine (reg) → (PC) ↑ (reg)

Single-operand group: OPR   dst

CLR(B)	CLeaR (Byte) 0 → (dst)
COM(B)	COMplement (Byte) ~(dst) → (dst)
INC(B)	INCrement (Byte) (dst) + 1 → (dst)
DEC(B)	DECrement (Byte) (dst) − 1 → (dst)
NEG(B)	NEGate (Byte) ~(dst) + 1 → (dst)
ADC(B)	ADd Carry (Byte) (dst) + (C) → (dst)
SBC(B)	SuBtract Carry (Byte) (dst) − (C) → (dst)

Single-operated group: OPR dst *(continued)*

TST(B)	TeST (Byte)
	0 − (dst)
ROR(B)	ROtate Right (Byte)
	rotate right one place with C
ROL(B)	ROtate Left (Byte)
	rotate left one place with C
ASR(B)	Arithmetic Shift Right (Byte)
	shift right with sign extension
ASL(B)	Arithmetic Shift Left (Byte)
	shift left with lo-order - 0
JMP	JuMP
	(dst) → (PC)
SWAB	SWAp Bytes
	bytes of a word are exchanged

Operate group: OPR

HALT	HALT
	processor stops; (RO) and the HALT address in lights
WAIT	WAIT
	processor releases bus, waits for interrupt
RTI	ReTurn from Interrupt
	↑ (PC), ↑ (PS)
IOT	Input/Output Trap
	(PS) ↓, (PC) ↓, (20) → (PC), (22) → (PS)
RESET	RESET
	an INIT pulse is issued by the CPU
EMT	EMulator Trap
	(PS) ↓, (PC) ↓, (30) → (PC), (32) → (PS)
TRAP	TRAP
	(PS) ↓, (PC) ↓, (34) → (PC), (36) → (PS)

Glossary[1]

absolute   Pertaining to an address fully defined by a memory address number or to a program which contains such addresses (as opposed to one containing symbolic addresses).

accumulator   A register in which numbers are totaled or manipulated, or temporarily stored for transfers to and from memory or external devices.

add   Restrictive: 2's-complement addition of binary numbers; general: any arithmetic addition.

address   Noun: a number which identifies one location in memory; verb: the process of directing the computer to read a specified memory location (synonymous with reference).

address modification   A programming technique of changing the address referred to by a memory reference instruction, so that each time that particular instruction is executed, it will affect a different memory location.

[1] Taken in part from "Hewlett-Packard Training Manual," Hewlett Packard, Cupertino, Calif. (1967).

address word A computer word which contains only the address of a memory location.

alter A modification of the contents of an accumulator or other register, e.g., clear, complement, or increment.

analog Pertaining to information which can have continuously variable values, as opposed to digital information, which can be varied in degrees no smaller than the value of the least significant digit.

AND A logical operation in which the resultant quantity (or signal) is true if all the input values are true and is false if at least one of the input values is false.

arithmetic logic The circuitry involved in manipulating the information contained in a computer's accumulators.

arithmetic operation Restrictive: a mathematical operation involving fundamental arithmetic (addition, subtraction, multiplication, division) specifically excluding logical and shifting operations; general: any manipulation of numbers.

Assembler A program for a computer which converts a program prepared in symbolic form (i.e., using defined symbols and mnemonics to represent instructions, addresses, etc.) to binary machine language.

base The quantity of different digits used in a particular numbering system. The base in the binary numbering system is 2; thus, there are two digits (0 and 1). In the decimal system (base 10), there are 10 digits (0 to 9).

base page The lowest numbered page of a computer's memory. It can be directly addressed from any other page (sometimes called *zero page*).

Basic Binary Loader A series of instructions which will load into memory programs prepared with absolute addresses, using defined input devices.

binary Denoting the numbering system based on the radix 2. Binary digits are restricted to the values of 0 and 1.

binary-coded decimal A coding method for representing each decimal digit (0 to 9) by specific combinations of 4 binary bits. For example, the 8, 4, 2, 1 BCD codes are commonly used, where 1 is 0001 and 9 is 1001.

binary point The fractional dividing point of a binary numeral; it is equivalent to the decimal point in the decimal numbering system (sometimes called *binal point*).

binary program A program (or its recorded form) in which all information is in binary machine language.

bistable Pertaining to an electronic circuit having two stable states, controllable by external switching signals; it is analogous to an on off switch.

bit A single digit in a binary number or in the recorded representation of such a number (by hole punches, magnetic states, etc.). The digit can have one of only two values, 0 or 1.

bit density A physical specification referring to the number of bits which can be recorded per unit of length or area.

bit-serial   One bit at a time, as opposed to bit-parallel in which all bits of a character can be handled simultaneously.

buffer   A register used for intermediate storage of information in the transfer sequence between the computer's accumulators and a peripheral device.

bus   A major electrical path connecting two or more electrical circuits.

carry   A digit, or equivalent signal, resulting from an arithmetic operation which causes a positional digit to equal or exceed the base of the effective numbering system.

character   The general term to include all symbols such as alphabetic letters, numerals, punctuation marks, mathematical operators, etc.; also, the coded representation of such symbols.

checkerboard   An alternating pattern of 0's and 1's stored in a computer for testing purposes.

clear   Reset; the binary 0 condition.

code   A system of symbols which can be used by machines such as a computer and which in specific arrangements has a special external meaning.

comparator   An instrument for comparing digitized measurements against preset upper and lower limits and for giving an indication of the comparison result.

compiler   A language translation program used to transform symbols meaningful to a human operator to codes meaningful to a computer. More restrictively, a program which translates a machine-independent source language into the machine language of a specific computer, thus excluding Assemblers.

computation   The processing of information within the computer.

computer (digital)   An electronic instrument capable of accepting, storing, and arithmetically manipulating information, which includes both data and the controlling program. The information is handled in the form of coded binary digits (0 and 1) represented by dual voltage levels, magnetic states, punched holes, etc.

computer word   See "word."

configuration   The arrangement of either hardware instruments or software routines when combined to operate as a system.

contents   The information stored in a register or a memory location.

core   The smallest element of a core storage memory module. It is a ring of ferrite material and can be magnetized in clockwise or counterclockwise directions to represent the two binary digits, 0 and 1.

CPU   See "processor."

current page   The memory page comprising all those locations which are on the same page as a given instruction.

data acquisition   The gathering, measuring, digitizing, and recording of continuous form (analog) information.

data reduction   The transformation of raw information gathered by measuring or recording equipment into a more condensed, organized, or useful form.

data word A computer word consisting of a number, a fact, or other information which is to be processed by the computer.

debug Check for and correct errors in a program.

decimal Denoting the numbering system based on the radix 10.

decrement To change the value of a number in the negative direction. If not otherwise stated, a decrement by 1 is usually assumed.

device An electronic or electromechanical instrument; most commonly implies measuring, reading, or recording equipment.

diagnostic Adjective: relating to test programs for detection of errors in the functioning of hardware or software, or the messages resulting from such tests; noun: the test program or message itself.

digital voltmeter An electronic voltage-measuring device which provides a readout in digital form on the instrument panel and commonly (essential for computer purposes) codes the measurement result in binary-coded decimal form as an electrical output.

direct memory access A means of transferring a block of information words directly between an external device and the computer's memory, bypassing the need for repeating a service routine for each word. This method greatly speeds the transfer process.

disable A signal condition which prohibits some specific event from proceeding.

disc storage A means of storing binary digits in the form of magnetized spots on a rotating circular metal plate coated with a magnetic material. The information is stored and retrieved by read/write heads positioned over the surface of the disc.

documentation Manuals and other printed materials (tables, listings, diagrams, etc.) which provide instructive information for usage and maintenance of a manufactured product, including both hardware and software.

double-length word A word, because of its length, which requires two computer words to represent it. Double-length words are normally stored in two adjacent memory locations.

driver An input/output routine to provide automatic operation of a specific device with the computer.

dump To record memory contents on an external medium (e.g., tape).

effective address The address of a memory location ultimately affected by a memory reference instruction. It is possible for one instruction to go through several indirect addresses to reach the effective address.

enable A signal condition which permits some specific event to proceed whenever it is ready to do so.

exclusive OR A logical operation in which the resultant quantity (or signal) is true if at least one (but not all) of the input values is true and is false if the input values are all true or all false.

execute   To fully perform a specific operation, such as would be accomplished by an instruction or a program.

exit sequence   A series of instructions to conclude operation in one area of a program and to move to another area.

Extend (or *Link*)   A 1-bit register which extends the effective length of the accumulator for certain additions and rotations.

fixed-point   A numerical notation in which the fractional point (whether decimal, octal, or binary) appears at a constant, predetermined position. Compare with floating-point.

flag bit   A signal, or the stored indication of this signal, which indicates the readiness of a peripheral device.

flip-flop   An electronic circuit having two stable states and thus capable of storing a binary digit. Its states are controlled by signal levels at the circuit input and are sensed by signal levels at the circuit output.

floating-point   A numerical notation in which the integer and the exponent of a number are separately represented (frequently by two computer words), so that the implied position of the fractional point (decimal, octal, or binary) can be freely varied with respect to the integer digits. Compare with fixed-point.

flowchart   A diagram representing the operation of a computer program.

format   A predetermined arrangement of bits or characters.

FORTRAN   A programming language (or the Compiler which translates this language) which permits programs to be written in a form resembling algebra, rather than in detailed instruction-by-instruction form (as for Assemblers).

FORTRAN *Library*   A collection of programs to provide the user with commonly used mathematical and formatting routines.

gate   An electronic circuit capable of performing logical functions such as AND, OR, NOR, etc.

hardware   Electronic or electromechanical components, instruments, or systems.

Hardware Diagnostics   A collection of programs designed to assist in the identification of hardware malfunctions.

high core   Core memory locations having high-numbered addresses.

inclusive OR   A logical operation in which the resultant quantity (or signal) is true if at least one of the input values is true and is false if the input values are all false.

increment   To change the value of a number in the positive direction. If not otherwise stated, an increment by 1 is usually assumed.

incremental magnetic tape   A form of magnetic-tape recording in which the recording transport advances by small increments (e.g., 0.005 in.), stopping the tape advancement long enough to record one character at the spot located under the recording head.

indirect address The address initially specified by an instruction when it is desired to use that location to redirect the computer to some other location to find the effective address for the instruction.

information A unit or set of knowledge represented in the form of discrete "words," consisting of an arrangement of symbols or (as far as the digital computer is concerned) binary digits.

inhibit To prevent a specific event from occurring.

initialize The procedure of setting various parts of a stored program to starting values so that the program will behave the same way each time it is repeated. The procedures are included as part of the program itself.

input Information transferred from a peripheral device into the computer. It also can apply to the transfer process itself.

input/output Relating to the equipment or method used for transmitting information into or out of the computer.

input/output channel The complete input or output facility for one individual device or function, including its assigned position in the computer, the interface circuitry, and the external device.

instruction A written statement, or the equivalent computer-acceptable code, which tells the computer to execute a specified single operation.

instruction code The arrangement of binary digits which tells the computer to execute a particular instruction.

instruction logic The circuitry involved in moving binary information between registers, memory, and buffers in prescribed manners according to instruction codes.

instruction word A computer word containing an instruction code. The code bits may occupy all or (as in the case of memory reference instruction words) only part of the word.

interface The connecting circuitry which links the central processor of a computer system to its peripheral devices.

interrupt The process, initiated by an external device, which causes the computer to interrupt a program in progress, generally for the purpose of transferring information between that device and the computer.

interrupt location A memory location whose contents (always an instruction) are executed upon interrupt by a specific device.

interrupt phase A predetermined state of the internal computer logic which causes the computer to suspend operation of a program in progress and branch to a specific service routine.

jump An instruction which breaks the strict sequential location-by-location operation of a program and directs the computer to continue at another specified location anywhere in memory.

label Any arrangement of symbols, usually alphanumeric, used in place of an absolute memory address in computer programming.

language   The set of symbols, rules, and conventions used to convey information either at the human level or at the computer level.

load   Put information into memory, a register, etc.; also, to put the information medium into the appropriate device (e.g., loading tape).

loader   A program designed to assist in transferring information from an external device into a computer's memory.

location   A group of storage elements (cores) in the computer's memory which can store one computer word. Each such location is identified by a number (address) to facilitate storage and retrieval of information in selectable locations.

logic diagram   A diagram that represents the detailed internal functioning of electronic hardware using binary logic symbols rather than electronic component symbols. (See "schematic diagram.")

logic equation   A written mathematical statement using symbols and rules derived from Boolean algebra; specifically (hardware design), a means of stating the conditions required to obtain a given signal.

logical operation   A mathematical process based on the principles of truth tables, e.g., AND, inclusive OR, and exclusive OR operations.

loop   A sequence of instructions in which the last instruction is a jump back to the first instruction.

low core   Core memory locations having low-numbered addresses.

machine   Pertaining to the computer hardware (e.g., machine timing, machine language).

machine language   The form of coded information (consisting of binary digits) which can be directly accepted and used by the computer. Other languages require translation to this form, generally with the aid of translation programs (Assemblers and Compilers).

machine timing   The regular cycle of events in the operation of internal computer circuitry. The actual events will differ for various processes, but the timing is constant through each recurring cycle.

macro instruction   An instruction, similar in binary coding to the computer's basic machine-language instructions, which is capable of producing a variable number of machine-language instructions.

magnetic-tape recording   A means of recording information on a strip of magnetic-coated material such that binary bits can be represented by reversals of the direction of magnetization.

magnitude   That portion of a computer word which indicates the absolute value of a number, thus excluding the sign bit.

math routine   A program designed to accomplish a single mathematical function.

media conversion   The transferal of recorded information from one recording medium (e.g., punched paper tape, magnetic tape, etc.) to another recording medium.

memory An organized collection of storage elements (e.g., ferrite cores) into which a unit of information consisting of a binary digit can be stored and from which it can later be retrieved; also, a device not necessarily having individual storage elements but which has the same storage and retrieval capabilities (e.g., magnetic discs).

memory cycle That portion of the computer's internal timing during which the contents of one location of core memory are read out and written back into that location.

memory module A complete segment of core storage capable of storing a definable number of computer words (e.g., 4,096 words). Computer storage capacity is incremental by modules and is frequently rounded off and abbreviated as 4K (i.e., 4,096 or approximately 4,000 words), 8K (8,192 or 8000), 16K, etc.

memory protect A means of preventing inadvertent alteration of a select segment of memory.

memory reference The address of the memory location specified by a memory reference instruction, i.e., the location affected by the instruction.

merge Inclusive OR.

microinstruction An instruction which forms part of a larger, composite instruction.

mnemonic An abbreviation or arrangement of symbols used to assist human memory, e.g., CLA means clear accumulator.

multilevel indirect Indirect addressing using two or more indirect addresses in sequence to find the effective address for the current instruction.

multiple-precision Referring to arithmetic in which the computer, for greater accuracy, uses two or more words to represent one number.

multiplexer A device for sequentially switching multiple signal sources to one measuring or recording instrument.

object program See "source program."

octal Denoting a numbering system based on the radix 8. Octal digits are restricted to the values 0 to 7.

octal code A notation for writing machine-language programs with the use of octal numbers instead of the binary system.

off-line Pertaining to the operation of peripheral equipment not under control of the computer.

one's-complement A number so modified that the addition of the modified number and its original value, plus 1, will equal an even power of 2. A 1's-complement number is obtained mathematically by subtracting the original value from a string of 1's and electronically by inverting the states of all binary bits in the number.

on-line Pertaining to the operation of peripheral equipment under computer control.

output   Information transferred from the computer to a peripheral device; also can apply to the transfer process itself.

overflow   A 1-bit register which indicates that the result of an addition has exceeded the maximum possible signed value. The addition result will therefore be missing one or more significant bits (H.P. 2100 family).

packed word   A computer word containing two or more independent units of information. This is done to conserve storage when information requires relatively few bits of the computer word.

page   An artificial division of memory consisting of a fixed number of locations dictated by the direct addressing range of memory reference instructions.

page zero   The memory page which includes the lowest-numbered memory addresses.

parity bit   A supplementary bit added to an information word to make the total of 1-bits be always either odd or even. This permits checking the accuracy of information transfers.

pass   The complete process of reading a set of recorded information (one tape, one set of cards, etc.) through an input device from beginning to end.

peripheral device   An instrument or machine electrically connected to the computer but which is not part of the computer itself.

phase   One of several specific states of the internal computer logic, usually set up by instructions being executed, to determine how the computer should interpret information read out of memory.

photoelectric reader   An input device which senses characters (on punched tape, cards, pages, etc.) by optical light strobe and detection circuits.

power-failure control   A means of sensing primary power failure so that a special routine may be executed in the finite period of time available before the regulated dc supplies discharge to unusable levels. The special routine may be used to preserve the state of a program in progress or to shut down external processes.

P register   The Program Counter register, i.e., the register which keeps track of (or counts) the stored locations of the instructions in a program being executed (H.P. 2100 family).

priority   The automatic regulation of events so that chosen actions will take precedence over others in cases of timing conflict.

process control   Automatic control of manufacturing processes by use of computer.

processor   The central unit of a computer system (i.e., the device which accomplishes the arithmetic manipulations), exclusive of peripheral devices; frequently (when used as an adjective) also excludes interface components, even though they are normally contained within the processor unit; thus processor options exclude interface (input/output) options.

program   A plan for the solution of a problem by a computer which consists of a sequence of computer instructions.

program listing   A printed record of the instructions in a program.

pseudoinstruction　A symbolic statement, similar to assembly-language instructions in general form, but meaningful only to the program containing it, rather than to the computer as a machine instruction.

punched tape　A strip of tape, usually paper, on which information is represented by coded patterns of holes punched in columns across the width of the tape. Commonly, there are eight hole positions (channels) across the tape.

read　The process of transferring information from an input device into the computer. Also, the process of taking information out of the computer's memory. (See "memory cycle.")

real time　Time elapsed between events occurring externally to the computer. A computer which accepts and processes information from one such event and is ready for new information before the next event occurs is said to operate in a real-time environment.

register　An array of hardware binary circuits (flip-flops, switches, etc.) for temporary storage of information. Unlike mass storage devices such as memory cores, registers can be wired to permit flexible control of the contained information, for arithmetic operations, shifts, transfers, etc.

relocatable　Pertaining to programs whose instructions can be loaded into any stated area of memory.

Relocating Loader　A computer program capable of loading and combining relocatable programs (i.e., programs having symbolic rather than absolute addresses).

reset　A signal condition representing a binary 0.

rotate　A positional shift of all bits in an accumulator (and possibly an extend bit as well) with the bits that are lost off one end of the accumulator "rotated" around to enter vacated positions at the other end.

routine　A program or program segment designed to accomplish a single function.

sampling　The process of taking a measurement of a signal existing at a measuring instrument's input during a short (sample) period. The length of the sample period is a predetermined function of the measuring instrument.

scanner　See "multiplexer."

schematic diagram　A diagram that represents the detailed internal electrical circuit arrangement of electronic hardware using conventional electronic-component symbols.

select code　A number assigned to input/output channels for purposes of identification in information transfers between the computer and external devices.

set　A signal condition representing a binary 1.

shift　Restrictive (arithmetic shift): to multiply or divide the magnitude portion of a word by a power of 2 using a positional shift of these bits; general: any positional shift of bits.

sign   The algebraic plus or minus indicator for a mathematical quantity; also the binary digit or electrical polarity representing same.

significant digit   A digit so positioned in a numeral as to contribute a definable degree of precision to the numeral. In conventional written form, the most significant digit in a numeral is the leftmost digit, and the least significant digit is the rightmost digit.

skip   An instruction which causes the computer to omit the instruction in the immediately following location. A skip is usually arranged to occur only if certain specified conditions are sensed and found to be true, thus allowing various decisions to be made.

software   Computer programs. Also, the tapes or cards on which the programs are recorded.

software package   A complete collection of related programs, not necessarily combined as a single entity.

source program   A program (or its recorded form) written in some programming language other than machine language and thus requiring translation. The translated form is the object program.

starting address   The address of a memory location in which is stored the first instruction of a given program.

statement   An instruction in any computer-related language other than machine language.

store   To put information into a memory location, register, or device capable of retaining the information for later access.

subroutine   A sequence of instructions designed to perform a single task, with provisions included to allow some other program to cause execution of the task sequence as if it were part of its own program.

symbolic address   A label assigned in place of absolute numerical addresses, usually for purposes of relocation. (See "relocatable.")

Symbolic Editor   A program which is used to add, delete, or correct select portions of any symbolic program.

symbolic file   A recorded collection of computer words, with a symbolic address assigned to each word.

truth table   A table listing all possible configurations and resultant values for any given Boolean algebra function.

two's-complement   A number so modified that the addition of the modified number and its original value will equal an even power of 2; also, a kind of arithmetic which represents negative numbers in 2's-complement form so that all addition can be accomplished in only one direction (positive incrementation). A 2's-complement number is obtained mathematically by subtracting the original value from an appropriate power of the base 2 (i.e., from 1_2, 10_2, 100_2, etc.) and electronically by inverting the states of all binary bits in the number and adding 1 (complement and increment).

user   The person or persons who program and operate a particular computer.

utility routine   A standard routine to assist in the operation of the computer (e.g., device drivers, sorting routines, etc.) as opposed to mathematical (library) routines.

waiting loop   A sequence of instructions (frequently only two) which are repeated indefinitely until a desired external event, such as the receipt of a flag signal, occurs.

word   A set of binary digits handled by the computer as a unit of information. Its length is determined by hardware design, e.g., the number of cores per location and the number of flip-flops per register.

working register   A register whose contents can be modified under control of a program. Thus a register consisting of manually operated switches is not considered a working register.

write   The process of transferring information from the computer to an output device; also, the process of storing (or restoring) information into the computer's memory. (See "memory cycle.")

Index

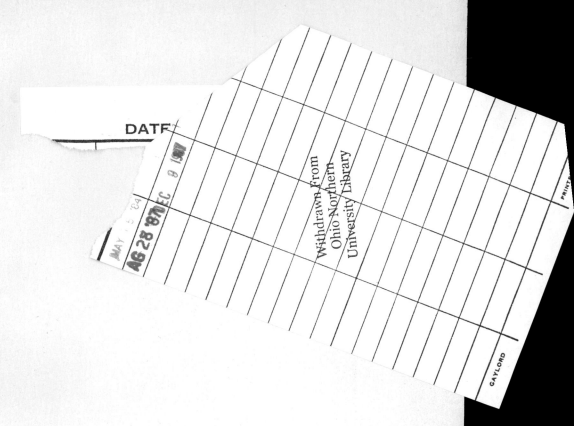